Pollen and Spores

Applications with Special Emphasis on Aerobiology and Allergy

Pollen and Spores

Applications with Special Emphasis on Aerobiology and Allergy

Shripad N. Agashe
Professor Emeritus in Botany
Bangalore University
Bangalore, India

Eric Caulton
Centre Director
Scottish Centre for Pollen Studies
School of Life Sciences
Napier University, Edinburgh, Scotland
United Kingdom

CRC Press
Taylor & Francis Group
Boca Raton London New York

CRC Press is an imprint of the
Taylor & Francis Group, an **informa** business
A SCIENCE PUBLISHERS BOOK

First published 2009 by Science Publishers

Published 2019 by CRC Press
Taylor & Francis Group
6000 Broken Sound Parkway NW, Suite 300
Boca Raton, FL 33487-2742

ISBN-13: 978-1-57808-532-3 (hbk)

Library of Congress Cataloging-in-Publication Data

Agashe, Shripad N.
Pollen and spores: applications with special emphasis on aerobiology and allergy/Shripad N. Agashe, Eric Caulton.
p. cm.
Includes bibliographical references and index.
ISBN 978-1-57808-532-3 (hardcover)
1. Palynology. 2. Pollen. 3. Respiratory allergy. I. Caulton, Eric. II. Title.

QK658.A353 2008
571.8'452--dc22

2008000562

Foreword

I feel very honoured to be asked to write a foreword to this important book which will undoubtedly take its place among the classic texts on pollen studies and aerobiology. The book has the distinct advantage of being written by two internationally renowned experts who have been able to draw on a wealth of experience, both from their own countries and also from their extensive knowledge of numerous countries throughout the world. The authors' enthusiasm and excitement for the subject area shine all through the book.

Previously the study of aspects of pollen and aerobiology by undergraduate and postgraduate students has been hindered severely by the fact that published material has been spread over a diverse range of disciplines including meteorology, ecology, botany, microbiology, allergy and forensic science. The text makes a major contribution towards improving this situation and fulfils a long standing need.

This work draws together numerous aspects of pollen studies, aerobiology and allergy into one accessible and well illustrated treatise. The comprehensive treatment has not only produced an extremely useful textbook but it has also resulted in an invaluable source of detailed references for researchers in the field. The timing of the publication is very appropriate. It coincides with a period of rapid increase in the number of courses in aerobiology which are offered in universities and as continuing professional development, such as in environmental health. This book will facilitate understanding and appreciation of the subject area whilst stimulating and inspiring students and researchers alike.

The breadth and depth of the content are particularly notable. The book achieves a synthesis of classic and traditional areas with newly developing applications. The early chapters place pollen studies and aerobiology within their historic contexts and cover the main aspects of aerobiology, whilst the later chapters deal with more novel applications and allergy. The authors are to be congratulated on accomplishing a skilful balance between the various themes presented.

The text meets and surpasses its objectives by providing a wealth of information within a clear and logical framework, supported by numerous references to published work for further reading. The breadth of content makes the book relevant to a wide audience of academics and professionals. It will be valued highly in many disciplines.

Professor Jean Emberlin
Director
National Pollen and Aerobiology Research Unit
University of Worcester
WR2 6AJ
UK

Preface

Palynology finds applications in various fields. Some of them are taxonomy, plant evolution, plant breeding programmes, biotechnology, microbiology of water, soil and air, the pharmaceutical industry, cosmetic industry, energy food industry, forensic science, aerobiology, allergy, epidemiology, meteorology, fossil fuel exploration and biodiversity.

On account of the above-mentioned applications, palynology has gained a lot of importance and is attracting different scientific disciplines. Published literature on the above aspects is widely scattered. Moreover undergraduate and postgraduate students are deprived of having 'One window system in palynology' in other words a textbook giving both the fundamental and applied information on pollen studies.

In view of the above, it is expected that the present comprehensive book on pollen studies and their applications will be useful to graduate, postgraduate and research students, and scientists in various industries in different countries. The book is also intended to serve as a source of reference to scientists in the various disciplines mentioned above. In order to accommodate both fundamental (basic) and applied aspects of pollen studies, and for easy understanding of the subject matter, the book is profusely illustrated.

Similar to other plant parts, pollen characters are so varied that the classification system of plants can be built up entirely on the basis of pollen morphology. An excellent and thought provoking book 'Pollen Grains' , their structure, identifications and significance in science and medicine published by R.P. Wodehouse in 1935 and reprinted in 1965, has been the main source of our inspiration to undertake this task of writing a comprehensive book concentrating on two important aspects: basic and applied. Several books and publications are available on the most basic aspects of pollen; however, applications of pollen though thoroughly exploited have not been compiled. Hence, the sole purpose of writing this book is to cover all aspects of pollen studies where our focus is always on undergraduate and postgraduate students. It is intended to inspire them to understand basic and applied aspects of pollen studies. Once they are exposed to these fascinating facts they can pursue any particular aspect for more detailed investigation.

The book will be also a good source of basic and applied information on pollen, which can be further exploited by researchers in various fields. It is up to user agencies to make use of the data in this book for exploiting respective fields. Pollen is aptly referred by some as GOLDEN DUST, extremely valuable on account of the golden yellow colour and their tremendous applications in science, industry and public health. No other plant part, even though extremely tiny in size is packed with so much information and power. It appears that honey bees, had realized the importance and utility of pollen earlier than man.

Any amount of written description will not be able to explain certain aspects of pollen which can be easily understood by illustrations. In view of this, the book is profusely illustrated with the primary intention of the students' proper understanding of structure, function and applications of pollen. The illustrations have been redrawn and modified in some cases from various existing publications. We are thankful to different sources for illustrations, which have been mentioned in the text wherever possible. We are especially grateful to Dr. T. S. Nayar, Dr. Cherian Panicker, Dr. K. Govind , Dr. B. E. Rangaswamy and Dr. M. P. Vidya for providing LM and SEM photomicrographs of some pollen and fungal spores. The senior author's wife Dr. Nirmala and daughter Sandhya were a great support and strength to us for initiating and completing this project besides preparing a number of illustrations and helping us in many ways in preparing the manuscript for this book.

Mr. B. R. Ramamurthy, the artist has redrawn many illustrations used in this book for which we are thankful. Data for compilation of some chapters were generated by the senior author during the Department of Science and Technology (D.S.T.) assisted project hence D.S.T.'s encouragement is thankfully acknowledged.

We are also thankful to the officials of Bangalore University, the Vice Chancellor, Registrar and Finance Officer , former colleagues and the Head of the Department of Botany, Bangalore University for providing the infrastructure and other facilities required for this project. There are so many scientists, botanists, clinicians and palynologists whose constant encouragement and interaction have helped us in completing this work. We thank all of them.

Several aerobiology research students of the senior author such as Drs. K. Manjunath, P. Sudha, Shashi Bala Sharma, K. V. Nagalakshamma, Meenakshi Chatterjee, Jacob Abraham, Soucenadin, Abdul Elfadil, Elizabeth Philip, H. S. Anuradha, M. P. Vidya, Binni, Mamlakatoi Hyderova, B. Pramila, have generated a lot of data which have gone into this book for which their efforts are thankfully acknowledged.

Grateful acknowledgements and thanks are due to the following persons for their encouragement, support and the provision of facilities for the work of the Scottish Centre for Pollen Studies, Napier University, Edinburgh, Professor Charles Bryce, Head of the School of Life Sciences, Miss Marina Mocogni and Mrs. Kim Kellet, Senior Technicians in the Biology and Biochemistry Sections respectively, and to David Caulton, son of Dr. Eric Caulton, Coauthor of this book, for his help with communications and advice throughout.

If postgraduate students for whom this book is primarily written, benefit from the palynological information incorporated, we will be really satisfied with our efforts.

Family members of Professor Shripad N. Agashe, wife Dr. Nirmala, daughters Sandhya and Swapna, Helen Caulton, wife of Dr. Eric Caulton and their son David Caulton have been directly or indirectly responsible for initiating and completing this book and to whom it is dedicated.

Bangalore, India
Edinburgh, United Kingdom, September 2008

Shripad N. Agashe
Eric Caulton

Contents

CHAPTER 1

Introduction

Palynology involves the study of pollen and encompasses the structural and applied aspects of pollen. Pollen grains are the male reproductive structures produced by the flowering plants (angiosperms) and gymnosperms (naked seeded plants). Palynology is a distinct branch of biology and is unique in many ways. According to the modern and wider definition of palynology, it also includes the study of spores produced by lower plants such as Algae, Fungi, Bryophytes and Pteridophytes.

According to geologists, the microfossils comprising pollen, spores, dermal appendages, cuticles, vascular elements, diatoms, desmids of plant origin and foraminifera, ostracods, microforaminifera are also studied within palynology. Scolecodonts of animal origin and certain microorganisms such as hystricospherids, dinoflagellates and acritarchs of doubtful origin are also included in palynological studies. Another unique feature of palynology is the fact that pollen and spores of both living plants and fossil plants are studied in detail.

Palynology, the science of pollen, gained a real impetus after the discovery of the microscope. This is logical as pollen grains are extremely tiny particles comparable to dust particles, which cannot be seen by the naked eye. The discovery of the microscope by Robert Hooke in 1665 was a landmark in the development of science, particularly palynology. Subsequent improvement in microscopy accelerated the study of pollen grains, especially the finer structure of the pollen wall and its myriad ornamentation patterns.

Pollen is ubiquitous in nature unlike other plant parts. It occurs buried deep in rocks, ground and surface water, and in air, both indoor and outdoors including the upper atmosphere. Besides, pollen finds its way through nasal and oral cavities to the digestive tract of humans and animals causing different degrees of discomfort. Pollen has an extremely long geological history, as it is well preserved in rocks as old as 400 m.y.

Pollen biology encompasses pollen production, its transfer to the stigma or pollination, and details of pollen-pistil interaction leading to fertilization and seed set. Any break in these sequential events affects seed and fruit set. Pollen biology studies are a prerequisite for any programme aimed at optimization and improvement of the yield of crop plants. Pollination ecology is also a part of pollen biology, which involves the study of various aspects dealing with efficient pollination. Pollen biotechnology is one of the techniques employed to study pollen biology for crop production and improvement. Pollen biotechnology is one of the most challenging areas of plant reproductive biology and plays an important role in crop improvement programmes.

The chapters included in the book are primarily devoted to two important aspects of palynology. It is well known that in order to exploit the applications of palynology in various fields, it is a prerequisite to have a thorough knowledge of different aspects of basic palynology. It is for this reason, that the first ten chapters comprise information on basic aspects of palynology such as pollen formation and pollen morphology of modern as well as fossil pollen and spores, current techninques and a general account of pollen physiology. These chapters are preceded by a chapter on a historical account of pollen studies, which to our mind is very appropriate and essential. Students of palynology should be aware of the origin and development of palynology up to the present. Many earlier scientists contributed significantly to palynology under adverse conditions, for example, limitations and availability of relatively simple microscopes. However, this historical account serves as a source of inspiration to learn more about pollen and exploit their potential applications.

The second and major part of this book comprises chapters on application of pollen studies in various fields such as agriculture, horticulture, plant breeding, enhancing honey production (melissopalynology), as an important tool in forensic science, reconstruction of past vegetation and environmental pollution and its effect on health particularly with reference to pollen allergy. Applications of airborne pollen and mould spores have been thoroughly explained in seven different chapters covering various aspects of aerobiology and allergy. This has been done on account of the significant role of aerobiological studies in allergy and immunology. Minor applications of pollen studies also include a brief account of copropalynology, which concerns pollen analysis of coprolites and other faeces of animals that throw light on past vegetation, feeding habits of animals of the present and past. A detailed account of most common aeroallergens and their source plants such as *Ambrosia* (Ragweed), grasses, and *Parthenium* has been given. A major chapter on 'Pollen Calendars: Global Scenario' includes information useful to aerobiologists and allergists from different parts of the world. This is followed by a comprehensive chapter on exploration of fossil fuels such as oil and coal.

Palynological information incorporated in the book is up-to-date to the extent possible and all the available sources, such as books; monographs, research articles have been referred for the compilation of information. An attempt has been made to include appropriate illustrations along with examples of common pollen sources.

CHAPTER 2

History of the Development of Palynology

Hyde and Williams coined the word 'palynology' in 1945 as a substitute for the science of pollen grains and spores. The term 'pollen' in Greek means 'flour'. Palynology is therefore derived from the Greek verb *'palynein'* meaning – to strew; to spread; to disseminate; to distribute in recognition of the fact that many pollen grains and spores are easily carried by the wind. However the systematic study of pollen and spores was initiated much earlier.

The function of pollen grains and their role in pollination and fertilization was known to ancient Assysrians. There is a large number of evidence in the form of illustrations and lithographs available to show that artificial pollination of Date Palms (*Phoenix dactylifera)* was a regular practice known to Assyrians and Babylonians in 700 B.C. **Figure 2.1** shows an eagle-headed human holding in one hand a bucket full of pollen and the other hand with male inflorescence dusting pollen on female flowers of date palm. This is also known from iconography dating as far back as 4000 – 3000 B.C. At the present time also artificial pollination of Date Palms **(Fig. 2.2)** is done as depicted in **Fig. 2.3** in which a Japanese botanist is seen in action.

The study of pollen grains began after the discovery of the microscope by Robert Hooke in 1665. He described this microscope in 'MICROGRAPHIA' published in London in 1665 some 350 years ago. Each notable improvement in the construction of the microscope has always been reflected with a corresponding relevance to the study of pollen morphology. Almost simultaneously Malpighi and Grew observed, and described pollen grains by using Robert Hooke's microscope.

Thus, Malpighi and Grew may be recognized as the co-founders of pollen morphology. Wodehouse (1935) in his book on **'POLLEN GRAINS, their structure, identifications and significance in science and medicine'** has included an excellent chapter on 'Historical Account of Palynology' which includes an exhaustive account of development of palynology from its

Fig. 2.1 Artificial pollination of Date Palm (*Phoenix dactilifera*) by an ancient Assyrian dated 4000-3000 B.C.: eagle headed human holding bucket full of pollen and dusting pollen on female flowers with a male inflorescence.

inception up to 1935. He discussed major developments in palynological knowledge in different centuries from the 13th to the 20th . Hence, we do not intend to repeat the same here. However, the main purpose of this account was to derive inspiration from the early founders and developers of the science of palynology and to understand and appreciate the efforts made by them in contributing to this science, many times under adverse working conditions. However, a brief life sketch and important palynological contributions of a few outstanding palynologists along with their significant achievements will be incorporated. Palynology is the scientific study of pollen and spores. It includes not only present day but also fossil examples. The study of fossilized plant material is part of the botanical discipline known as palaeobotany.

Palynology is not only a scientific discipline in its own right, but also a subdiscipline within pollen analysis and aerobiology. Pollen and spores are frequently dispersed from their points of origin and carried on wind

Fig. 2.2 Present day Date Palm (*Phoenix dactilifera*).

currents thus forming a component, the biospora of particulate matter in the air. Pollen and spores may be considered as 'organic pollutants' in air.

Pollen grains are microspores, which carry the male genetic component gametes of the plants, which produce them. Pollen grains are essentially part of the sexual reproductive process, spores, by contrast, may be the product of a sexual process, or may be vegetatively produced, the latter having the function of spreading the species which produce them–an aspect known as vegetative or asexual reproduction. Palynology as a modern scientific discipline is not old although the interest in and study of small biological particles such as spores and their possible role in the life of plants, extends back to at least classical times.

Fig. 2.3 Artificial pollination of Date Palm (*Phoenix dactilifera*) by a Japanese scientist at the present time.

The scientific study of pollen, both structure and function, has its origin in the early part of the 20th century. The establishment of a methodology to record both quantatively and qualitatively, the occurrence of pollen was first published in 1916 by Lennart von Post, a Swedish botanist (von Post, 1916). This method of illustrating the spore content of peat cores, later ice and submarine seabed cores, has become a standard method involving not only pollen and spores but also micro faunal and mineral deposits.

Contemporary pollen, which descends under gravity or is washed down by rainfall is called 'pollen rain'. All surfaces receive the descending pollen, which become buried in soil or sink to the bottom of lakes, rivers, seas and soil. Pollen in the air is also recovered from the groomed surfaces of fauna and can be extracted from the faeces enabling the scientist to determine diet and or habitat (Caulton 1988; Caulton *et al.*, 2005). Fossilized dung of bats inhabiting in caves has revealed similar indicator pollen of diet from the Pleistocene period.

PROFESSOR GUNNAR ERDTMAN

He was one of the foremost palynologists of the world in the 20th century and rendered valuable service to the science of palynology, particularly in pollen morphology. He started his work in pollen analysis in the footsteps of Professor Lennart von Post, the inventor of the method of pollen analysis of peat bog for reconstruction of past vegetation. Erdtman worked in the summers of 1918-1920 and 1922 as a member of the peat bog investigation team of the Geological Survey of Sweden. He contributed significantly to trends of post-glacial forest history of Western Europe. In 1930-1931 he earned the Rockfeller Scholarship to work on pollen analytical work in Western Canada and visited several parts of the U.S.A., including California and Arizona.

Later, Erdtman concentrated on experimental work on pollen, which resulted in 1934 (Erdtman 1934) in the invention of the Acetolysis technique for the preparation of pollen for microscopic studies.

In 1950 at the International Botanical Congress in Stokholm, Sweden, he organized the first International Palynological Meeting. At the 1954 International Botanical Congress in Paris, for the first time, a special section was devoted to palynology with Professor Gunnar Erdtman as its President.

Swedish Natural Science Research Council in 1948 established the famous Palynological Laboratory at Stokholm, Bromma and later at Solna. Prof. Erdtman was appointed as its Director. In addition, he also served as Professor of Geological Palynology at the University of Stokholm. He published more than 250 original research papers and several books dealing with palynology. One outstanding book entitled 'Pollen Morphology and Plant Taxonomy' was published in 1952, the other important book published by him in 1969 was 'The Handbook of Palynology' which has served as a standard manual for a long time.

Professor Gunnar Erdtman
1897 - 1973

Professor Erdman's other significant publications include; 'Grana Palynologica' (9 Vols. 1954-1909), 'An Introduction to a Scandinavian Pollen Flora' Vol. I in 1961, II in 1963 and 'World Pollen Flora'. He presented papers and conducted short-term courses in palynology in more than 60 universities all over the world.

CHAPTER 3

Pollen Formation, Development of the Pollen Wall and Tapetum

GENESIS OF POLLEN

Pollen is considered as the male gamete in flowering plants or angiosperms and gymnosperms through which genetic information is transmitted to the offspring. To the naked eye pollen grains appear mainly in the form of a yellow or cream coloured powder, which look alike, but they are quite different in their wall pattern. In fact each pollen grain bears a species-specific wall surface pattern, which helps immensely in identification and classification. These varied pollen wall patterns exhibit the beautiful art of nature. This becomes more apparent if the pollen walls are observed under SEM.

Typically, pollen has two cells known as the vegetative cell and generative cell. The latter gives rise to two sperm cells, which fuse with egg cells of the ovule during double fertilization. The vegetative cell comprises the bulk of the pollen cytoplasm, which is responsible for the development of the pollen tube **(Fig. 3.1)**.

Unlike animals, sperm cells in plants are not motile and hence they have to be transferred towards the egg cells of ovules for fertilization. This

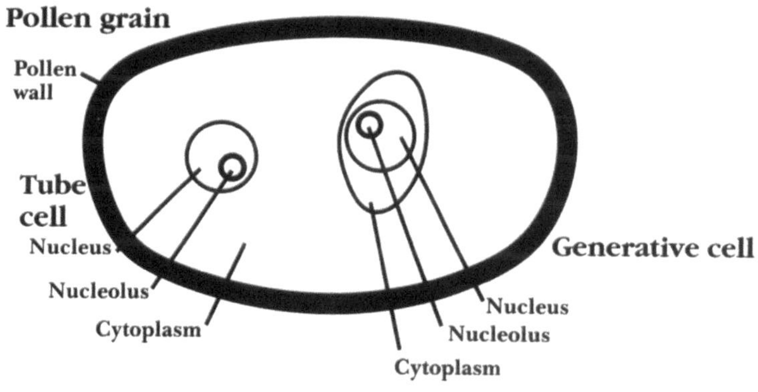

Fig. 3.1 Structure of pollen contents.

is usually done by pollen tubes, which arise during germination of pollen on the stigma. When the pollen germinates, the pollen tube emerges through one of the apertures (pores) in the pollen grain wall.

The pollen tube acts as a channel of transport for the sperm cells to the embryo sac of the ovule to achieve fertilization. During fertilization one sperm cell fuses with the egg to form the diploid zygote and the other sperm cell fuses with two haploid polar nuclei to give rise to the endosperm, which serves to nourish the developing embryo during seed development. This is termed as double fertilization, which is characteristic of angiosperms only and not gymnosperms.

There is a tremendous variation in the size and shape of pollen grains. The range of size variation of pollen grains in terms of diameter is from 5-200 μm with the average size being 30 μm. They vary considerably in the shape, which may be mostly spherical, oval or cubic, hexahedral, fibrous etc.

Fine structure of pollen grain walls revealed under SEM appears quite different from the one observed under the light microscope, for example a pollen grain of rice (Poaceae), Caryophyllaceae and Chenopodiaceae however appear as a vessel with a smooth wall surface. However, under SEM the black spots are clearly demonstrated as granulate concaves and plugs are identified at the aperture opposite the end from which the pollen tube elongates.

The structure of a flower, a sexual reproductive organ of a plant is shown in the **Fig. 3.2**. From the palynological point of view, the most

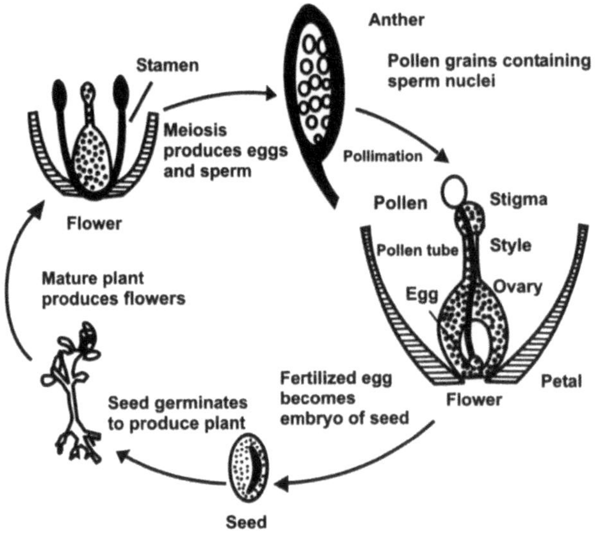

Fig. 3.2 Life cycle of a flowering plant showing the structure of a flower.

important part in the flowers is the stamen or male reproductive organ consisting of pollen sacs or anthers borne on elongated stalks or filaments.

POLLEN FORMATION

The male part of flowers is known as the androecium, which consists of structures called stamens. Each stamen comprises two parts: a stalk or filament, which terminates in the pollen-bearing structures, the anthers.

Pollen formation occurs in the anthers. Each anther is four lobed, each lobe is known as a loculus. The anther structure is composed of three parts: an outer wall, a lining layer of nourishing cells, the tapetum and a central mass of sporogenous tissue. The outer wall comprises relatively large cells with thin walls to allow loss of moisture at the onset of pollen release, anthesis, the cell wall which adjoins the ends of each lobe is thickened to resist increasing tension set up during the onset of pollen release anthesis. The tapetum or nourishing layer comprises a tissue in which, during its development is the depository of starch granules. These serve to provide essential nutriment and energy for the ongoing development of the innermost cell mass, the sporogenous tissue. The sporogenous tissue will eventually form tetrads of pollen grains (microspores) after a series of mitotic divisions followed by a final meiotic division, which produces each member of each tetrad having half of the parental chromosome number (n or haploid state).

Production of Pollen

Pollen grains or microspores are produced from pollen mother cells (microsporocytes), within the anthers (microsporangia or pollen sacs) of the flower. During its development two types of cells are differentiated within the anther; reproductive or sporangenous cells which give rise to pollen grains and the non-reproductive cells, which form the tissue layers such as epidermis, cortical and tapetal cell layers. During the formation of the pollen grains (microgametophyte) two distinct successive developmental phases are recognized, they are the microsporogenesis and microgametogenesis, which are illustrated in **Fig. 3.3**. The pollen mother cells are poorly attached to each other by plasma connections. Later, pollen mother cells assume a spherical shape and get detached followed by a thick callose layer around them. The pollen mother cells (2n) undergo meiosis to give rise to four haploid microspores (n).

During this phenomenon, a callose wall separates the members of the tetrad from one another, which is continuous with the callose surrounding the entire tetrad as seen in the illustration **(Fig. 3.4)**.

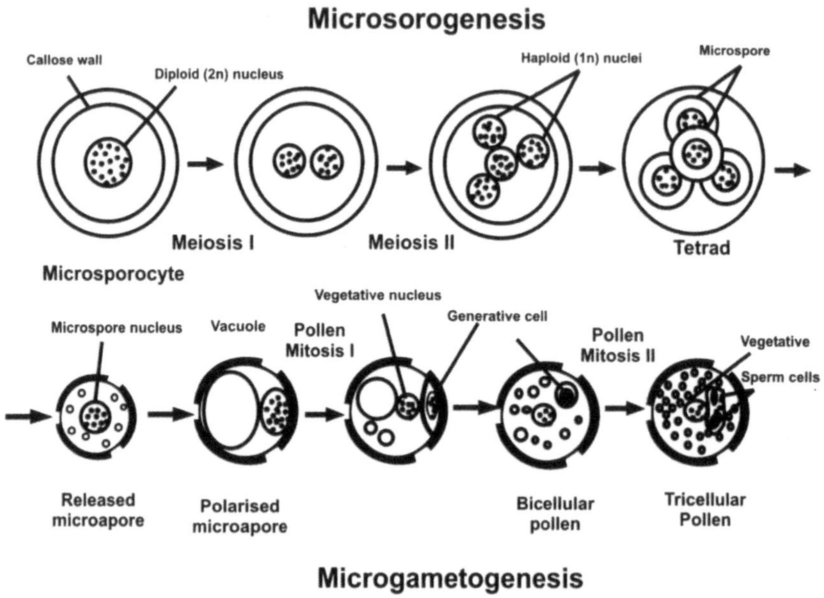

Fig. 3.3 Microsporogenesis and microgametogensis.

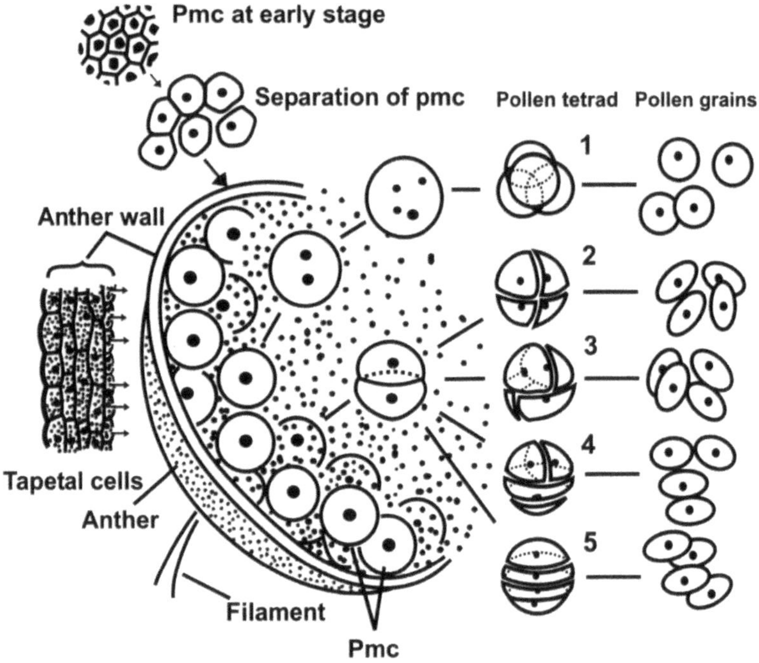

Fig. 3.4 Genesis of pollen grains.

Usually these four microspores or pollen grains remain attached as tetrad in various ways such as tetrahedral, tetragonal, linear (e.g. *Mimosa pudica*), rhomboidal (e.g. *Anona muricata*), decussate (e.g. *Magnolia grandiflora*), square depending on the orientation of meiotic spindle axes and the related cleavage planes.

Pollen grains get separated from tetrad and single pollen are called monads. In some plants two-pollen remain attached and are referred as dyads. In a few plants, for example *Calluna* and *Typha* 4 pollen remain attached and are released from anthers as tetrads. In plants like *Acacia* 16 pollen remain attached and are referred as polyads. In Asclepiadaceae and Orchidaceae numerous pollen are attached in units known as pollinia, which are basically polyads.

The Role of Tapetum in Pollen Development

The role of tapetum ascribed to the development of exine of the pollen grains as well as their nutrition is not very clear. Investigators such as Ubisch (1927), Maheshwari (1950), Johri and Tiagi (1952). have stated that tapetum provides the material for the development of the exine. They came to this conclusion having located similar types of granules at the place of degenerated tapetum and also surrounding the developing microspores. Regarding the role of tapetum in nutrition, the general inference comes from the rich cytoplasmic contents of the tissue and also from the fact of invasion of its cytoplasm between the growing pollen grains in some cases.The following facts observed in some experiments carried on male sterile plants support both these findings.

Origin of Exine

On the basis of experimental work on plants of *Sesamum* and *Gossypium*, Singh (1965) concluded that the development of microspores to form normal and healthy pollen grains follows the tapetal degeneration and the resultant product is deposited on the pollen. The development of exine of the pollen grains is dependent in some way on the behaviour of the tapetum, but the exact mechanism has yet to be found out.

Nutritional Role of Tapetum

On the basis of Feulgen positive reactions in anthers at various stages the increase in DNA in the generative cell is due to its supply from the degenerating tapetum. It has been made abundantly clear that the pollen grains thrive and develop further on degenerating tapetum but the exact nature of this dependence is so far not clearly demonstrated.

SPOROPOLLENIN

It is one of the most complex chemical substance present in the outer wall of the spores and pollen grains. This substance is resistant to many physical and chemical forces. This complex is highly resistant to decay. Sporollenin appears to be a polymerized cyclic alcohol related to suberin predominant in cork, callus and cutin. The durability of pollen walls is mainly due to the presence of this chemical substance. In fact, the durability of pollen walls and hence preservation of pollen grains and spores millions of years ago is the fundamental basis of the science of palynology.

It was first observed and named 'sporonin' by John, in 1814, and characterized by Berzelius in the 1830s. In addition to the walls of pollen and spores, the other palynomorphs such as dinoflagellates, acritarchs also have sporopollenin in their walls.

Sporopollenin is probably the most inert organic compound known. It resists acetolysis. Heslop-Harrison worked on the origin of sporopollenin and indicated that tapetum in anthers of flowers is responsible for producing and depositing sporopollenin in the exines of pollen. Chemically, it has been suggested that sporopollenin is a copolymer of betacarotine, a xanthothyl such as antheraxanthin, and fatty acids.

Sporopollenin is often referred as the 'cousin' of rubber as they share many common features such as elasticity, sensitivity to oxidation and alkalinity, general durability. However, the chemical nature of rubber differs from sporopollenin.

As mentioned earlier, sporopollenin is present in exine (outer wall) of the spores and pollen. In some fern spores, exine is enveloped by an additional wall referred as perine or perispore. Generally, aquatic flowering plants have very little sporopollenin or it is totally absent in the pollen wall. Perhaps this is the reason why the pollen of aquatic plants are not preserved in the form of fossils. There are three types of sporopollenin containing elements associated with spores and pollen. These are: viscin threads, elaters and ubisch bodies or orbicules.

The evolutionary history of sporopollenin indicates that the oldest sporopollinious acritarichs occur in Precambrian rocks, which are 1.2 to 1.4 billion years old. Sporopollenin perhaps played a significant role in these organisms with regard to protection of protoplasm against ultraviolet radiation. The green algae are presumably responsible for the development of sporopollenin and its introduction into the armament of higher green plants. Its main function in the higher plants appears to be protection against oxidation and desiccation. The natural colour of sporopollenin is pale yellow, but with thermal maturation the colour deepens through dark yellow through orange, reddish brown, finally to black. The specific gravity of the substance is about 1.4.

It is known that enzymes mostly do not affect sporopollenin. Hence, pollen and spore exines pass through most animal guts (including humans) unaltered though the contents of the pollen grains are digested. The oldest sporopollenin containing palynomorphs are sphaeromorphs acritarchs, over one billion years old have been described from the erstwhile Soviet Union and other parts of the world.

CHAPTER 4

Pollen Morphology

Pollen morphology is one of the most important and fundamental branches of palynology. It will not be an exaggeration if pollen morphology is referred to as the mother of palynological studies. Proper identification of pollen and spores both of living and fossil plants is a prerequisite for exploiting their applications. Pollen morphology is the principle tool used for correct identification. The slightest error in identification leads to an erroneous conclusion.

A recent example in this context which can be cited is concerned with the identification of airborne pollen in Bangalore. Due to lack of proper background knowledge of pollen morphology, *Cassia* pollen were shown to be the second most abundant pollen in Bangalore's atmosphere in a publication by the Asthma Research Society (1979). However, a systematic aeropalynological survey (Agashe *et al.*, 1994) carried out by sound pollen morphological knowledge, proved that Bangalore's atmosphere was rich in *Casuarina equisetifolia* pollen, which, since it looks superficially similar to *Cassia* pollen, was earlier identified incorrectly.

Various features of pollen are studied in pollen morphology. Some of these pollen morphological characters are: symmetry, size and shape, pollen wall, exine stratification, ornamentation, furrows/grooves and apertures. The last three characters seem to be the most important pollen morphological characters useful in basic identification and classification of pollen. A number of terms have been coined to describe various pollen morphological characters. At times, confusion arises as the literature abounds with respect to different terminologies used by different palynologists. However, in the present account the almost worldwide-accepted terminology proposed by Gunnar Erdtman, will be used. The pollen morphological descriptions will be adequately supported by suitable illustrations and examples.

MORPHOLOGY OF ANGIOSPERM POLLEN

Pollen grains, when mature and dehisced, show a range of size, shape and even colour. Colour requires the pollen to be seen en masse, for example the pollen of Willow (*Salix*) is yellow, Dandelion (*Taraxacum*) is orange and Willow Herb/Fireweed (*Chamaenerion*) is grey. Such a colour aspect is of value to beekeepers in identifying pollen source. Most pollen however are yellow when seen en masse on a white background. The range of colours represented en masse is well illustrated by Hodges (1975).

The majority of pollen grains range between 20-30 µm (microns*). Some are very small, for example Forget me not (*Myosotis*) whose pollen measures 18 µm. The largest category where size may exceed 100 µm is exemplified by Rosebay Willow herb/Fireweed (*Chamaenerion*) whose large triangular pollen measures c. 150 µm.

The shape of pollen grains is, like size, variable: round, oval (flattered or elongated), long or triangular, semicircular, boat shaped. Others may have several sides (flat or rounded). It is important to bear in mind when examining pollen grains with a view to their identification, that they are the products of a biological system, which is subject to variation, that a degree of care is sometimes needed to interpret what is seen.

The wall (exine) of pollen grains have apertures – Poplar (*Populus*) is exceptional in lacking them as shown in **Fig. 4.16 –A-C** which take the form of simple holes (pores) and furrows (colpi). Pores and furrows may merge or be irregular in appearance.

The exine surface can vary in its structure considerably. It can be smooth, have granules, be striped (striated), have a mesh or network, small holes or pits, or appear dotted (which are the bases of spines more fully seen in the side view.The exine when viewed from the edge, may appear thin or be composed of two or more layers, which may be separated by thick, thin or beaded rods.

Grains are usually simple, separating from the quartet during development in the pollen sac, each having its own complete exine. Some pollen grains are released during dehiscence as quarters – compound grains for example in the Ericaceae and Juncaceae, families. The appearance of the cytoplasm may be as glass (hyaline or granular). Some plant families, for example Asteraceae, show a great variety of forms: grains with large spaces (fenestrae) such as Dandelions (*Taraxacum*), Hawkweeds (*Hieracium*) and Sowthistles (*Sonchus*), whilst others have spines of varying number and length, for example (*Senecio*). Again, in the Asteraceae family, Mugwort (*Artemisia vulgaris*) has a three-lobed grain with long rods (tectum) within each lobe separating the exine layers – this grain can be confused with those of Privet (*Ligustrum*) and grains of Cruciferae, such as Oil Seed Rape (*Brassica napus*), which are similarly three-lobed. Pollen of the large family of grasses, Poaceae, have the simplest structure, which is more or less round, thin walled with one single pore. Pollen of cultivated cereal grasses

* Pollen grains are measured in microns. (1 micron = 0.001 mm) magnification for measuring size is usually x 400 or x 1000 under the light microscope.

are usually much larger than their wild relatives – being tetraploid or polyploid genetically.

It is unlikely that genetically modified pollen (GM) can be distinguished from the natural non-GM, morphologically. GM treatment is of no significance to the palynologist working with pollen of wild plants as opposed to pollen of cultivated crop plants.

Pollen grains are reproductive cells (male gametophytes). These specialized cells are provided with an extremely hard outer wall (exine) and an inner softer cellulose wall (intine) surrounding the cytoplasm with the vegetative and generative cells (nuclei), and organelles.

The pollen grains may be single (monads) which is most common, or two-pollen united (diads) for example *Scheuchzeria palustris* or united in fours (tetrads), or many, for example, multiples of fours (polyads). In Orchidaceae, Asclepiaceae and a few other families the grains are united in club-like masses (pollinia) **(Fig. 4.1)**.

With tetrads as a starting point we may discern polarity and symmetry. There are different kinds of tetrads, tetrahedral, rhomboidal, tetragonal, linear, etc. In Dicolyledons, the tetrahedral type (two planes) dominates, in Monocotyledons the tetragonal or rhomboidal type occurs (one plane only) **(Figs. 4.2 and 4.3)**. The tetrad configuration (different types of tetrads) is often linked to the microspore divisions, for example tetrahedral tetrads (simultaneous), and one-plane tetrads (successive). A polar axis, equatorial axis, polar diameter and equatorial diameter, distal pole, proximal pole, distal face and proximal face are distinguished in a pollen grain as shown in **Figs. 4.4** and **4.5**. With reference to this, it is possible to describe the location of various morphological features of the pollen grains and spores. Pollen grains or spores may have similar poles (isopolar), almost similar (subisopolar), or dissimilar poles (heteropolar), e.g. *Lycopodium* **(Fig. 4.6)**.

Symmetry

Pollen or spores may be symmetric or asymmetric. Asymmetric grains have no planes of symmetry and are either fixiform (with fixed shape, which is the common case) or nonfixiform (without fixed shape, very rare). Symmetric grains may be of two types: radiosymmetric (radial) grains have more than two vertical planes of symmetry, or, if provided with but two such planes, always with equilong equatorial axes. Bilateral spores are more or less flattened having two vertical planes of symmetry but in contradiction to the radiosymmetric spores with two such planes, the equatorial axes are not equilong. However, sometimes it is difficult to determine the symmetry as for example in *Crypteronia* and *Isoglossa*. Most Dicotyledons are radially symmetrical, whereas most Monocotydons and primitive Dicotyledons are

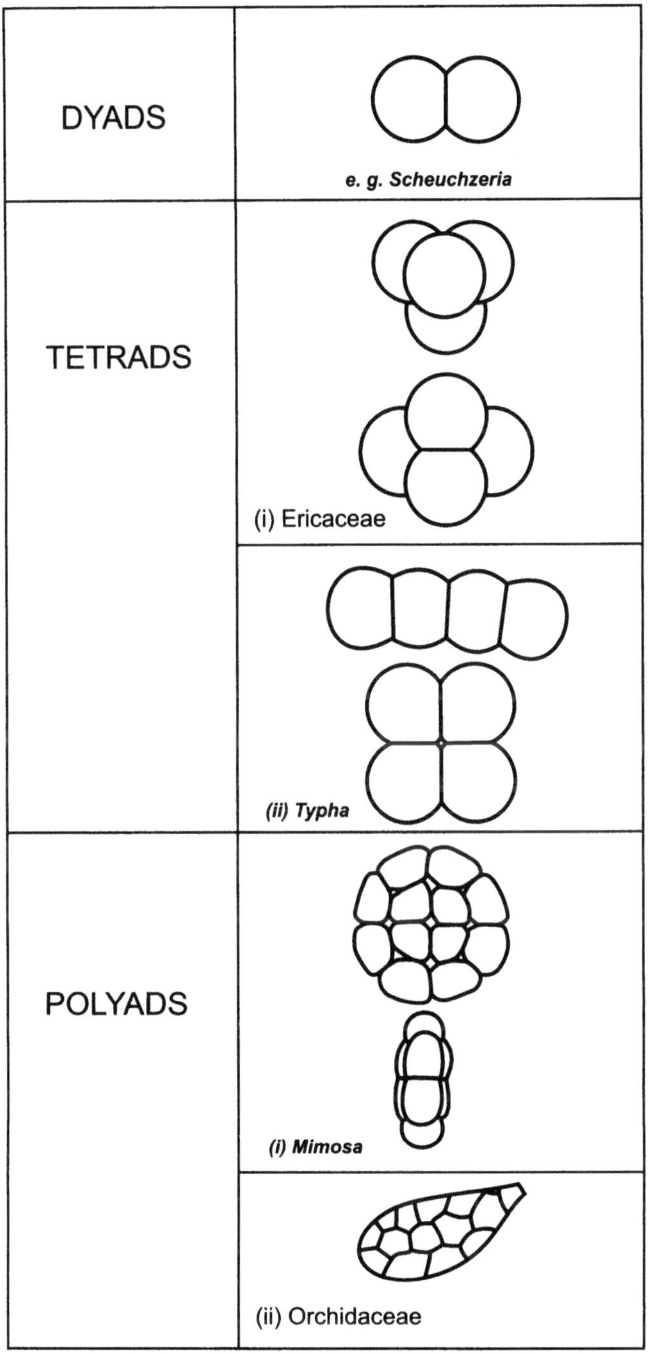

Fig. 4.1 Types of pollen associations at the time of their release from anthers.

A. Tetrahedral B. Tetragonal

Fig. 4.2 Modes of pollen association in tetrads.

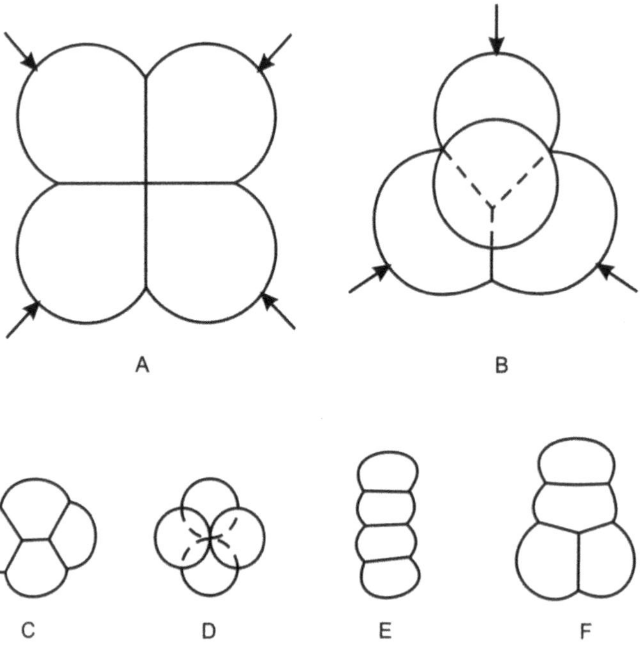

Fig. 4.3 Different types of pollen tetrads.

bilateral. Bilateral spores are also commonly produced in some Pteridophytes.

Size and Shape

Size is important since structural differences are sometimes inadequate for distinguishing species, and size becomes a reliable criterion. In *Picea* for example, such measurements have aided species identification. Pollen grains of angiosperms range from 5-200 μm in diameter. However, most grains seem to fall in a range of 25 μm-100 μm in living angiosperms.

Very few pollen grains have 5 μm as maximum diam. Families such as Boraginaceae (*Myosotis*), Piperaceae, Crypteroniaceae Cunoniaceae have

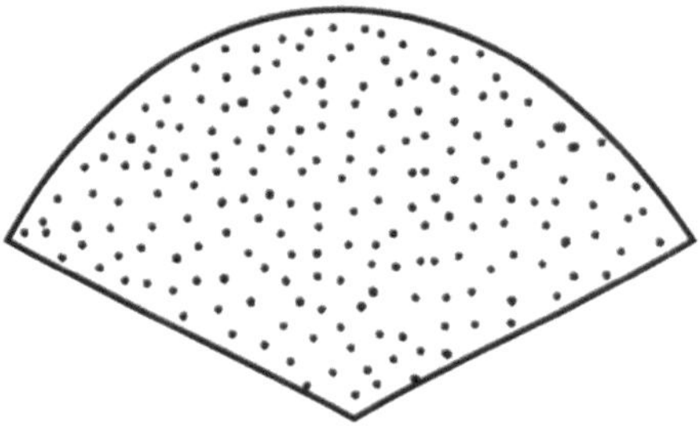

Fig. 4.4 Part of pollen showing distal pole and distal face, i.e., face away from the centre of the tetrad.

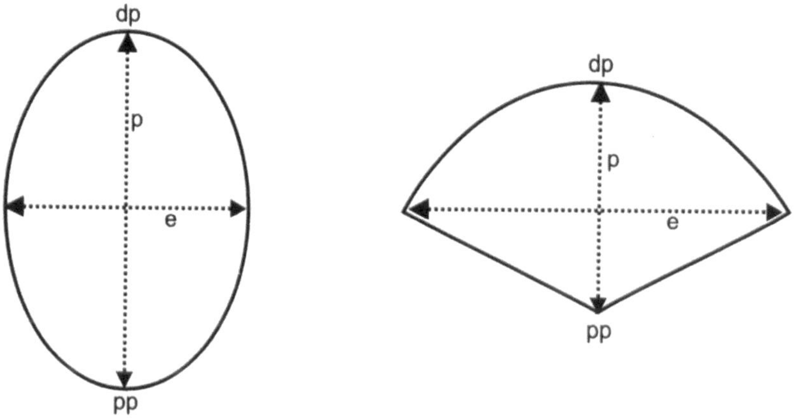

Fig. 4.5 Pollen and spore showing polar axis (p), equatorial axis (e), distal pole (dp) and proximal pole (pp).

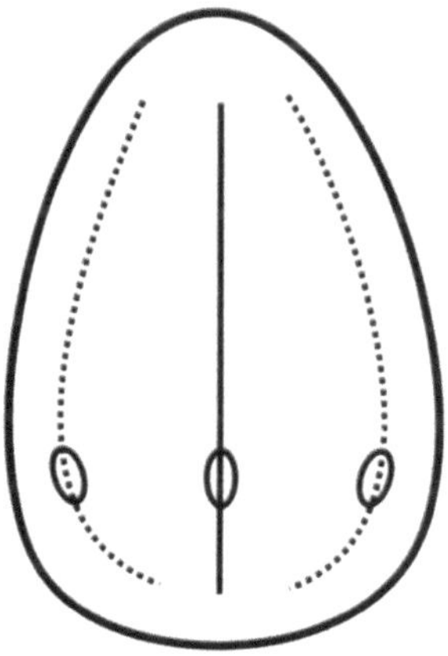

Fig. 4.6 Heteropolar: Pollen in which the distal and proximal faces of the exine are different either in shape, ornamentation or apertural system as in *Echium vulgare* (Boraginaceae).

pollen of lowest size ranges. However, pollen grains in excess of 200 µm in diam. are recorded in certain species of Dispsaceaeae, Nyctaginaceae, Oenotheraceae; Onagraceae Malvaceae, Cucurbitaceae. Probably the largest (diam. up to 350 µm) pollen grains are those of *Cymbopetalum odoratissimum*, a member of Annonaceae, (Walker 1971). Most of the studied monocots according to Cranwell have grains between 15-80 µm. In marine angiosperms, *Zostera marina* has tubular pollen exceeding 2,500 µm in length and 3-4 µm in diameter. *Amphibolis* has tubular pollen exceeding up to 5 mm long (Ducker *et al.*, 1978).

There often appears to be a direct relation between size and number of pollen produced per anther. For example, in *Mirabilis* only 32 very large pollen occur per loculus, whereas as many as 50,000 pollen grains are produced per loculus in *Borago*. In *Rumex acetosa* 30,000 grains are produced per stamen. It also appears that, as a rule the largest grains are produced by ephemeral flowers lasting only a day. Generally pollen of anemophilous plants are 'small' and those of entomophilous plants are 'large.'

Spores of certain Pteridophytes are extremely large, e.g., *Selaginella exaltata* megaspores have a diameter of about 1.5 mm. In Carboniferous deposits, megaspores measuring up to about 6-7 mm have been encountered as in *Triletes giganteus*.

SHAPE OF THE POLLEN GRAINS

The shape of pollen is often an important pollen morphological character. The shape of pollen varies in different views. The outline in polar view or Amb (L. Ambitus: short form Amb) is circular, triangular, square, pentagonal, rounded, three-lobed or in other geometrical shapes. Pollen and spores also differ considerably in their contours, walls **(Figs. 4.7 and 4.8)** and apices **(Fig. 4.9)**. As early as 1943, Erdtman suggested certain terms to describe shapes of pollen grains based on the ratio of polar axis to equatorial axis P : E. In the equatorial view the ratio between the polar and equatorial diameters multiplied by 100 gives an indication of the shape. The following terms are used to describe the shape of the pollen grains. Here P refers to polar diameter and E refers to equatorial diameter.

Fig. 4.7 Modes of contours of pollen walls in polar view.

STRAIGHT CONVEX CONCAVE

Fig. 4.8 Modes of pollen side walls in polar view.

ANGULAR ROUNDED TRUNCATED

Fig. 4.9 Types of pollen apices in polar view.

SHAPES OF THE GRAIN (FIGS. 4.10 – A-F)

a. Peroblate – P/E X 100 = < 50
b. Oblate – P/E X 100 = 70 – 75
c. Suboblate – P/E X 100 = 89 – 100
d. Prolate spheroidal – P/E X 100 = 101 – 114
   Subprolate- P/E X 100 = 115 – 133
e. Prolate – P/E X 100 = 134 – 200
f. Perprolate – P/E X 100 = > 200

SIZE CLASSES

Walker and Doyle (1975) have simplified the following six classes of size of pollen grains based on diameter or length of the longest axis.

1) Minute grains – <10 μm
2) Small grains – 10-24 μm
3) Medium sized grains – 24-49 μm
4) Large grains – 50-99 μm
5) Very large grains – 100-199 μm
6) Gigantic grains – > 200 μm

In bilateral grains, pollen are plano – convex, concavo – convex, or biconvex in lateral view. In other views, different shapes such as ellipsoidal, lenticular, oval or other types are often present.

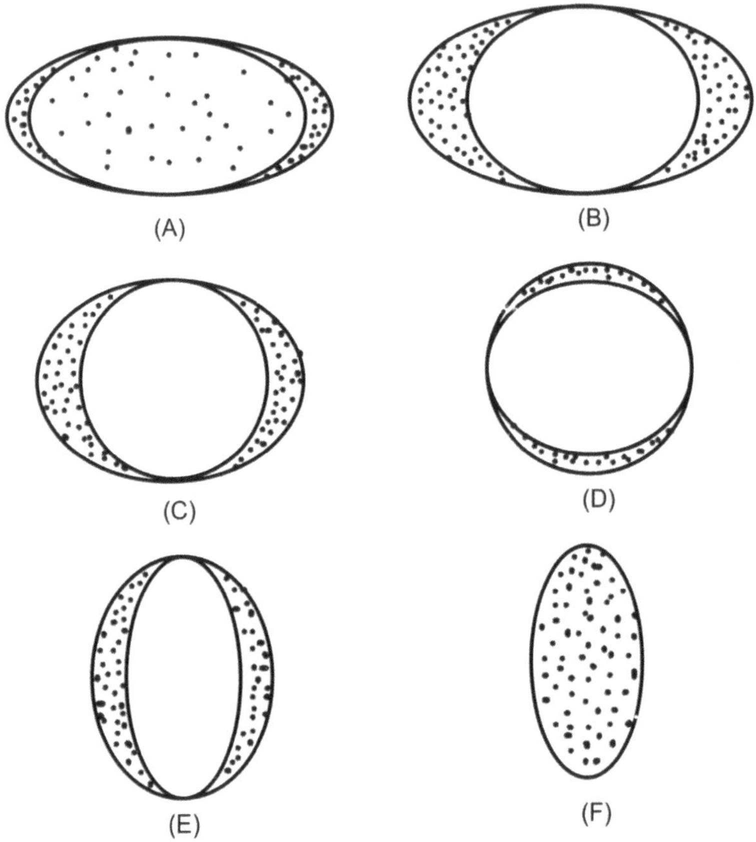

Fig. 4.10 A-F- Various shapes of pollen based on P/E × 100, A. Peroblate, B. Oblate, C. Suboblate, D. Prolate spheroidal, E. Prolate, F. Perprolate.

THE POLLEN WALL

The pollen grain wall is one of the most remarkable structures in plants. Each species of flower produces pollen with a uniquely structured wall. This uniqueness of the wall enables the identification of the parent plant when viewed under the microscope. When viewed optically with the naked eye, a mass of a specific pollen type may show a distinct colour, for example some members of the Asteraceae have orange pollen whilst some Onagraceae have grey pollen when seen en masse. Most pollen however are shades of yellow. This colour feature is of significance in honey analysis, melissopalynology (see Chapter 12). Microscopically, the wall may appear, smooth, spiny, furrowed, patterned, pitted, etc. Furrows (colpi) and pores may occur separately or in combination of various numbers. A complex terminology has been developed to describe wall ornamentation, furrows and pores.

The pollen grain wall (sporoderm) comprises two layers, the outer (sculptured) layer is called exine and an inner layer is the intine. The exine structure, described above is composed of a complex of substances collectively known as sporopollenin. All structures evolve in relation to function and persist as refinements by the process of natural selection. The pollen wall can be studied directly—ontogeny. By means of experimentation its function(s) may be determined. The evolutionary development of the pollen wall however can only be deduced hypothetically – phylogeny. The ontogeny of the wall has been touched on above. There are several functions attributable to the pollen wall: The physical contents – cytoplasm within – are adequately protected as a living, viable entity capable of germination under appropriate conditions; the pores provide means of water transfer and substances involved in pollen dispersal and pollen stigma (female receptive part of a flower) interactions are contained within. Sporopollenin is synthesized by the tapetum (Blackmore and Ferguson, 1986) and the exine prior to the completion of the synthesis and deposition of sporopollenin, serves as a communication surface for the selection and transport of supplies up to the germination stage described sporopollenin as extraordinary molecular complex. Sporopollenin protects the living internal contents from the harmful effects of the environment, for example, radiation hazards, due to its ability to absorb UV radiation. The lipid component acts as a seal being impermeable to water thereby protecting the grain against dessication (Thanikaimani 1986). Sporopollenin appears to be consistent in its properties throughout all taxa.

Pollen grain ontogeny in both gymnosperms and angiosperms have features in common; the initiation of the pollen mother cells from sporogenous central tissue; the ultimate meiotic divisions to produce a tetrads of haploid cells and the development of the sporopollenous exine wall prior to the development of the cell wall (intine).

POLLEN WALL

The pollen wall is often compared to mammalian skin and hence it is at times known as sporoderm, its main function being protection against desiccation and mechanical injury. The pollen wall in angiosperms and gymnosperms is most durable and resistant due to its chemical constituent in the form of sporopollenin. It is a complex chemical substance, which withstands physical and chemical reactions including strong acids.

DEVELOPMENT OF THE POLLEN GRAIN WALL

As early as 1911, Rudolph Beer noted the sculpturing elements present on pollen while they were still within the pollen mother cell wall. The timing of the appearance of the sculpturing elements related to the appearance of fibrils that extended from the nucleus to the cell wall while the sculpture formed. Recent studies have elucidated the following stages of pollen grain wall formation.

As soon as the cells of the tetrad are defined and separated by a callose wall, a cellulose layer forms between the plasmalemma of the pollen and the cellose. The cellulose completely surrounds the pollen grain wall except where apertures develop. At the site of apertures, an area of endoplasmic reticulumn moves to the plasmalemma, which in turn lies against the callose tetrad wall. The association of the endoplasmic reticulumn with the aperture area prevents cellulose from forming over areas with which it is associated. The apertures are formed at the three areas where each grain comes in contact with the other grains in the tetrad. In the cellulose layer, probacula or procolumellae form, which are composed of lipoprotein. Thus, before sporopollenin begins to be deposited, the aperturation and columellae patterns are established.

After the probacula form, extensions of their bases extend and connect the probacula to form the Nexine 1 (foot layer). At this time sporopollenin is deposited in the probacula. This is called the primexine stage. Once the pollen grain is released from the tetrad, the pollen grain expands and the primexine stretches and becomes thinner. As the primexine stretches more sporopollenin is deposited. It is at this stage that the tectum and foot layer become developed **(Fig. 4.11)**.

The nexine 2 (endexine) and the cellulose intine begins slightly before the separation of pollen from the tetrad. Nexine 2 is built up as sporopollenin lamellae are produced at the plasma membrane. The origins of sporopollenin differ in different parts of the pollen wall. The primexine and nexine 2-sporopollenin probably comes inside the haploid spore. After

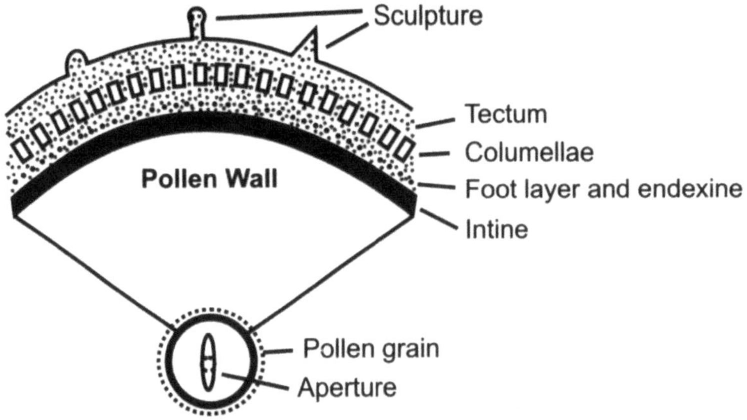

Fig. 4.11 Pollen wall stratification.

the pollen tetrad has divided, when the callose wall breaks up the source of sporopollenin is thought to be with the tapetum, which surrounds the pollen sac. Within the cells of the tapetum, one finds Ubish bodies, which are believed to be carotenoid precursors of sporopollenin. As the tapetum disintegrates, lipids and proteins that form the pollenkitt are released.

CHEMICAL COMPOSITION OF POLLEN WALL

It has been experimentally proved that the exine is exceedingly hard and resistant. It can be treated by strong acids or bases without being destroyed, or heated up to almost +300°C .

Oxidation processes and certain biological organisms (Phycomycetes, Actinomycetes and Bacteria) may corrode and destroy the exine. Pollen and spores originally embedded in sediment or peat, maintain their structure ±intact throughout the geological ages (from Palaeozoic and onwards). It is still possible to study their morphology and use them to date various geological strata. Chemically seen the exine is composed of sporopollenin, a substance resembling lignin and formed through polymerization of hydrocarbons, carotenoids and/or carotenoid esters (pine pollen: C_{90} H_{158} O_{44}). Certain recent green algae, e.g. *Chlorella*, and fossil ones from Devonian appear coated with sporopollenin substances. In ferns an extra-exinous layer (perine) often occurs. The spore wall (including the outer layer, perine) of fern and moss-spores *resembles* that of pollen as to resistance, whilst that of fungal spores is chemically different.

POLLEN WALL STRATIFICATION

Basically the pollen wall is divided into two layers. The inner layer is called as intine and the outer layer is referred as exine **(Fig. 4.12)**. The intine though less resistant to acids is an essential layer composed of cellulose and pectins. It forms the wall of the growing pollen tube and storage functions. Exine is more resistant, hard, sclerodermatous and highly variable layer in its structural characters. Functionally it is a nonessential layer, but most important for pollen morphology on account of its variability. Erdtman preferred to distinguish a pollen wall into the inner layer intine and outer layer as sclerine composed of the inner layer exine and the outer layer perine, which is thin, membranous, less resistant layer present prominently in some moss and pteridophyte spores. Perine is destroyed during acetolyses. Perine is absent in most of the angiospermous pollen and hence sclerine is actually synonymous with exine **(Table 4.1a)**.

An important part of Sclerine is Sexine (S stands for Sculpture) which is actually Sculptured exine. It is further divided into inner Endexine and an outer ectosexine. The inner non-sculptured part of the exine is termed as

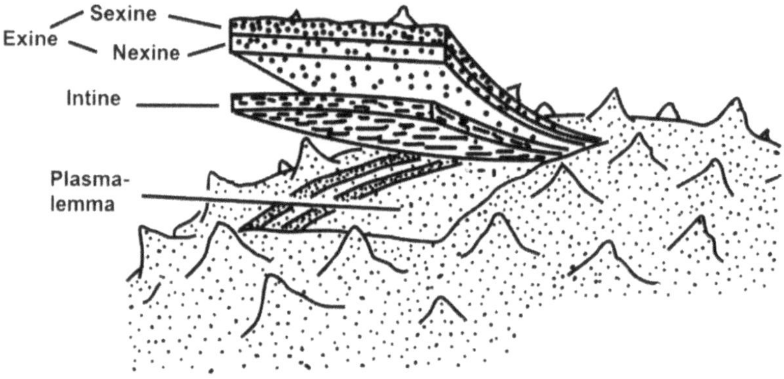

Fig. 4.12 Generalized structure of pollen wall showing different layers.

Table 4.1a *Sporoderm stratification, Erdtman 1952*

<table>
<tr><td rowspan="6">SPORODERM</td><td rowspan="5">SCLERINE</td><td colspan="3">PERINE</td><td rowspan="3">SCULPTINE</td></tr>
<tr><td rowspan="4">EXINE</td><td rowspan="2">SEXINE</td><td>ECTOSEXINE</td></tr>
<tr><td>ENDOSEXINE</td></tr>
<tr><td rowspan="2">NEXINE</td><td>ECTONEXINE</td><td rowspan="2">NEXINE</td></tr>
<tr><td>ENDONEXINE</td></tr>
<tr><td colspan="5">INTINE</td></tr>
</table>

Sexine = from S in Sculptured exine
Nexine = from N in Non-sculptured exine.

Nexine (N stands for non-sculptured part of exine). Nexine is divided into outer thick, not very refractive layer known as ectonexine, and inner; more refractive layer endonexine. Erdtman differentiated one more innermost layer of nexine thus classifying nexine into Nexine 1, Nexine 2 and Nexine 3 from inside to outside **(Figs. 4.13 and 4.14)**.

Further classification of exine along with its morphological characteristic features is discussed under a separate pollen morphological category namely exine stratification. In a way pollen morphology is essentially exine morphology. Exine is often compared with fingerprints on account of its highly variable characters which are of great diagnostic value.

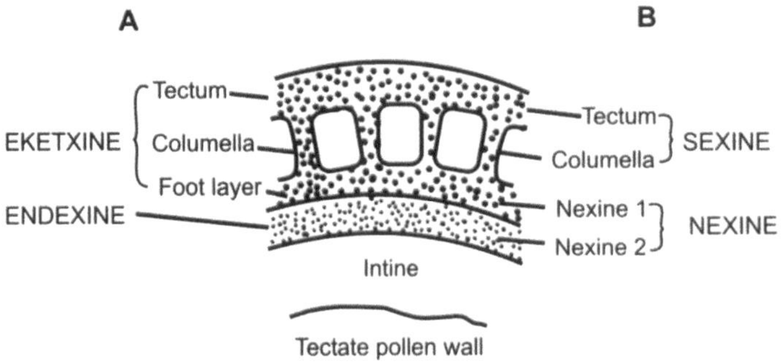

Fig. 4.13 Structure of tectate pollen wall: Pollen wall with tectum which is formed due to fusion of sculpture elements of sexine such as columellae and forming a roof over endexine.

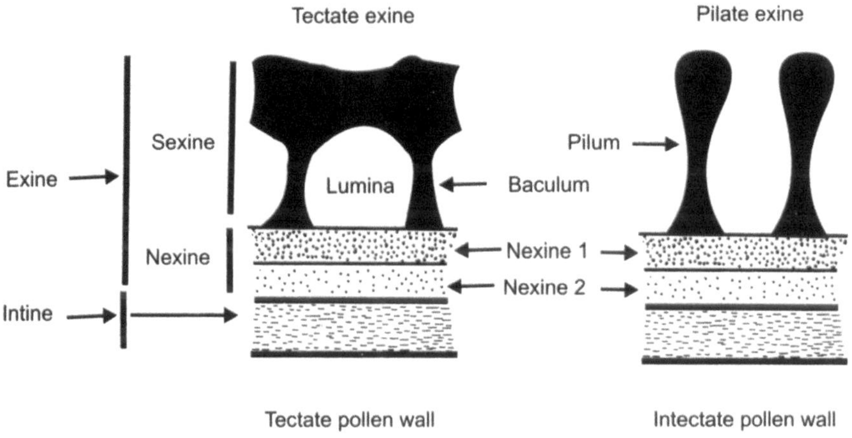

Fig. 4.14 Exine stratification to show tectate and Intectate (pilate) types.

THE STRUCTURE OF EXINE OF POLLEN WALL

Among the flowering plants one of the basic elements of the sexine appears to consist of small drumstick shaped rods (pila), projecting at right angles from the endexine surface.

Each pilum has a head (caput) supported by a rod like baculum the capita forming the upper part of sexine (ektosexine) The bacula form the lower part of sexine (endosexine). If the capita amalgamate or hypothetically a layer of some sort should be formed on the top of pila, coalescing with or enveloping the capita while leaving the bacula free, a tegillum (tectum or small roof) is formed. The layer below the tectum is referred as Infratectum, which may be granular, columellar alveolar or structureless as shown in **(Fig. 4.15)**.

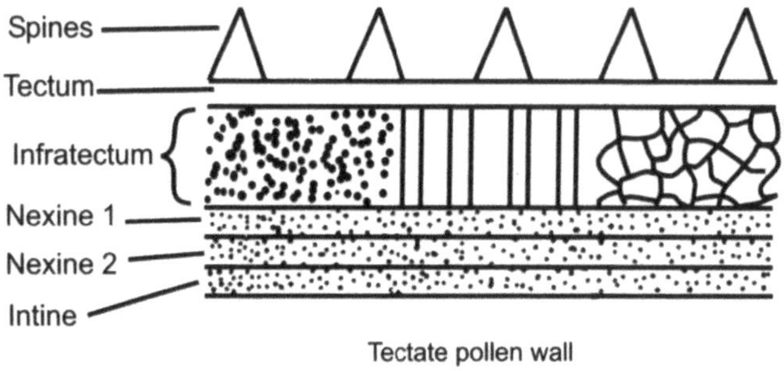

Fig. 4.15 Tectate pollen wall showing details of Infratectum: pl. Infratecta, adj. Infratectate, a general term for the layer beneath the tectum, which may be granular, columellar, alveolar or structureless.

The exine shows morphological features (comparable to fingerprints) that are of high diagnostic value. In simple pollen, the wall is composed of only two layers an inner uniformly structured endexine and a single layered outer ektexine that may have external sculpture or may be smooth.

The tectate (roofed) wall is made up of three or more layers: the endexine, columellae or granules fused at their apices to form a 'roof', and sculptured elements deposited on this layer. In contrast, in some pollen grains the wall will be without tectum and is referred as intectate or atectate, wherein the sculpture elements such as baculae or columellae, etc., are free from each other without forming a roof over endexine **(Fig. 4.16-B)**. In addition, several illustrations showing the structure of Tectate, Semitectate and Intectate pollen walls with suitable examples have been incorporated in **Figs. 4.16-A-a to 4.16-A-o.**

L – O ANALYSIS OF THE POLLEN WALL

It is common knowledge that the structure of a pollen wall under the microscope looks different if the adjustment of microscope focus, particularly the distance from the mounted pollen and objective lens of the microscope is changed. Under a different focus the same pollen wall feature exhibits

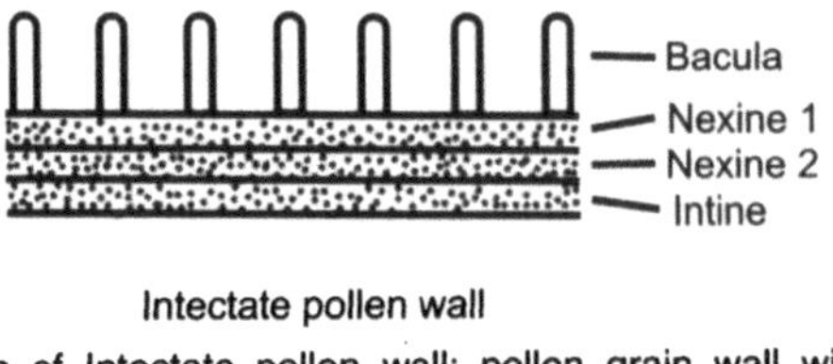

Fig. 4.16B Structure of Intectate pollen wall: pollen grain wall without a tectum. The sculpture elements of sexine are free from each other without forming a roof over endexine, e.g. *Viscum* (Loranthaceae), *Ilex* (Araliaceae).

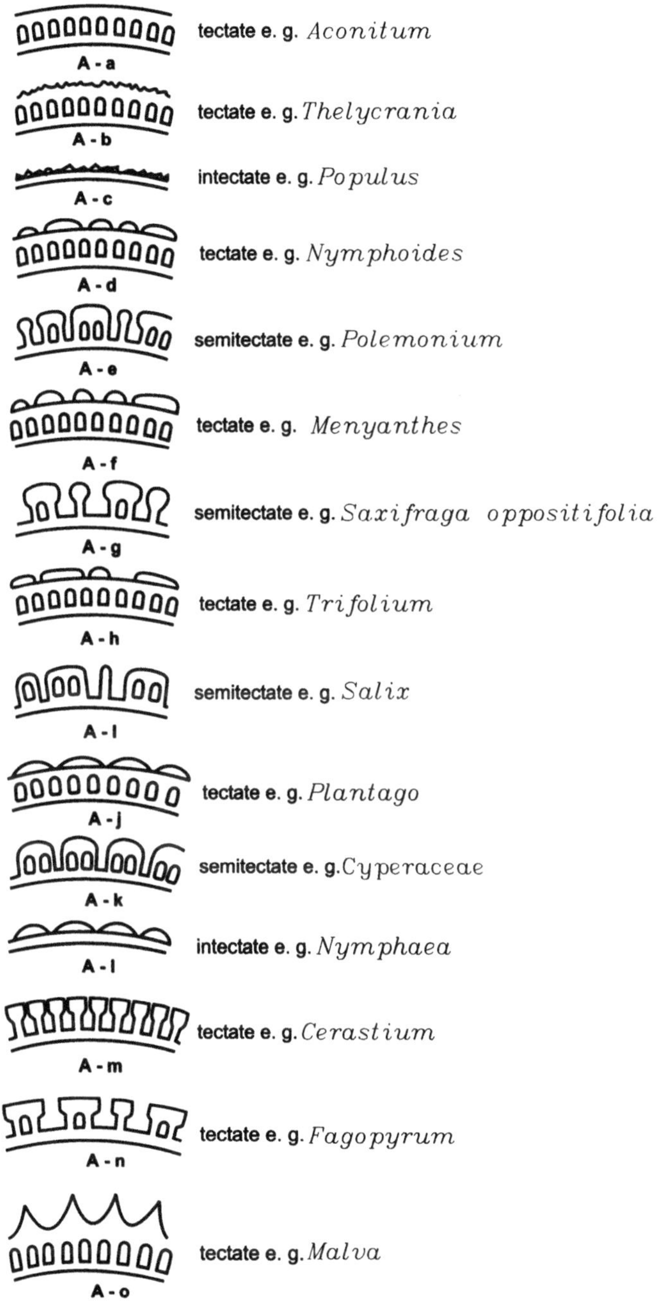

Fig. 4.16 A-a-o Structure of Tectate, Semitectate and Intectate pollen walls with suitable examples.

different patterns such as bright or dark. This analysis is referred as L – O analysis (adapted from H. Welcker 1885 for diatoms) where L stands for Lux or light or bright and O stands for obscuritas or darkness (Erdtman, 1956). In glycerine-jelly mounted pollen, elements projecting from exine are bright first and dark then, holes (perforations) appear dark first and then bright. The L–O analysis should preferably be done under oil immersion. By using this analysis routinely, the pollen wall morphology can be properly described. This can be explained with the help of the following example and illustration of L–O analysis of a sexine of pollen wall, which is described as punctitegillate, spiniferous.

A sexine like that shown in **Fig. 4.16-c** (E: a) exhibits different patterns at different adjustments of the microscope. At high adjustment, small white islands produced by the spinules are seen (E: b: 1). When focusing at a slightly lower level the same islands turn dark (E: b: 2-3). At medium adjustment very small dark islands appear, produced by the puncta in the tegillum (E: b: 4). They later become very bright (E: b: 5). At a lower adjustment numerous small white islands, caused by the baculae appear and later likewise turn dark (E: b: 6, 7). These patterns are referred to as S, T and I Patterns respectively (S: supra tegillar, T: tegillar, I: infra tegillar).

EXINE SCULPTURE

In the pollen grains most endexine layers are uniform in structure and are undifferentiated. The ektexine (sexine) usually is not uniform structurally; differentiation often develops in the form of irregularities on the exterior, which are known as sculpture patterns. Sections of the pollen wall illustrating the common types of sexine sculpture elements and sculpture

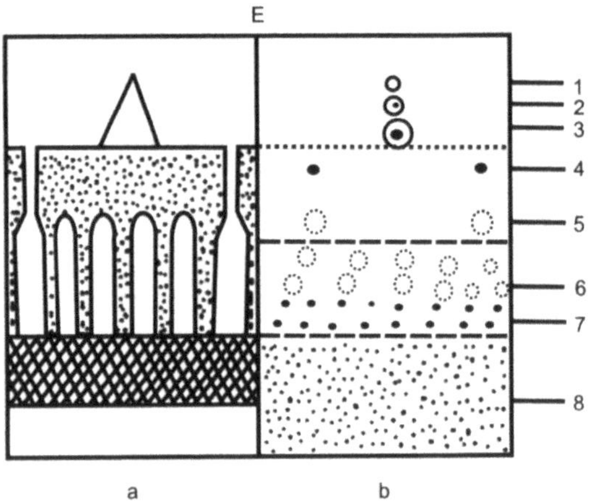

Fig. 4.16c L-O analysis of pollen wall.

patterns are depicted in **Figs. 4.17-a-e and Figs. 4.18-a-l** respectively. Combinations and intergradations of these forms are common.

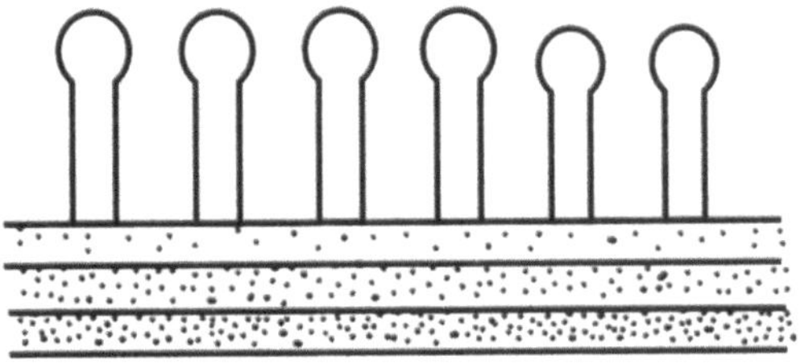

Fig. 4.17a Pila: sing. Pilum, adj. Pilate They are sexine elements usually standing directly on the nexine, consisting of a rod-like base or stock (columella) and a swollen apical part head (caput).

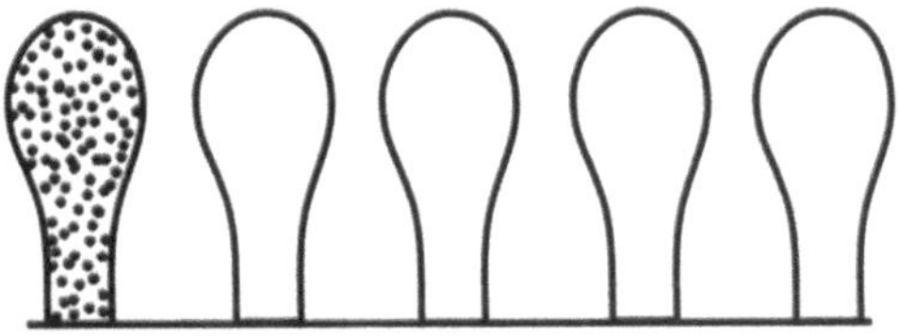

Fig. 4.17b Clavae: sing. clava, adj. Clavate Club shaped elements of the sexine/ektexine that are higher than 1 μm, with diameter smaller than height and thicker at the apex than the base.

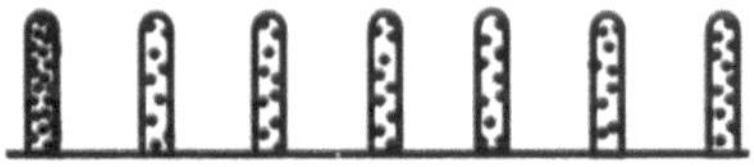

Fig. 4.17c Bacula; sing. Baculum, adj. Baculate Cylindrical, free standing exine element more than 1 μm in length and less than this in diameter. e.g. *Raistria sactora*.

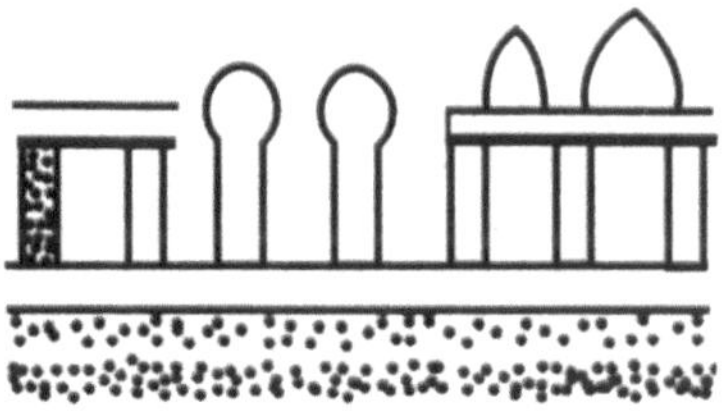

Fig. 4.17d Columellae; sing. Columella adj. Columellate. Rod-like elements of the sexine/ektexine, either supporting a tectum or a caput. The difference between a baculum and a columella in current usage is, that a baculum is always free standing element of sculpturing, whereas a columella is part of the structure.

Fig. 4.17e Gemmae: sing. Gemma adj. Gemmate. Sexine elements which are constricted at base, higher than 1 μm, and that has approximately same width as their length.

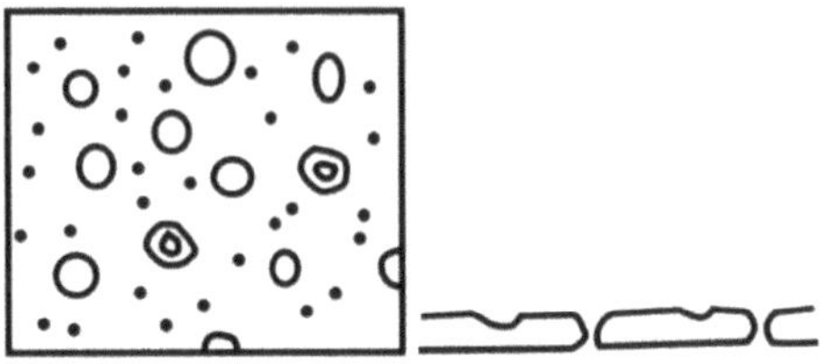

Fig. 4.18a Foveolate Syn pitted: Wall with foveolae (sing. Foveola). This wall ornamentation consists of more or less rounded depressions or lumina more than 1 μm in diameter. The distance between foveole is greater than their breadth.

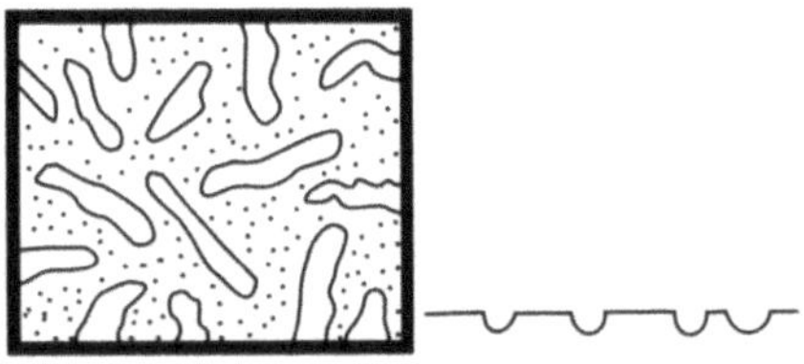

Fig. 4.18b Fossulate syn. Canaliculate: Wall with fossulae (sing. Fossula). Wall ornamentation consisting of an elongated, irregular grooves in the surface.

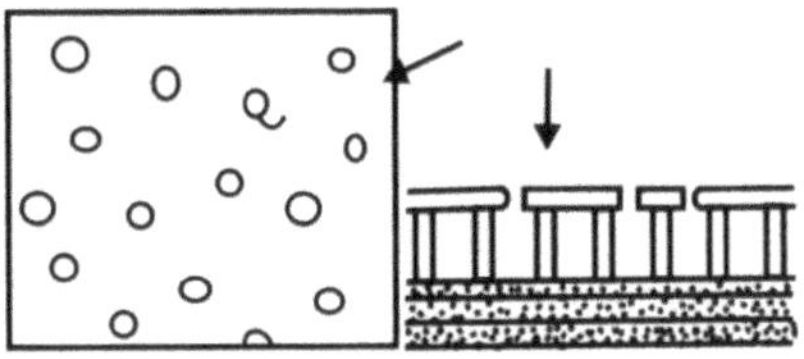

Fig. 4.18c Punctate: (sing. Punctum, pl. puncta). It is characterized by rounded or elongated tectal perforations, less than 1 μm in length or diameter.

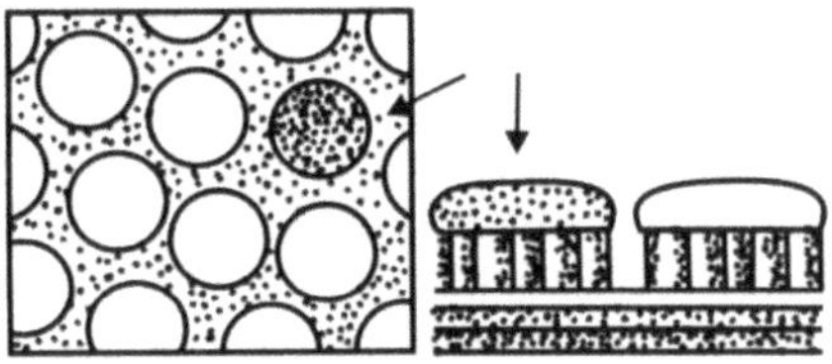

Fig. 4.18d Areolate: A feature of pollen wall ornamentation (sculpturing) in which the sexine/ektexine is composed of circular or polygonal areas separated by grooves which form a negative reticulum. e.g. *Phyllanthus* (Euphorbiaceae).

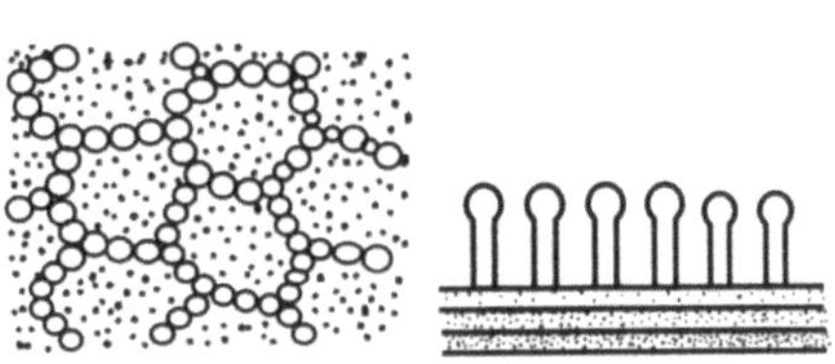

Fig. 4.18e Retipilate: Pollen wall with reticulum formed by rows of pila instead of muri e.g. *Callitriche* (Callitrichaceae).

Fig. 4.18f Microreticulate: A reticulum which is characterized by the consistent presence of one porate aperture in each lumen. e.g. *Froelichia floridana* (Amaranthaceae).

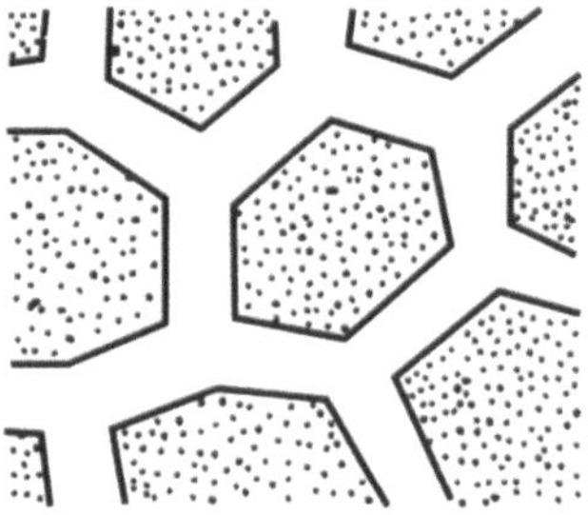

Fig. 4.18g Brochate (Reticulate): (sing. Brochus, pl. brochi). It consists of one lumen of a reticulum and half of the width of the adjacent muri. It is a type of reticulate pattern. It resembles (sing. reticulum pl. reticula) a network – like pattern consisting of lumina or other spaces wider than 1 µm bordered by elements narrower than the lumina.

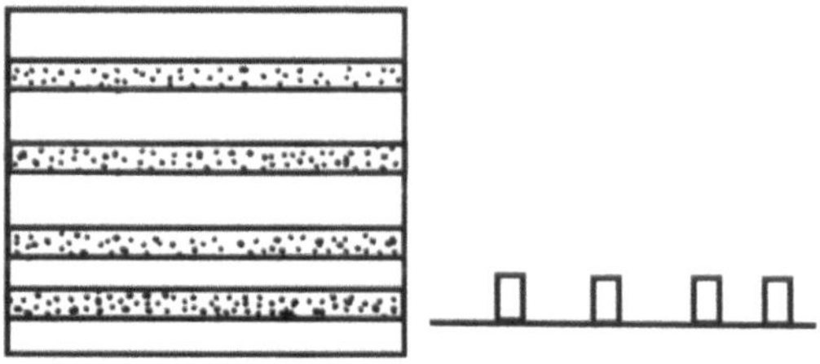

Fig. 4.18h Cicatricose: It has more or less parallel ridges that are narrower than the spaces separating them.

Fig. 4.18i Striate: The pollen wall with narrow ridges forming the muri in striate pattern.

Fig. 4.18j Corrugate syn. Rugulate: It has a wrinkled texture. Here the elevated projections are ridges with regular or irregular radial humps or bulges.

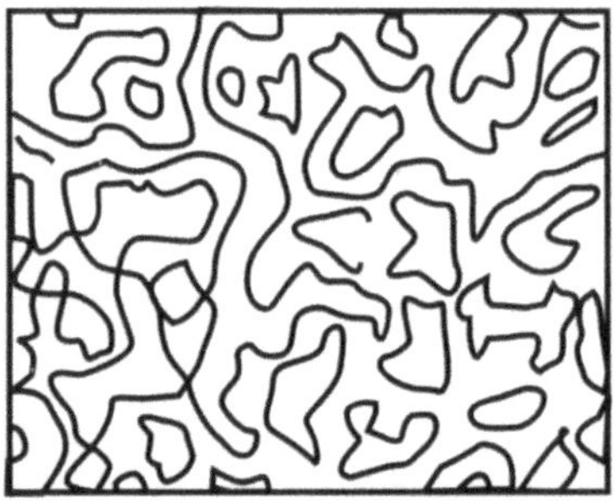

Fig. 4.18k Ornate: A reticulate ornamentation consisting of broad, curved muri and lumina that are often anastomosing. e.g. *Ceiba aesculifolia* (Bombacaceae).

Fig. 4.18l Croton pattern: A characteristic type of ornamentation comprising rings of five or six (sometimes more) raised, often triangular, sexine elements arranged around a circular area usually formed by capitate columellae (pila). e.g. *Croton, Jatropha* (Euphorbiaceae).

The sculpture elements and patterns found on Palaeozoic (and recent) spores are essentially the same as those found on pollen grains. In an attempt to bring out uniformity in the description of Palaeozoic spores an international committee was organized. In the committee's report, recommendations regarding terminology were made. These terms do not entirely coincide with the sculptural terms applied to pollen grains. The sculpture types applicable to both pollen and spores are described below and some of them are illustrated in surface view and optical section **(Figs. 4.19 – a-t)**.

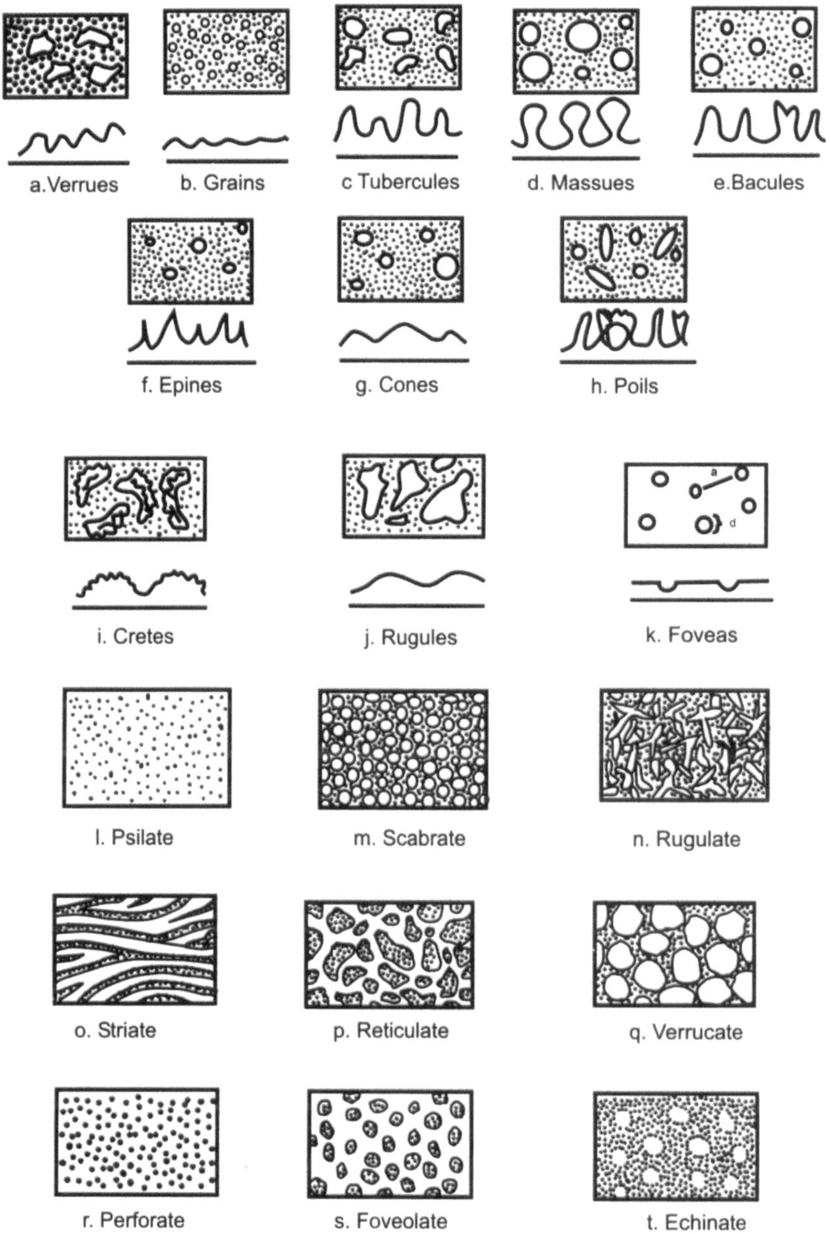

Fig. 4.19a-t Showing various sculptural elements and patterns in surface view and optical section. a. Verrues b. Grains c. Tubercules d. Massues e. Bacules f. Epines g. Cones h. Poils i. Cretes j. Rugules k. Foveas l. Psilate m. Scabrate n. Rugulate o. Striate p. Reticulate q. Verrucate r. Perforate s. Foveolate t. Echinate.

Chagrenate: From Turkish, Gaghri – untanned leather. Shiny smooth, translucent.

Psilate: From Greek *psilos* – smooth. A mat like surface they may even possess small depressions less than 1 µm in diameter as in *Betula.*

Scabrate: From Latin *scaber* – flecked. Very small projections less than 1 µm in diameter, as in *Artemisia* or *Quercus.*

Verrucate: From Latin *verruca* – a wart. A rounded projection broader than high which is not constricted at base.

Gemmate: from Latin *gemma* – a bud. A spherical projection with constricted base, comparable to gemma cups of Bryophytes and in *Juniperus.*

Baculate: From Latin *baculum* – a rod. Rod like projections whose greatest diameter is less than the height as in *Nymphea.*

Pilate: With pilar elements, which have blunt drumstick-shaped projections with a swollen top and short shaft.

Clavate: From Latin *Clava* – a club. A club like projection whose apex is greater in diameter than the base as in *Ilex.*

Echinate: From Latin *echinatus* – prickly, spiny projections tapering from base to a sharp point, as in *Taraxacum* (Dandelion). Spinules are defined as processes < 3 µm in length and differentiated from spines which are the processes > 3 µm in length.

Rugulate: From Latin *ruga* – a wrinkle. Irregularly distributed elongated ridges or wrinkles, as in *Nymphoides peltatum* or *Selaginella deusa* megaspore.

Striate: From Latin *Stria* – striped. More or less parallel ridges that are wider than the spaces separating them as in *Menyanthes.*

Cicatricose: From Latin *cicatrix* – a scar. More or less parallel ridges that are narrower than the spaces separating them, as in *Anemia phyllitidis.*

Reticulate: From Latin *rete* – a net. Ridges forming a net. The walls (Muri) narrower than the diameter of the spaces (Lumina); as in negative surface of *Illicium.*

Corrugate: From Latin *Corrugatus* – wrinkled. Projections are ridges with regular or irregular radial humps or bulges, as in *Riccia natans.*

Foveolate: From Latin *Fovea* – a pit. Circular pits whose diameter is less than the diameter of the ridges separating them, as in *Lycopodium phyllanthum* and *Ledum.*

Fossulate: From Latin *fossa* – a ditch. Possessing a negative reticulum. Surface with irregular grooves, as in *Pteris tripartia* or *Ledrum palustre.*

Canaliculate: From Latin *canalis* – a channel, canal. More or less parallel cavities or channels narrower than the bastions between them as in *Anemia simii.*

In the sexine, the internal rod like bacules constitute structure while any external elements located upon the tectum constitute sculpturing. Thus, a distinction is made in such pollen between internal or structural components and external supratectal sculpturing elements.

THE STRUCTURE AND SCULPTURE OF POLLEN EXINES

The terminology applied to the various subdivisions of the exine is somewhat varied as shown in the table from Erdtman, Kupriojova, Faegri **(Table 4.1b).**

Table 4.1b *Comparison of exine classification terminology used by different palynologists (Van Campo 1966).*

TOMSOVIC 1960			KUPRIJANOVA 1956			FAEGRI 1956			ERDTMAN 1952		
EXINA	SEXINA	EKTOSEXINA	EXINA	EKTEXINA	SEXINA	EXINE	EKTEXINE	TECTUM	EXINE	SEXINE	EKTOSEXINE
		ENDOSEXINA						COLUMELLAE			ENDOSEXINE
		BASOSEXINA						FOOT LAYER		NEXINE	EKTONEXINE
		MESEXINA			MEXINA						MESONEXINE
		NEXINA			ENDEXINA			ENDEXINA			ENDONEXINE

PRINCIPLE TYPES OF APERTURES AND TYPES OF POLLEN

Apertures

Apertures are generally thin walled areas in the outer pollen wall or exine through which the pollen tube usually emerges at the time of germination. The pollen grains may be aperturate or nonaperturate or inaperturate (Atreme, Gr.:Treme=aperture), nonaperturate spores are known as alete.

The function of the exine, like that of the skin of animals, is the protection of the organism from injury by external agencies such as excessive desiccation, destruction by light and mechanical injury.

In addition, exine performs two other functions, which are the special properties of pollen grains. These are to provide for the emergence of the pollen tube at the time of fertilization and to accommodate changes in volume as the grain takes up and gives off moisture, which it readily does in response to the ever changing humidity of the atmosphere that it encounters as a free living organism. This function is also called Harmomegathic function, a mechanism accommodating changes in the volume of the semi-rigid exine. Normally, for the emergence of pollen tubes germ pores are provided and furrows are used for volume change accommodation. Thus, the apertures may be referred as safety valves at volume changes. On the basis of shape the apertures are termed as furrows (colpi) or pores **(Figs. 4.20 and 4.21)**.

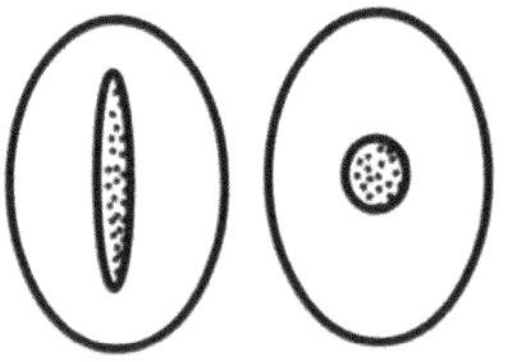

Fig. 4.20 Monoaperturate pollen: pollen grain with single colpus (Monocolpate) or single pore (Monoporate).

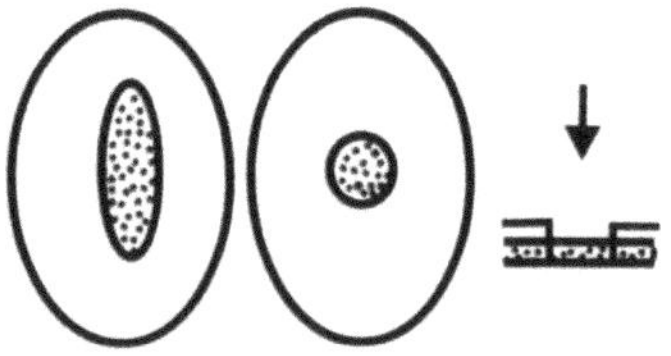

Fig. 4.21 Colpus membrane and pore membrane.

The apertures, both colpi and pores are always covered with a membrane which may be smooth or differently ornamented. However, these features are extremely small which cannot be seen clearly in a light microscope but are obvious if observed under Scanning Electron Microscope (SEM). The membranes are psilate (smooth), granulate (provided with granules), crustate (thickly beset with coarse granules), etc.

Germinal Furrows

The form and character of the germinal furrows are normally strictly phyletic, tending to be constant throughout families and other large groups. However, their number and arrangement are controlled to a large extent by their internal environment, which in turn is determined by the number and arrangement of the grains as they are formed from the pollen mother cells, and may therefore be different even in grains from the same anther. If the germ pores are absent in pollen, the furrow functions directly as germ pore as an exit for the emergence of pollen tube. External factors also affect the germinal furrows e.g. wind pollination may completely banish the germinal furrows.

Furrows

These are similar to boat shaped depressions in the exine, the ektexine being much reduced but with the endexine less affected (Furrow or Colpus, Pl: Colpi; adj: Colpate). Unlike pores, furrows do not completely penetrate the exine and consequently, if pores are lacking, the pollen tube must force its growth through the covering membrane of the exine at germination.

Characters of the Furrows

The characters of the furrows, such as shape, size, distribution, number are of the greatest value in the identification and classification of plants.

Pore

The other type of aperture is called a pore, which is typically an isodiametric, circular opening. This aperture is short with length and breadth ratio of 2:1 or 1:2. If the pores are elongated, the ends are rounded as in *Hibiscus*. Exine structure at the region of pores is of various types, which has been depicted in **Figs. 4.22 – A – D.**

Fig. 4.22 Exine structure at the region of pore (Redrawn from Moore and Webb 1978). A. Pore with a costa (thickening of the nexine) e.g. Poaceae B. Pore where sexine separates from nexine to form a vestibulum e.g. *Betula* C. Pore with an annulus formed by a slight thickening of the sexine. The nexine is absent in the vicinity of the pore e.g. *Myrica*. D. Pore with an operculum (thickening of the middle of the aperture membrane) and an annulus (Sexine thickening) e.g. *Plantago lanceolata*.

There is great variation in the number and position of pores in the pollen wall. The pollen, depending on the number of pores are referred as monoporate, triporate or polyporate **(Fig. 4.23)** pores are variously situated in the pollen and accordingly are described as angulapertureate **(Fig. 4.24)**

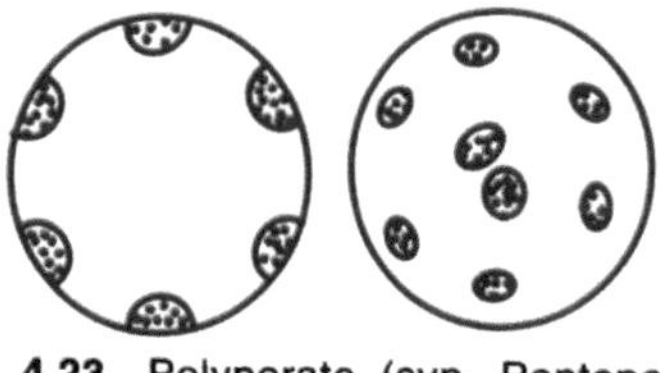

Fig. 4.23 Polyporate (syn. Pantoporate and zonoporate) pollen showing many pores distributed throughout the pollen wall. E.g. *Amaranthus spinosus*.

Fig. 4.24 Angulapertureate: pollen with equatorial aperture situated at the angles of the outline in polar view as in *Corylus* (Betulaceae).

if pores are situated at the angles of the pollen wall; lobate if situated in the indentations between the lobes of pollen wall **(Fig. 4.25)**. In pollen of *Corylus* a lens shaped structure is observed beneath the pore **(Fig. 4.26)**. In *Betula* a prominent protruding thickening of the exine exists around a pore, which is termed as aspis **(Fig. 4.27)**.

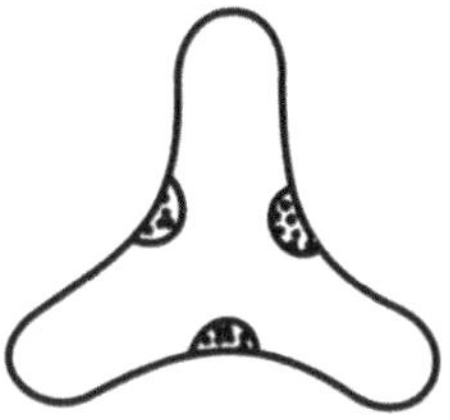

Fig. 4.25 Lobate pollen grain with equatorial aperture in the indentations between the lobes.

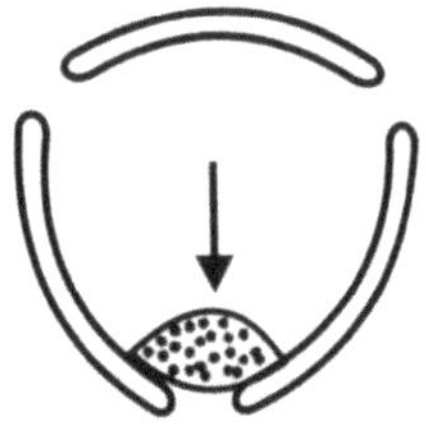

Fig. 4.26 Pollen showing oncus, a lens – shaped structure, which occurs beneath the apertures e.g. *Corylus* (Betulaceae).

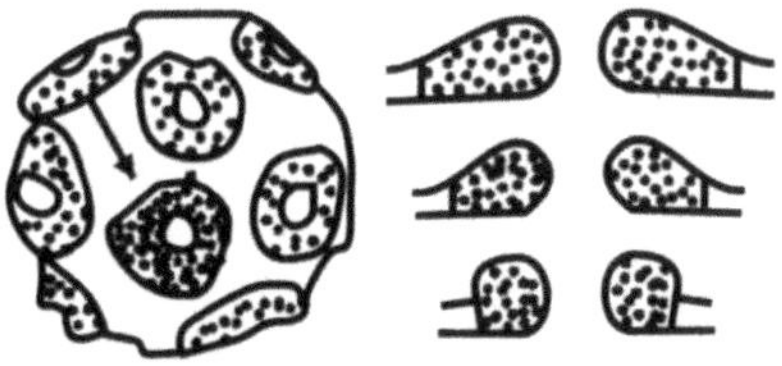

Fig. 4.27 Aspis: A prominently protruding thickening of the exine around a pore as in *Betula*, (Betulaceae) *Dorstenia* (Moraceae).

The grains with pores are referred as porate. In certain cases the pore is closed by an operculum (lid) and it is often surrounded by an annular area (annulus) as in pollen of some grasses **(Figs. 4.28 a, b, c)**.

A thickening of measurable bulk and clearly defined of an aperture membrane is called operculum. In some grains the central or median part of the aperture membrane shows distinct granules. These heavily covered thickened membranes of apertures are called opercula (sing. operculum) and are distinct from the rest of the exine surrounded by a very thin membrane as in Nymphaeceae. Beautiful opereula are found in the furrows of many Asteraceae pollen grains.

In many grasses, opercula are easily lost when the grain expands with moisture. In simple terms a covering of the apertural membrane is called an operculum. In grass pollen, the operculum appears as a half-raised lid, which may get lost altogether in the airborne pollen. The operculum of *Plantago lanceolata* seems surrounded by a moat looking like a doughnut.

Polyporate pollen grains occur in Ranunculaceae, Cactaceae, Phytolacaceae, Nyctaginaceae, Portulacaceae, Basselaceae, Polygonaceae, Amaranthaceae and Chenopodiaceae. It is generally believed that polyporate grains appear to be derived from tricolpate pollen (Savitsky and Martynyuk 2001).

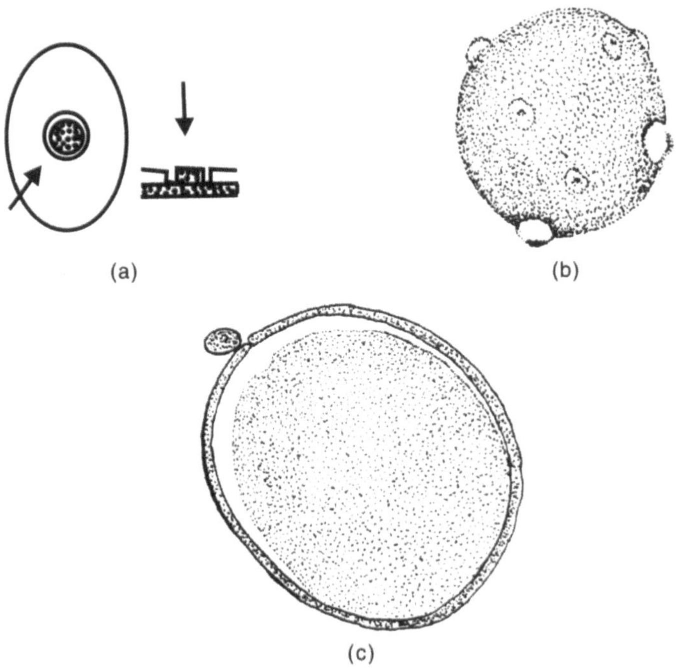

Fig. 4.28 Pollen with operculum (pl. Opercula, adj. Operculate). A distinctly delimited sexine / ektexine structure which covers part of an ectoaperture and which is completely isolated from the rest of the sexine.

Sulcus

It is a longitudinal aperture in the distal face of the grain and usually with the distal pole in its centre, with the ratio of length to breadth > 2 as in *Lilium bulbiferum* and *L. longiflorum*. The pollen in this case is referred as sulcate **(Fig. 4.29)**. Pollen of some palms are characterized by having two sulci. They are referred as disulcate pollen as in *Chamaerops humilis* **(Fig. 4.30-a)** and in *Metroxylon* **(Fig. 4.30-b)**.

Furrows or colpi

The difference between furrows and pores seems to be purely morphological. Phylogenetically, furrows are apparently the primitive form; pores must have developed later by the contraction of furrows. Transitional forms also occur.

Monocolpate grains generally occur in Monocotyledons such as in *Lilium longiflorum* as well as many primitive dicotyledons and in Ranales. Majority of the Dicotyledons produce pollen with three equidistant furrows referred

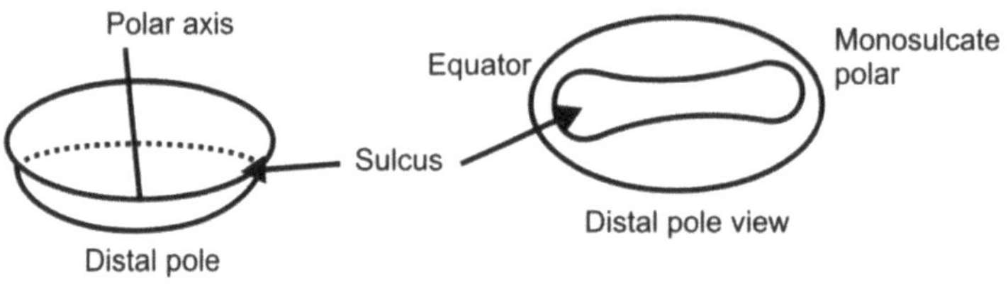

Fig. 4.29 Sulcate pollen in equatorial and polar view.

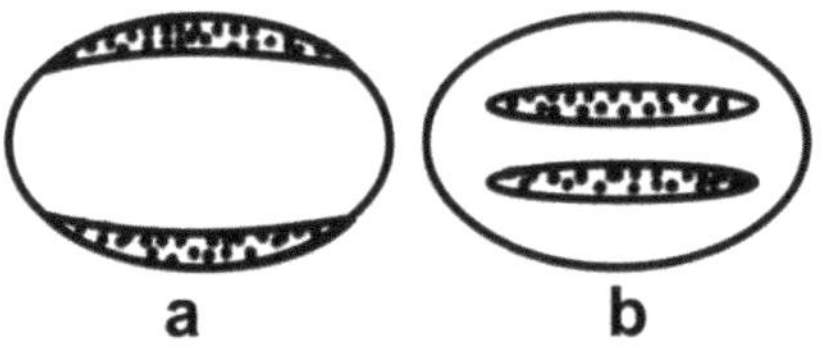

Fig. 4.30 Disulcate pollen: a. distal disulcate pollen in *Chamaerops humilis* (Palmae) b. equatorial disulcate pollen in *Metroxylon salomonense* (Palmae).

as tricolpate **(Fig. 4.31)** as in *Prunus mume* (Japanese apricot). In *Pardoglossum* (Boraginaceae) pollen with simple apertures like colpi and pores alternate with each other round the equator and are referred as Poro-colpate pollen **(Fig. 4.32)**. In certain pollen a structure called margo exists. This is an area of exine around an ectocolpus that is differentiated from the remainder of the sexine **(Fig. 4.33)**. There appears to be an evolutionary trend with regard to the position of aperture. Proximal apertures are present in mosses, ferns and pteridosperms. There seems to be a shift of apertures to distal position in monocots, which culminates in assuming equatorial position in dicotyledons.

Fig. 4.31 Polar view (amb) of tricolpate pollen in which polar axis is directed forwards the observer.

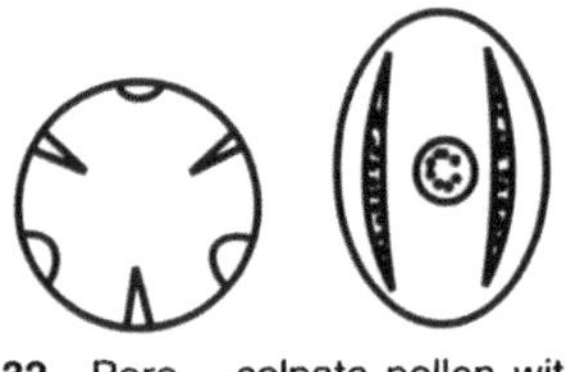

Fig. 4.32 Poro – colpate pollen with colpi alternating with pores round the equator e.g. *Pardoglossum* (Boraginaceae).

With reference to the apertures the pollen surface may be divided into two, apocolpium, which is a region at the pole of colpate grains delimited by lines connecting the apices of the colpi in radially symmetrical grains **(Fig. 4.34)** and apoporium **(Fig. 4.35)** which is the corresponding term for porate grains. The areas between colpi are called mesocolpia **(Fig.4.36-a)**, and between pores, mesoporia **(Fig. 4.36-b)**.

Fig. 4.33 Margo: (Pl. Margines, adj. Marginate). An area of exine around an ectocolpus that is differentiated from the remainder of the sexine, either in ornamentation or by dfference in thickness.

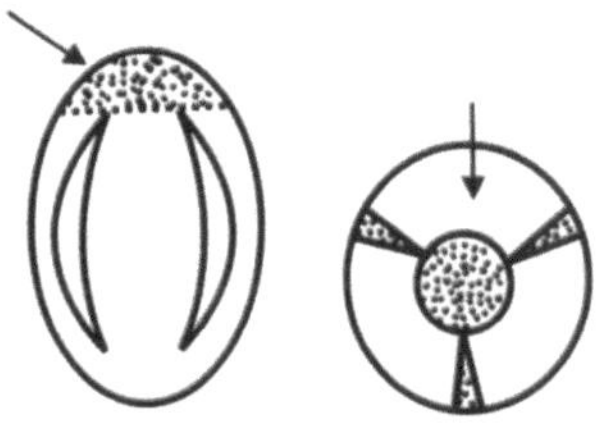

Fig. 4.34 Apocolpium: A region at the pole of a zonocolpate pollen grain delimited by lines connecting the apices of the colpi.

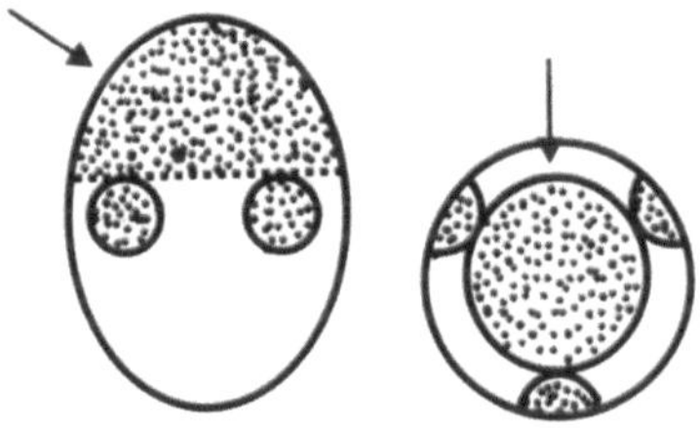

Fig. 4.35 Apoporium: An area at the pole of a zonoporate pollen grain that is delimited by a line connecting the borders of the pores.

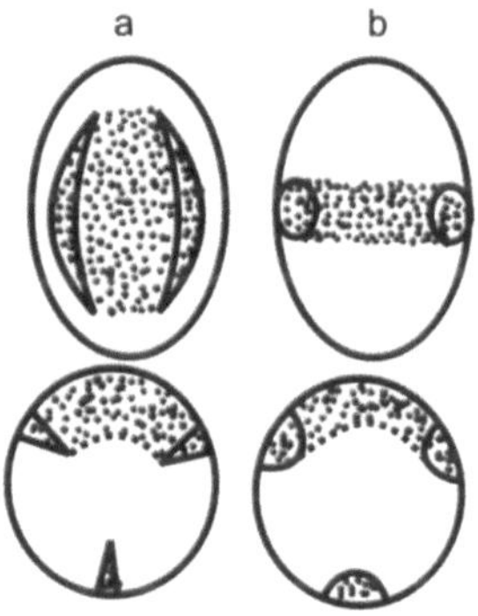

Fig. 4.36 Mesocolpium a and Mesoporium b. The area of pollen grain surface delimited by lines between the apices of adjacent colpi or the margins of adjacent pores.

The apertures described above are referred as simple apertures. In contrast, the compound apertures are a combination of furrows and pores or pores within pores.

COMPOUND APERTURES OR COMPOSITE APERTURES

They consist of an outer part (marginal) either colpoid or poroid and an inner central part usually referred as an OS = Latin meaning mouth (sing: os; plural: ora). Like a mouth it can vary in outline from circular to transversely or longitudinally elongated. The ora are referred as lalongate (**Fig. 4.37**) if they are elongated laterally at right angles to the colpus or lolongate (**Fig. 4.38**) if elongated meridionally or longitudinally.

Fig. 4.37 Lalongate: Compound aperture with shape of a transversely elongated endoaperture as in *Filipendua* (Rosaceae).

Fig. 4.38 Lolongate: It describes the shape of a longitudinally elongated endoaperture as in *Rumex* spp. (Polygonaceae).

Colporate grains contain a furrow with a pore in the centre for example *Artemisia* (**Fig. 4.39**). In case of colporate grains (colp-orate) the outer (ectoporium) and inner (endoporium) faces are incongruent. Frequently, the pore in a colporate grain extends beyond the furrow margin. In some cases, the pores may extend laterally (parallel to the equatorial plane and coalesce, thereby being too long to be termed as pores). These equatorially elongated pores may be called 'transverse furrows' as seen in pollen of Umbelliferae.

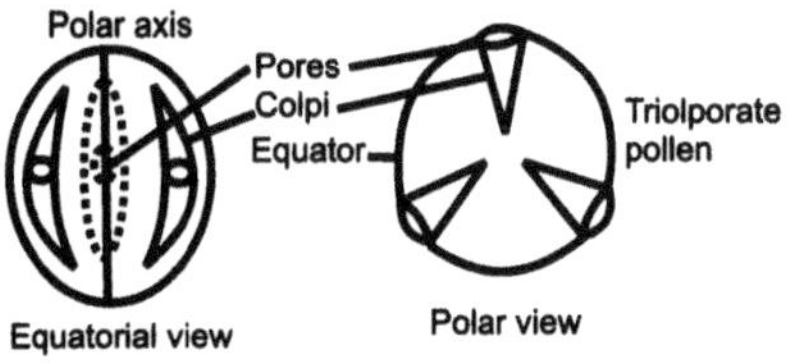

Fig. 4.39 Tricolporate pollen in equatorial and polar view.

Pororate

If the grains have an outer pore (ectoporium) and an inner one (endoporium – os) the grains are called pororate (**Fig. 4.40**). Thus, colporate and pororate pollen with compound apertures differ in ora or endoaperture (**Figs. 4.41 and 4.42**).

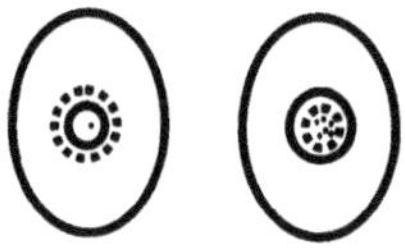

Fig. 4.40 Pororate pollen.

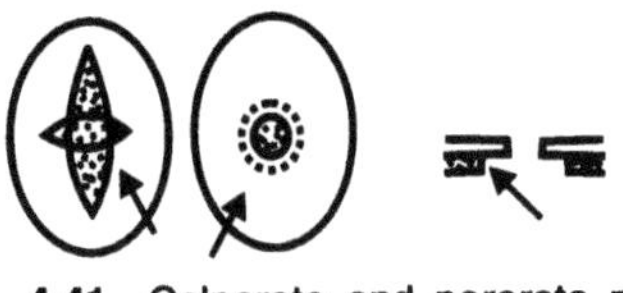

Fig. 4.41 Colporate and pororate pollen showing Os (pl. Ora adj. Orate) syn: endoaperture, meaning mouth.

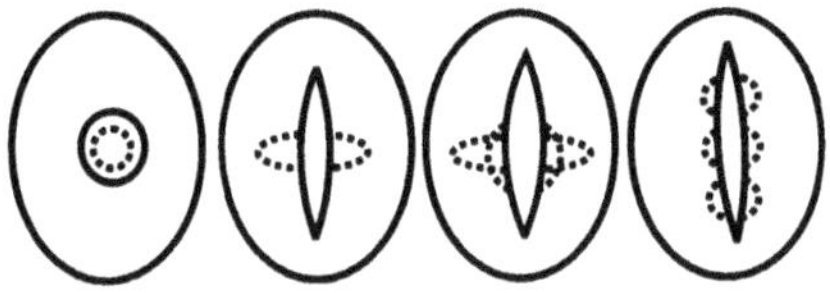

Fig. 4.42 Compound apertures with two or more components that are situated in more than one wall layer.

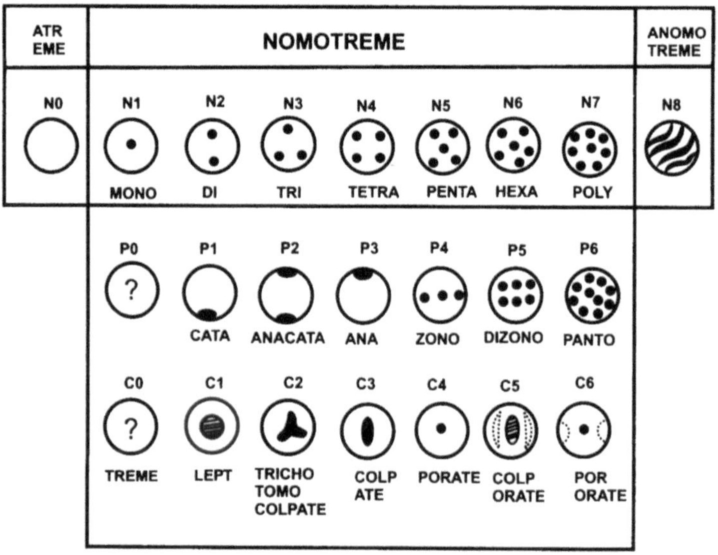

Fig. 4.43 NPC system of classification proposed by Erdtman in 1969.

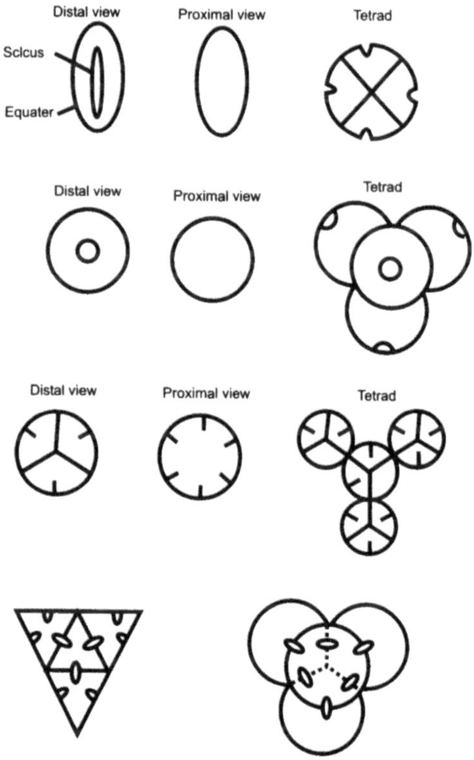

Fig. 4.44 Various types of apertures in distal and proximal views.

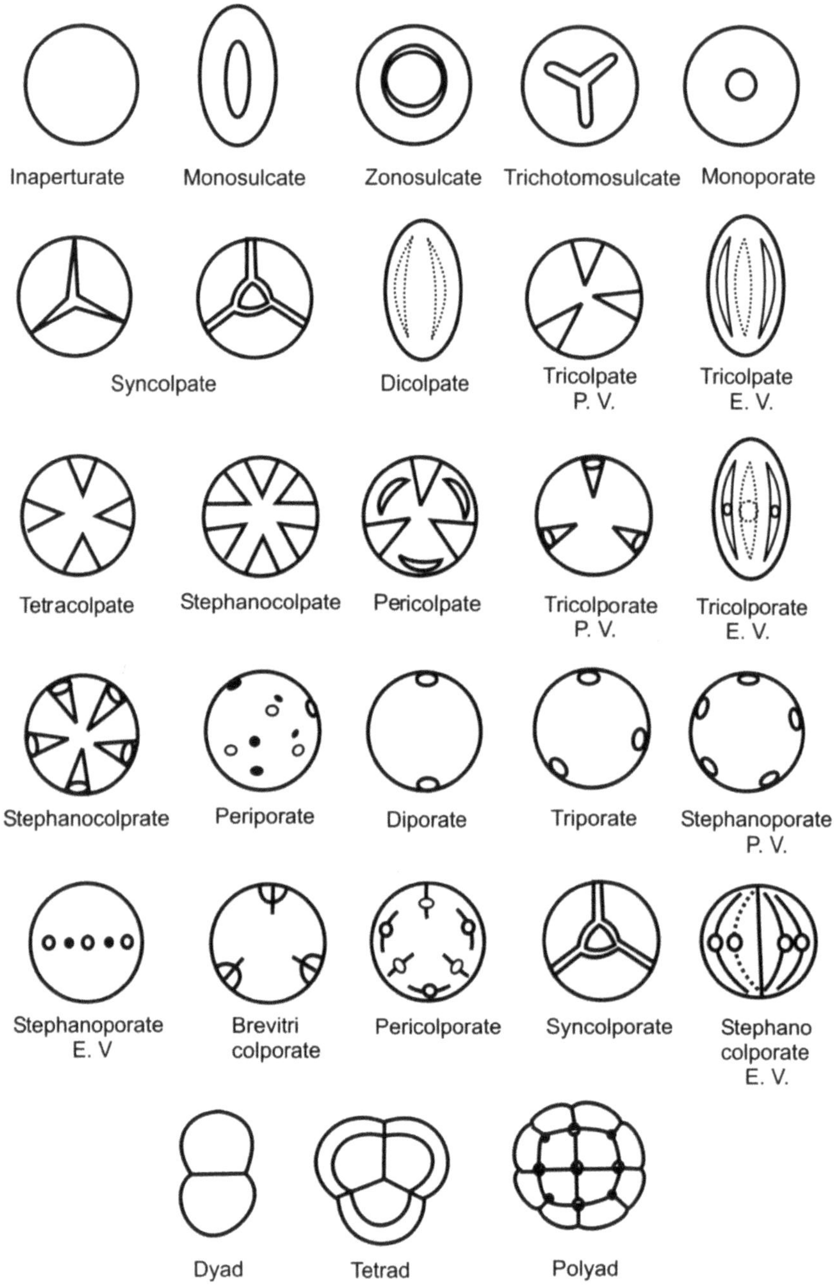

Fig. 4.45 Different types of pollen based on features of apertures.

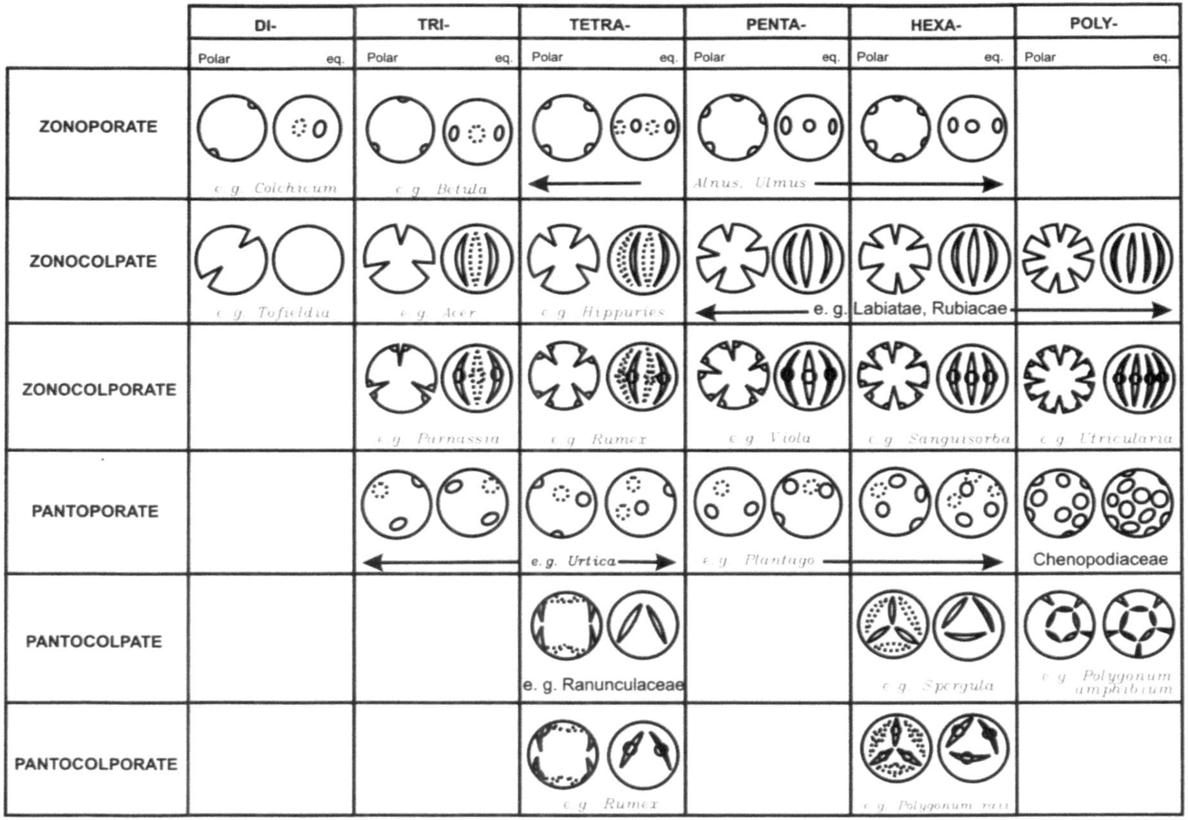

Fig. 4.46 Types of pollen wall apertures.

Table 4.1c Summary of the aperture terminology

Group	*Aperture type*	*Aperture condition*
BRYOPHYTES	Laesura (pl.laesurae; adj.-lete)	monolete, trilete (cata-)
	Leptoma (pl-leptomata; adj- -lept)	-lept (cata-)
	No apertures	alete, inaperturate
PTERIDOPHYTES	Laesura (pl.laesurae; adj.- lete)	monolete, trilete (cata-)
	No apertures	alete, inaperturate
GYMNOSPERMS	Sulcus (pl. sulci; adj. Sulcate)	monosulcate (ana-)
	Leptoma (pl. leptomata; adj.-lept)	–lept (ana)
	No apertures	inaperturate
ANGIOSPERMS	Sulcus (pl. sulci; adj. Sulcate),	monosulcate (ana-)
Monocots	cf.colpus	trichotomosulcate (ana-)
	Sulcus (branched)	meridionosulcate (ring-like)
	Meridionosulcus (syn. Nonasulcus)	
	Sulculus (pl- sulculi; adj. Sulculate),	Sulculate
	cf. colpus:	
	Zonasulculcus	Sulculate (ring-like,
	Ulcus (pl.ulci;adj. Ulcerate), cf. porus	Equatorial)
	Foramen (pl. foramina; akj. Forate),	Ulcerate (ana-)
	cf. panto- porate	Forate (panto-)
	Spiral aperture(s)	spiraperturate
	No aperture(s)	inaperturate
Dicots	Colpus (pl.colpi; adj. Colpate), cf.	colpate (equatorial, panto-)
Simple	Sulcus	trichotomocolpate
Apertures	Colpus (branched)	porate (equatorial, panto-)
	Porus (pl. pori; adj. porate), cf.	colporate
	Ulcus, foramen	
Compound	Colporus pl. colpori; adj.	
Apertures	Colporate), i.e.	
	Colpus + os (endopore)	
	Porus + os (endopore)	pororate
	Spiral aperture(s)	spiraperturate
	No aperture(s)	inaperturate

The apertures may be irregular-shaped or like a spiral. The apertures may be located at the proximal pole (fern-moss-spores), at the distal pole (Gymnosperms, Monocotyledons and primitive Dicotyledons), at the *equator* (Dicotyledons) or scattered ±regularly over the surface that is *panto-position* (Monocotyledons and Dicotyledons). Erdtman and Straka (1961) proposed a classification system for pollen called NPC-system **(Fig. 4.43)** based on number (N), position (P), and character (C) of apertures. The following prefixes may be used: *ana-* (distal), *cata-* (proximal), *Panto-* (spread over the surface).

In order to bring clarity in the understanding of various types of apertures in distal and proximal views and pollen types based on various apertural features, summarized illustrations have been included in **Figs. 4.44, 4.45** and **4.46** respectively.

CHAPTER **5**

Pollen Morphological Description of Some Plants

A: POLLEN MORPHOLOGY OF CERTAIN TROPICAL ANGIOSPERMOUS PLANTS

Tropical angiospermous flora of India, very rich in families, genera and species is highly diversified. Similarly the pollen morphology of these plants shows great variation. It is not possible to describe pollen morphology of many of these plants in this volume. However, the Mimosaceae (Mimosoids) family is chosen for a detailed pollen morphological description for two reasons. The plants belonging to this family are cultivated widely and some of them grow naturally and abundantly in different regions of India and other tropical areas. From the student's point of view the pollen material for detailed pollen morphological study is easily accessible. The pollen of this family vary from monads to polyads. There is tremendous variation in the pollen morphology of these plants. Significant contribution on this theme has been recently made by Cheriyan Panicker (2002).

It is appropriate to mention here that detailed pollen morphology of several plants is discussed in the chapters on aerobiology as these pollen have great relevance to allergic manifestations. However, pollen illustrations of some tropical angiospermous plants, particularly abundant in the Western Ghats have been incorporated in this chapter **(Figs. 5.1 a-p)**.

Pollen Morphology of Mimosoids

The following pollen morphological descriptions of some mimosoids are accompanied by **Fig. 5.2 a-r**

***Acacia auriculiformis* A. cunn.**

It is commonly known as Australian Acacia or golden shower plant. In India the plant was introduced from tropical Australia. The trees without spines grow up to 10-15 m high. The branchlets are glabrous and phyllodes are elliptic, falcate, thick coriaceous, glabrous and palminerved. The trees produce drooping spikes of golden yellow flowers. Stamens are numerous

a. *Anona squamosa*, Anonaceae – pollen grains are in Dyads with foveolate – striate sculpturing.

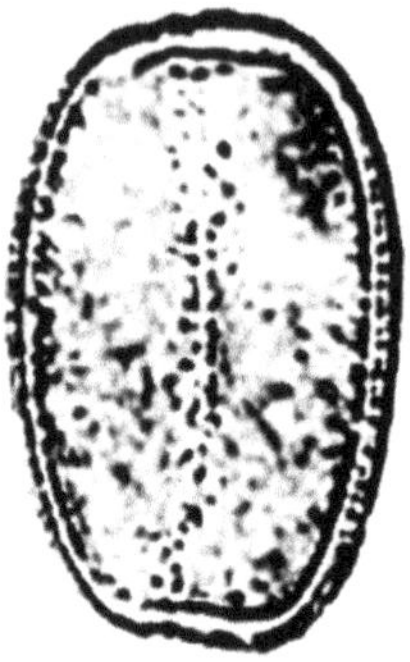

b. *Glosiosa superba – Liliceae* – Peroblate monocolpate pollen.

c. *Gloriosa superba – Liliaceae* – pollen exine showing reticulate sculpturing.

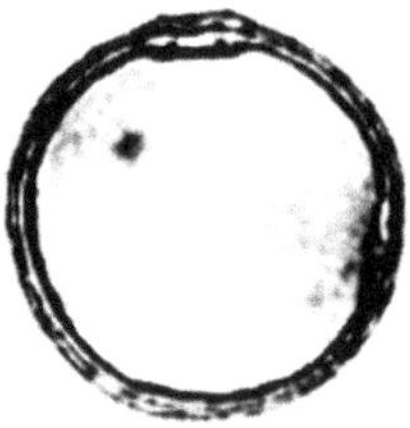

d. *Dendocalamus strictus* – Poaceae – pollen spherical, monoporate, oblate spheroidal spherical, monoporate, costate.

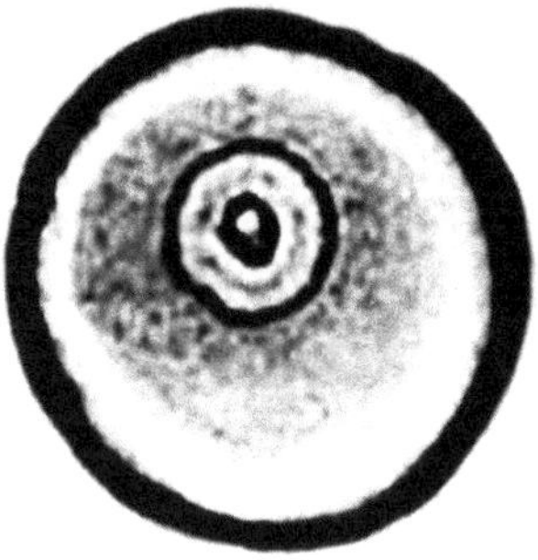

e. *Dendocalamus strictum*, Poaceae – pollen showing single pore and psilate sculpturing.

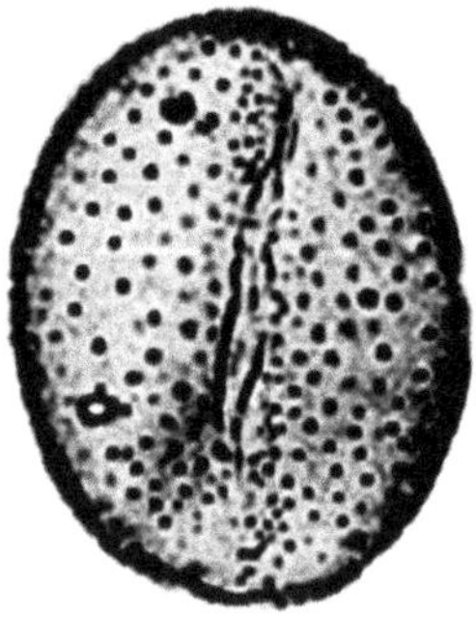

f. *Crinum pretense*, Amaryllidaceae pollen grain subprolate, dizonocolpate, exine sculpturing echinate.

g. *Crinum pretense*, Amaryllidaceae trizonosyncolpate pollen with echinate sculpturing.

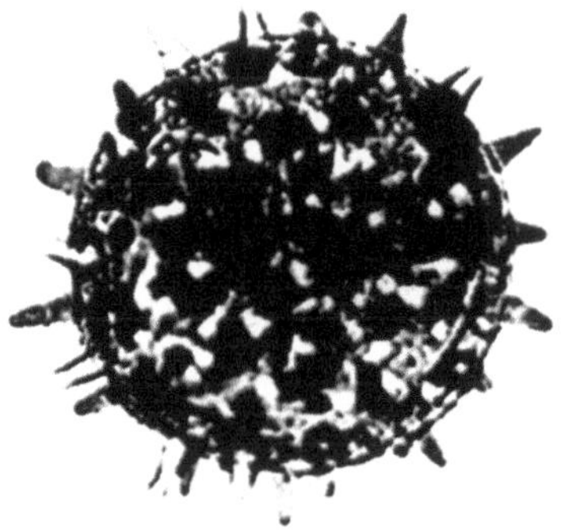

h. Urena lobota – Malvaceae – pantoporate pollen with circular to oblate pori, echinate sculpturing, echinae lanceolate in outline, tips obtuse.

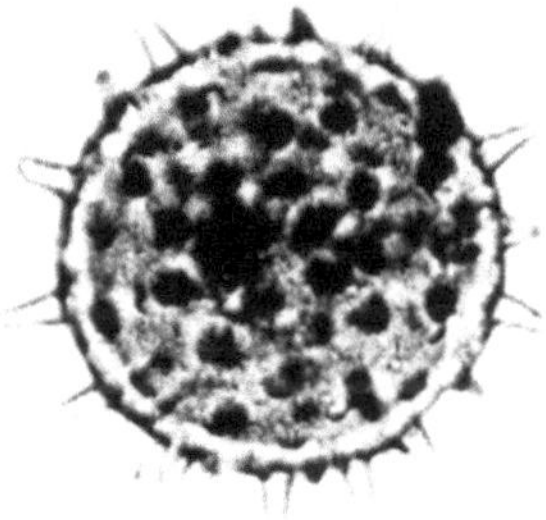

i. *Hibiscus hirtus* – Malvaceae – pollen oblate spheroidal, pantoporate, exine with echinate sculpturing, with convex bulbous protuberances.

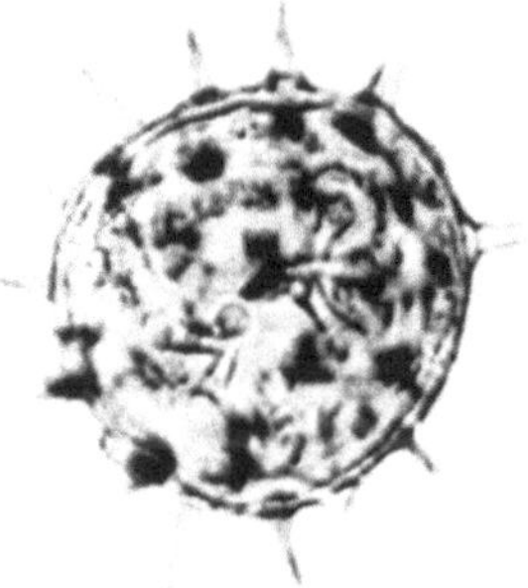

j. *Abutilon indicum*, Malvaceae – pollen oblate spheroidal, trizonocolporate, colpi very faint, narrowly elliptic. Exine with echinate sculpturing, echinae with acute tips and swollen bases.

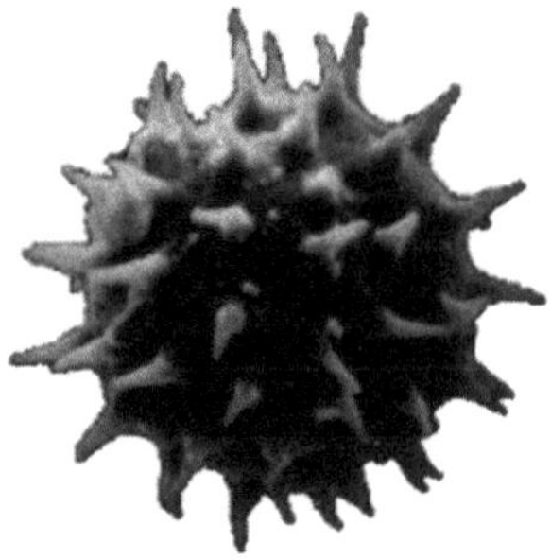

k. *Hibiscus lobatus* – Malvaceae – pantoporate pollen with circular to oblate pori, echinate sculpturing, echinae lanceolate in outline, tips obtuse.

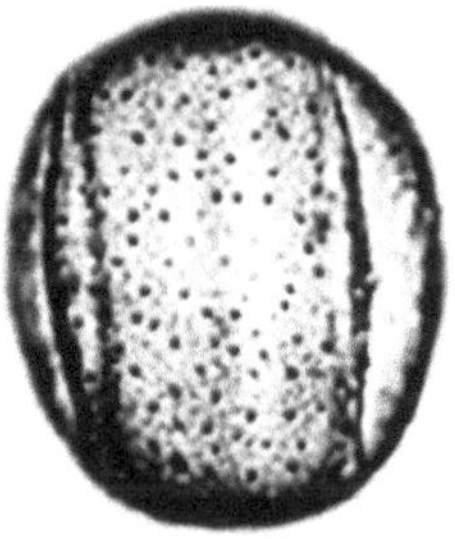

l. *Cereus dayamii* – Cactaceae – Pollen oblate spheroidal, trizonocolpate, with echinate sculpturing.

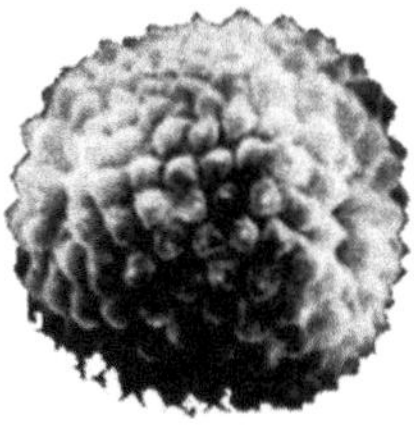

m. *Xanthium strumarium* – Asteraceae – enlarged view of exine sculpturing.

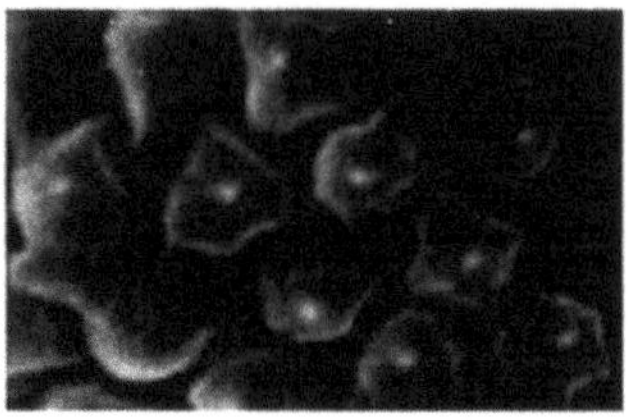

n. *Xanthium strumarium* Asteraceae – pollen grain suboblate, trizonocolporate, Exine 4.25 μm thick, sculpturing scabrate or microechinate, echinae situated on polygonal cushion bases.

o. *Citrullus lanatus* – Cucurbitaceae – prolate pollen, trizonocolporate, exine with reticulate sculpturing, lumina irregularly polygonal, with margo.

p. *Cissampelos pareira*, Menispermaceae – Pollen prolate spheroidal, polar outline triangular, trizonocolporate, colpi narrowly elliptic, operculate. Exine with reticulate sculpturing, lumina irregularly trigonal to variously polygonal.

Fig. 5.1a–p Pollen morphology of tropical angiospermous plants, particularly abundant in the Western Ghats, India.

with 3-3.5 mm long filaments, which are basaly united. The flowering period is from July to February. The mode of pollination is amphiphilous.

It is the second largest pollen producer in the family Mimosaceae. About 120 anthers are produced per flower. The catkin bears 115 flowers with 120 anthers producing 8 polyads containing 16 pollen. Thus, about 128 pollen grains are produced per anther and 15,360 per flower and 17,66,400 per inflorescence.

The pollen grains in this plant are clustered in spherical polyads with 16 grains in each, arranged in a regular pattern that is 8 grains in the centre arranged four in two layers each, one layer directly above the other which are surrounded in turn by 8 peripheral grains. Apertural condition in individual pollen grains is characterized by three pores, which are not so well defined. SEM of these grains shows exine surface with the central square shaped depressions bounded by raised undulate margins. These depressions and margins have low pock-mark depressions **(Figs. 5.2 a, b)**.

The pollen grain is squarish with a reticulate exine surface, packed close to each other like bricks. Polyads vary in diameter from 41.2 to 45.3 μm, whereas the size of individual pollen grains varies from 8-11 μm across.

***Acacia dealbata* link**

Commonly called Silver wattle, consists of unarmed trees 12-15 m in height, with silvery pubescent branchlets bearing bipinnate leaves, which are grown as avenue trees. The plant is native of Northeast Australia, which was introduced and grown at higher altitudes in the Nilgiris. The tree produces flowers in axillary and terminal panicled globular heads 3-4 mm across. Flowers are minute, scented and yellow in colour. The flowers have numerous stamens with 1-1.5 mm long filaments.

The anthers produce polyads with 16 pollen units adhering to each other. The pollen are identical and tetragonal in shape. Pollen grains are triporate with deep rectangular depressions and clearly defined rounded edges. The exine surface is pockmarked and reticulate. Polyad size ranges from 38.12 to 41.36 μm **(Fig. 5.2c)**.

Adenanthera pavonina

The species is a native of India, Sri Lanka, Thailand and China. The trees of this species are known to grow to 8-12 m in height with glabrous branchlets and two pinnate leaves. Small yellow coloured flowers are produced in axillary raceme, which may be 12-20 cm long. The androecium has 10 stamens with free filaments and gland-crested anthers. 16-celled polyads produced in anthers are spherical. Polyads vary in size from 41-43.9 μm and individual pollen grains measure 12.67 to 16.67 μm across. The individual pollen grains in polyad are irregularly arranged. Occasionally polyads get separated into tetrads during acetolysis. The boundary of the grains in polyads is smooth but at its centre, exine materials are deposited regularly giving a rough appearance. The grains are trizonoporate with exine characterized by rugulate sculpturing **(Figs. 5.2-d, e)**.

***Albizia falcataria* Linn**

The plants of this species are natives of Guinea and Solomon isles. They grow up to 15 m high. They bear creamy white flowers in terminal or axillary panicles 15 cm long. The androecium comprises numerous stamens with 1 cm long filaments united to form a 3 mm long staminal tube.

The anther produces 16-celled polyads measuring 66.26 to 70.42 μm in which the grains are arranged in a regular pattern. Individual pollen grains have a cushion like swollen surface without depressions. The central area of the pollen grain wall is marked by distinct rugulate ornamentation, gradually smoothening towards the margins. Pollen grains are 3-zonoporate with sparsely present micropunctae **(Figs. 5.2f, g, h)**.

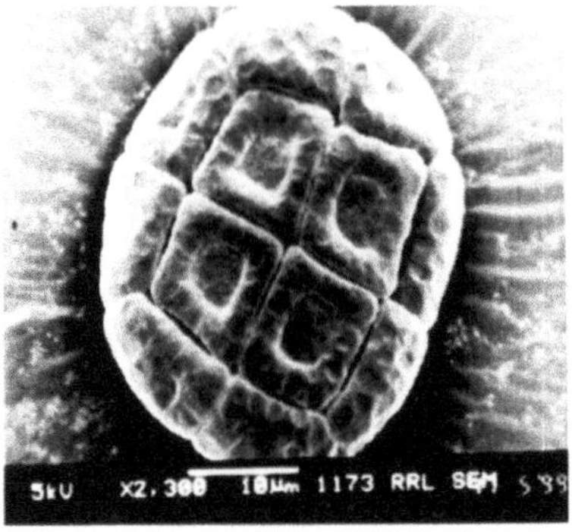

Fig. 5.2a & b *Acacia auriculiformis*
a. SEM of polyad showing surface details

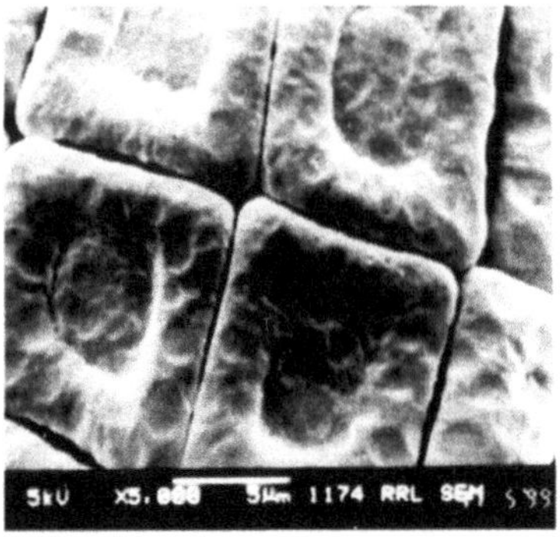

b. Enlarged SEM view of polyad showing exine ornamentation, major and minor surface depressions.

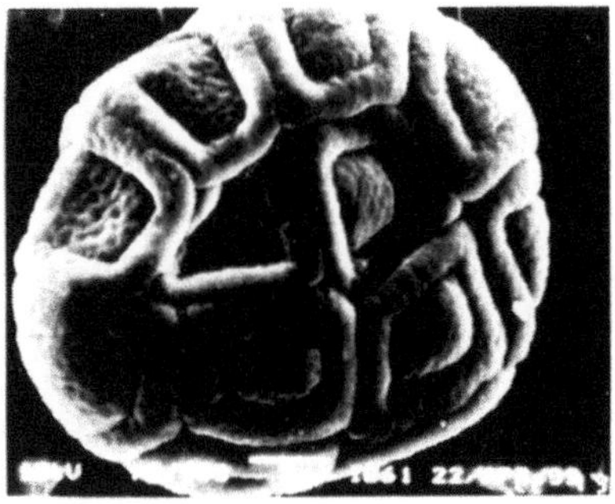

Fig. 5.2c *Acacia dealbata* SEM of polyad showing pollen arrangement.

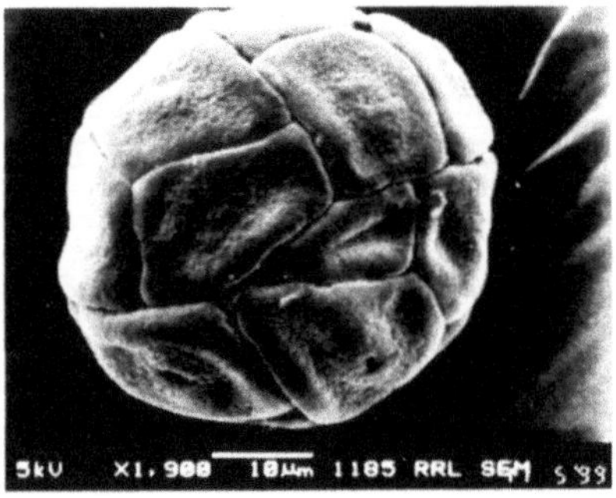

Fig. 5.2d *Adenanthera pavonina* SEM of polyad showing irregular arrangement of pollen.

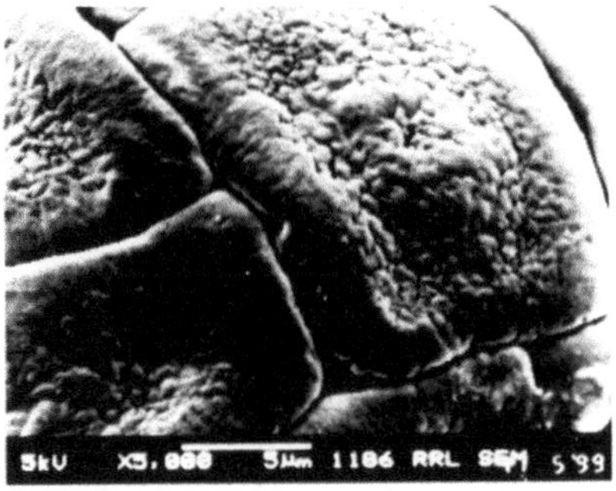

Fig. 5.2e *Adenanthera pavonina* Enlarged SEM view of part of polyad showing the rugulate pattern of exine ornamentation. Broad interspaces of pollen are seen clearly.

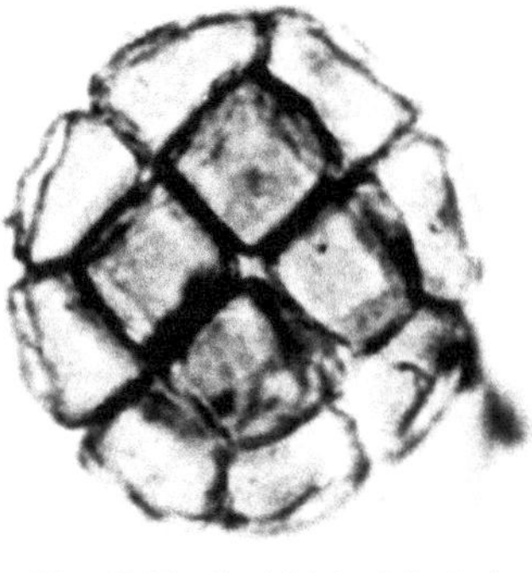

Fig. 5.2f,g,h *Albizia falcataria*
(f) LM photomicrograph of polyad.

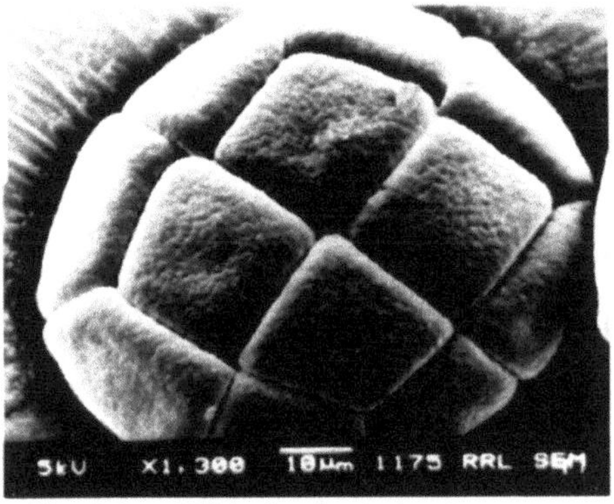

Fig. 5.2g SEM of polyad showing regular rectangular pollen with cushion like surface.

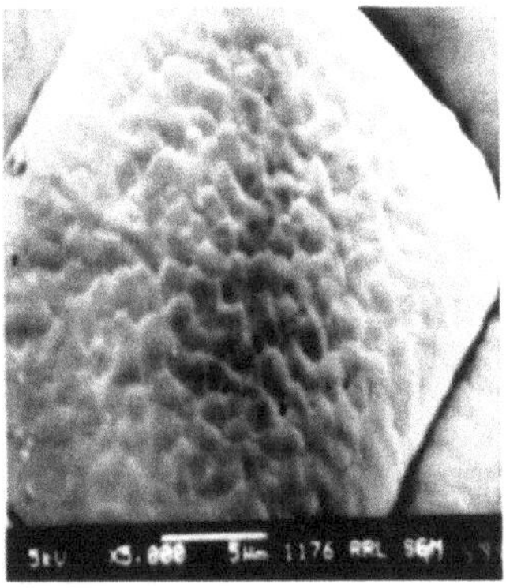

Fig. 5.2h Enlarged SEM of a pollen in polyad showing central rugulate ornamentation, gradually smoothening towards the margins.

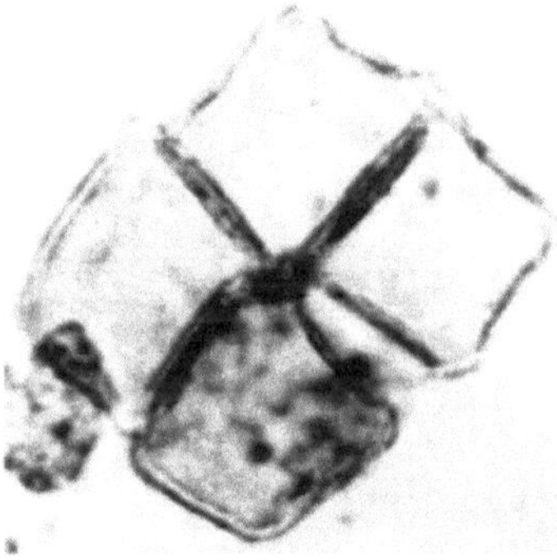

Fig. 5.2i *Albizia lebback* LM of polyad showing a group of four pollen grains separated during acetolysis.

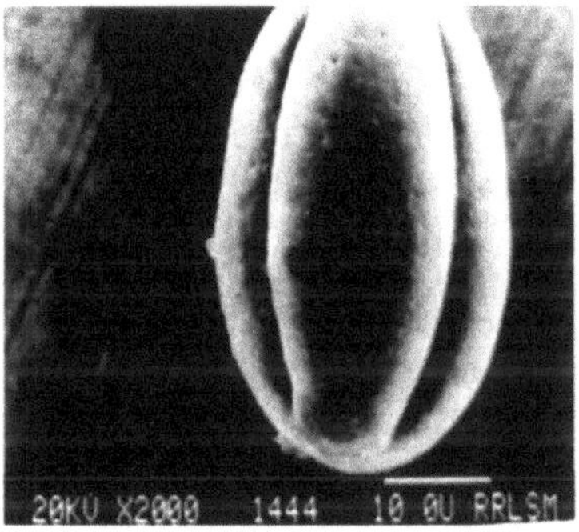

Fig. 2.5j,k,l *Leucaena leucocephala* j. SEM of equatorial view of pollen showing two elongated colpi.

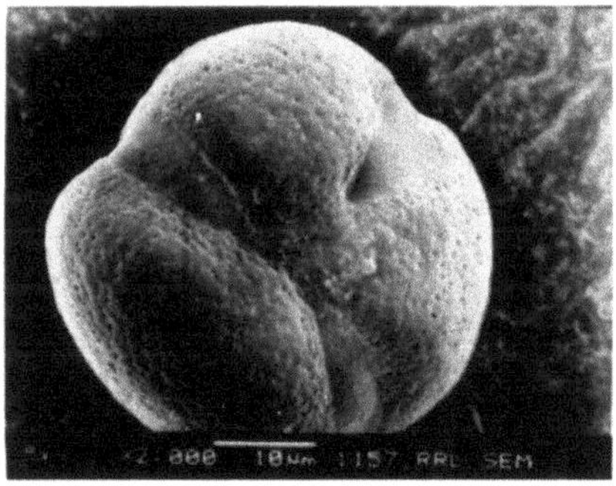

Fig. 5.2k SEM of polar view of pollen showing three broad furrows terminating in rounded ends.

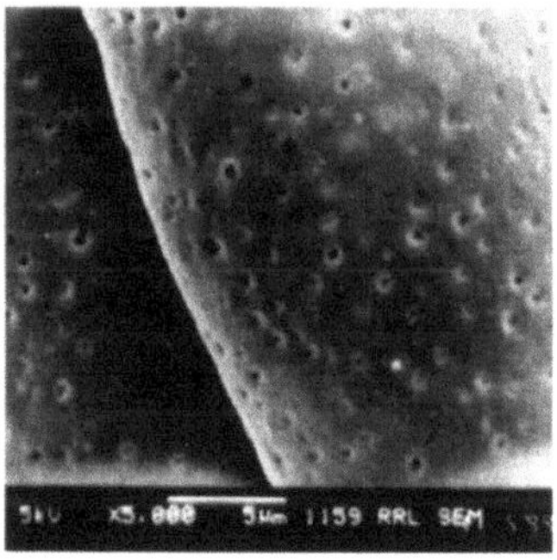

Fig. 5.2l SEM showing scrobiculate exine pattern.

Fig. 5.2m & n *Mimosa invisa* var. *invisa*
m. SEM of pollen tetrad.

Fig. 5.2n Enlarged SEM showing tuberculate exine sculpturing.

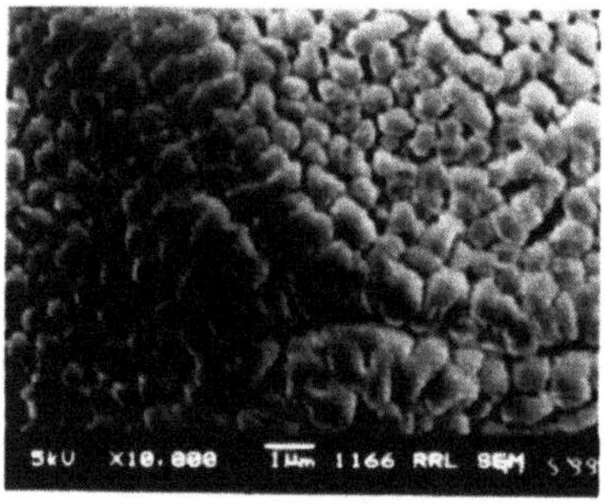

Fig. 5.2o *Mimosa pudica* SEM picture of portion of a tetrad showing compound tubercles fused along the lateral ridges.

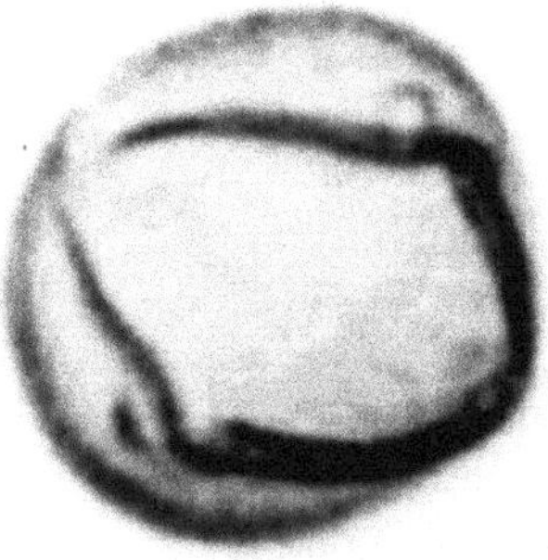

Fig. 5.2p and q *Prosopis juliflora* LM photomicrograph of equatorial view showing details of colpus.

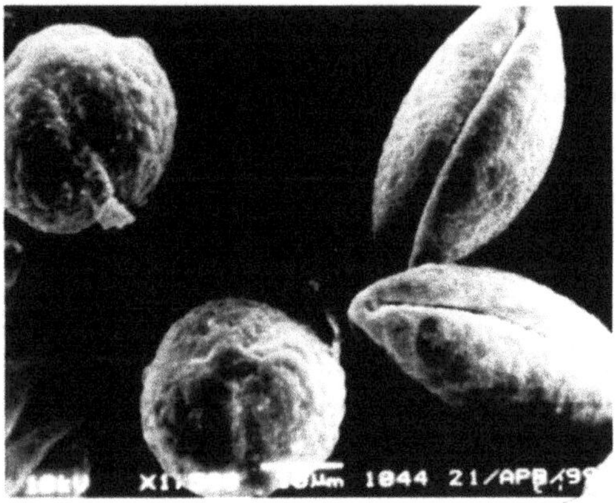

Fig. 5.2q SEM photomicrograph of pollen grains in equatorial and polar views.

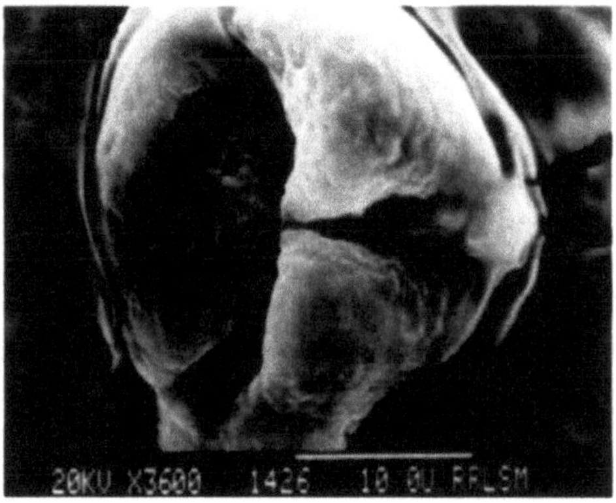

Fig. 5.2r SEM photomicrograph showing enlarged polar view of pollen.

Figs. 5.2 a to r Pollen morphology in SEM of some Mimosoids.

***Albizia lebback* Linn**

The plants of this species are natives of India, Sri Lanka, South East Asia and South China. They are usually grown as avenue trees for shade and have scented greenish white flowers. The trees can grow as tall as 20 m. The flowers are in axillary, solitary or 2-3 umbellate heads with numerous stamens, which are monoadelphous with 3 mm long filaments, which are united below to form up to 5 mm long staminal tube.

Pollen grains, which have been clinically proved to be allergenic are clustered in spherical polyads with 16 grains which are tetergonal. Polyads when subjected to acetolysis may separate into pollen tetrads, which measure in 64.2 to 66.3 μm in diameter. Individual pollen grains are 19 - 21 μm in diameter. The pollen wall consists of exine, which is indistinctly foveolate with low laminar and lowly raised muri. Pollen are three zonoporate without depression **(Fig. 5.2i)**.

***Leucaena leucocephala* Linn**

The plants of this species though natives of South America are widely grown in the tropics. The leaves of these plants are used as fodder for cattle. The plants represent small trees growing to a height of 4-6 m. They produce creamy white flowers in axillary dense globose heads; solitary or in clusters of 2-3, only 10 stamens are produced per flower. The plants flower profusely from July to March.

The pollen grains produced in anthers are monads, their polar view shows them to be spheroidal or triangular. They show tricolpate apertural condition where colpi are broad and taper to rounded ends. Pollen are 3-zonocolporate wtih scrobiculate exine ornamentation **(Figs. 5.2j, k and l)**. Individual scrobicule with punctate lumen are surrounded by a thick annulus. Pollen grains vary from 57.26 to 59.42 μm in length in equatorial view whereas in polar view they range between 39.4 and 42.4 μm in diameter.

Mimosa invisa* mart.*var.invisa

The plants of this species are natives of South America, but now grow profusely as 'weeds' in India. They are straggling shrubs with an angled stem, pubescent-hirsute and armed with downwardly pointed prickles. Flowers are borne in inflorescence, which is axillary, solitary or paired globose heads 1-1.2 cm in diameter. The flowers are characterized by a minute calyx and corolla with four pink petals united at the base. The androecium consists of eight stamens having 4-5 mm long filaments.

Anthers have pollen tetrads, which are spherical or prolate spheroidal. In the pollen tetrad both tetrahedral and isobilateral pollen arrangements are observed. Exine ornamentation is tuberculate, pores are ill defined **(Figs. 5.2m, n)**. Pollen tetrads are 19.6 to 23.2 μm and individual pollen grains are 10.2 to 11.2 μm in diameter.

***Mimosa pudica* Linn**

These spreading herbs with prickly stems are natives of South America but were introduced and naturalized in India. The plant is referred to as a 'sensitive plant' or 'Touch me not' on account of their leaves, which are sensitive to touch when 2-3 pairs of pinnae fold in. Flowers are in axillary heads solitary or in clusters of 2-3, with 2-3 cm long peduncles. Flowers are pink and 0.6 mm across. The calyx tube is 1 mm long and four lobed. Petals are four in number. Stamens are four in number having 6 mm long filaments.

Pollen are produced in tetrads and released to the air in the same condition after dehiscence of the anther. Both tetrahedral and isobilateral arrangements are seen in the tetrads which are 7.6 to 9.2 μm in diameter, whereas individual pollen grains are 3.9 to 4.5 μm in diameter. Exine ornamentation is characterized by tubercules. The central area of individual pollen grains is marked by tubercles, merging into the lateral ridges with completely fused tubercles **(Fig. 5.2o)**.

***Prosopis juliflora* sw**

Though it is native of North and Central America, it is cultivated and grows profusely in semiarid regions of both the northern and southern regions of India and the Middle -East. It is often introduced for afforestation. The armed trees grow up to a height of 8 m. Branchlets are glabrous with 1-1.5 cm long axillary spines. Leaves are bipinnate with 1 - 3 pairs of pinnae. Each pinna bears 15 - 18 pairs of leaflets, which are oblong, inequilateral and obtuse at both ends.

The specific name Juliflora is derived from the Latin word juliflorous meaning blooming with catkins, which are axillary, solitary or in clusters of 2-4 and 8-12 cm long. The creamy flowers in the catkin have campanulate calyx and five white petals, which are joined below to form a tube. Stamens are 10, free having 4-4.5 mm long filaments and glandular anthers.

The pollen grains are in monads. They are spheroidal or triangular in shape, radially symmetrical, subisopolar with 2-zonocolporate aperature. Exine is punctuated, scrobiculate, formed into islands of various sizes and shapes. The valleys between islands have punctuated scrobiculate ornamentation. Colpi have a circular operculum at the region of the ora. Pollen grains range from 23.6 to 27.1 µm in polar view and 26.6-28.2 µm in equatorial view **(Figs. 5.2p, q, r).**

B: POLLEN MORPHOLOGY OF CERTAIN ANGIOSPERMOUS PLANTS FROM TEMPERATE REGIONS

A brief description of a few allergically important plants and their pollen, common in temperate regions has been given below. On the basis of similarity in pollen morphological characters they have been grouped in 5 groups and the same are illustrated in the form of line drawings in **Figs. 5.2, 5.3, 5.4, 5.5, 5.6 and 5.7.**

In addition, pollen morphology of a few common temperate plants has been described below. These descriptions are supported by photomicrographs in **Figs. 5.8, 5.9, 5.10, 5.11, 5.12, 5.13 and 5.14.**

Quercus robur – Oak

Oak trees are very common in the deciduous forests of Central Europe. The pollen grains are mostly suboblate, tricolpate measuring 20-30 µm. The pollen is mildly allergenic.

Corylus – Hazelnut

In Europe *Corylus avellana* and *C. carpinifolia* are the chief contributors to airborne pollen spectrum. *Corylus avellana* is represented by a high shrub with a broad crown and several almost parallel stems. The plants are monoecious with drooping yellow male catkins and small inconspicuous female inflorescences with brightly red pistils **(Fig. 5.8)**. The plant produces hazelnuts as fruits in autumn. The pollen grains are sub-oblate to spheroidal, 20-25 µ in diameter , triporate with large onci. The pollen are allergenic with known cross-allergenicity with *Betula* and *Alnus*.

Alnus – Alder

The plants of this genus are found growing predominantly in north western Europe. The trees are monoecious with flowers arranged in unisexual inflorescences. The plants flower during March-April and May. Sometimes trees start flowering during December (Christmas time). The pollen grains measure 20-30 µm in size having a sub-oblate form with five pores around the equator, intine showing small clear onci **(Fig. 5.9)**. These pollen are highly allergenic, showing cross reactivity with the pollen of *Betula*.

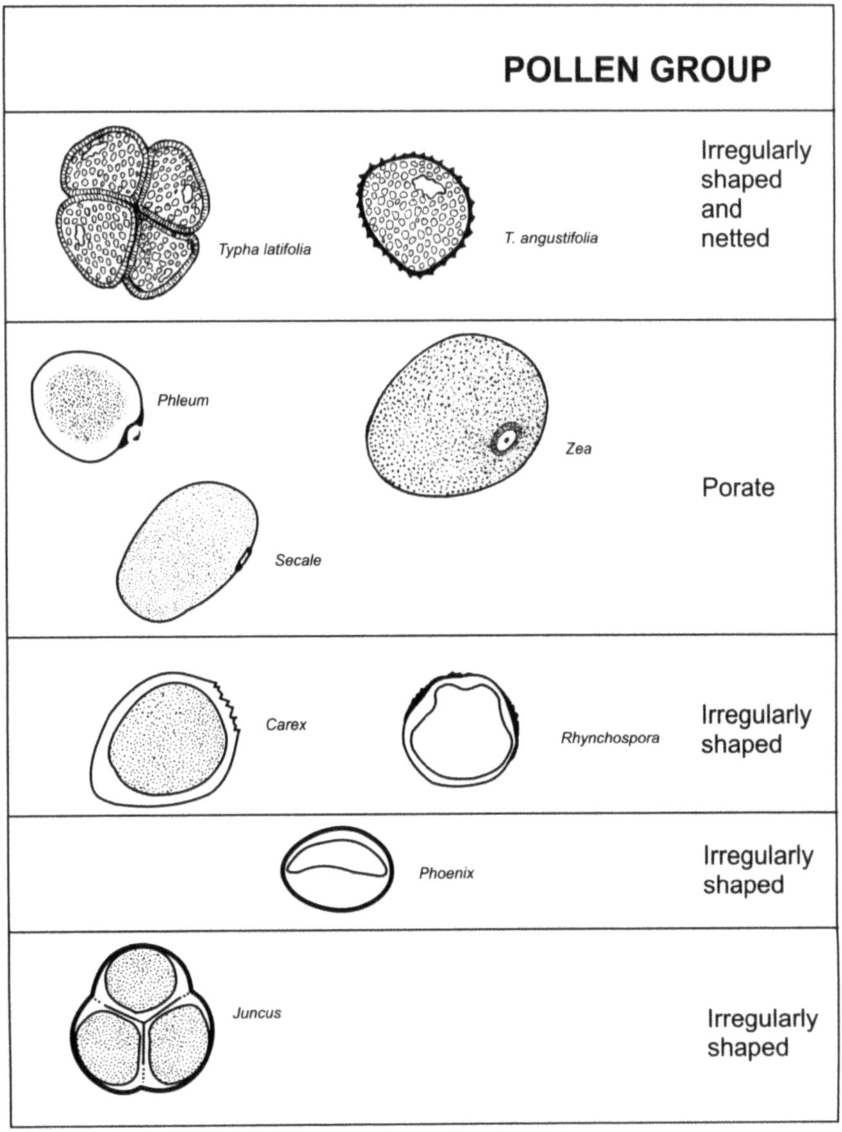

Fig. 5.3 Pollen Group: Irregularly shaped and netted, *Typha latifolia, T.angustifolia.* Porate: *Phleum, Zea, Secale.* Irregularly shaped: *Carex, Rhynchospora, Phoenix, Juncus.*

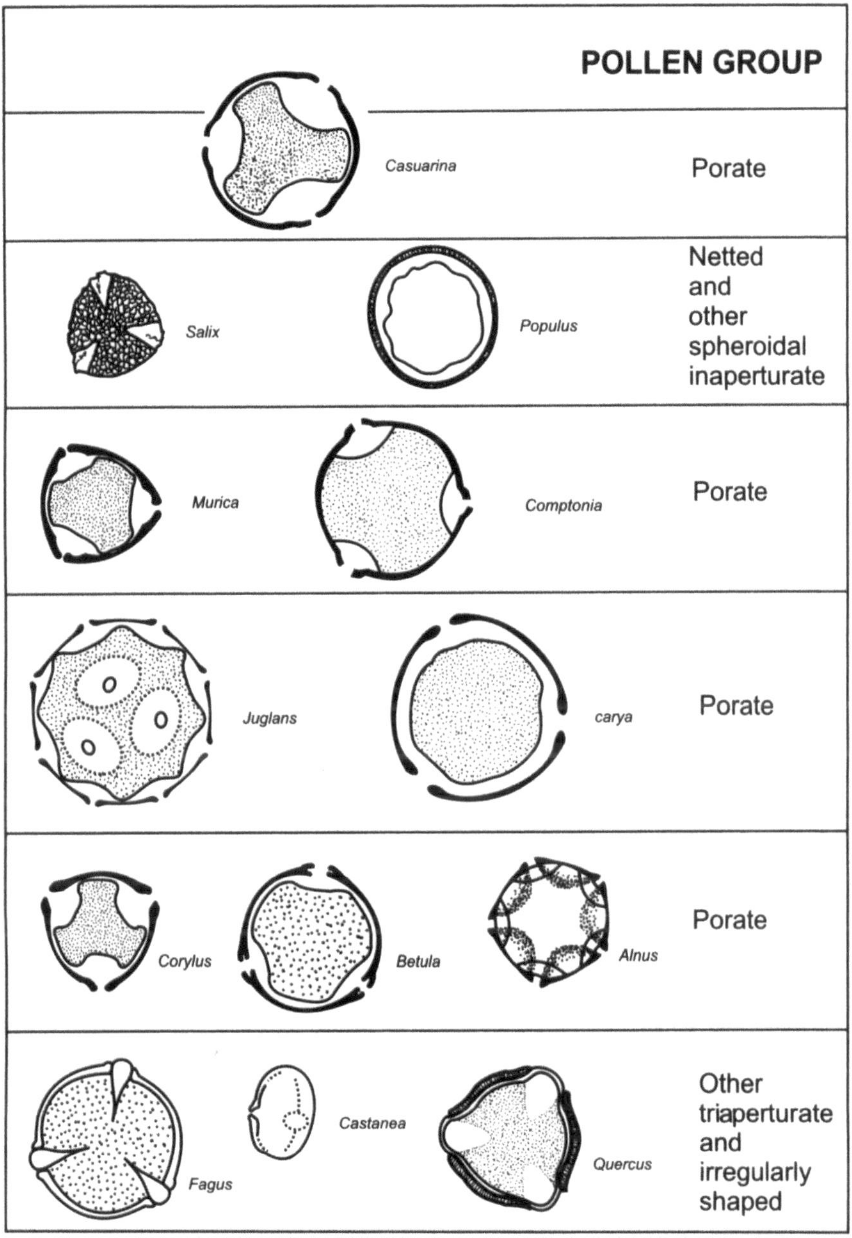

Fig. 5.4 Pollen Group: *Porate: Casuarina, Murica, Comptonia, Juglans, Carya,* Corylus, Betula, Alnus. Triaperturate and irregularly shaped: *Fagus, Castanea, Quercus.* Netted and other spheroidal inaperturate: *Salix, Populus.*

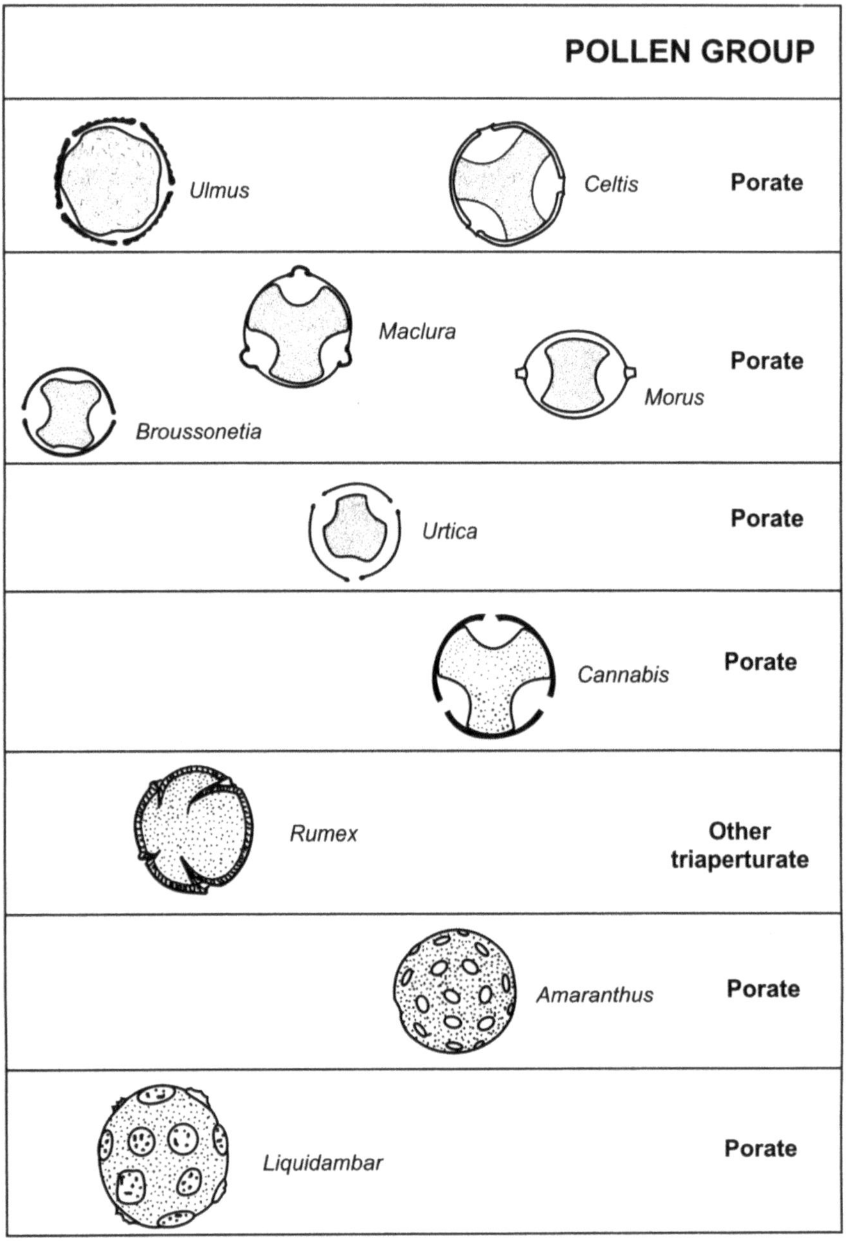

Fig. 5.5 Pollen Group: Porate: *Ulmus, Celtis, Broussonetia, Maclura, Morus, Urtica, Cannabis, Amaranthus, Liquidamber.* Other Triapertuate: *Rumex.*

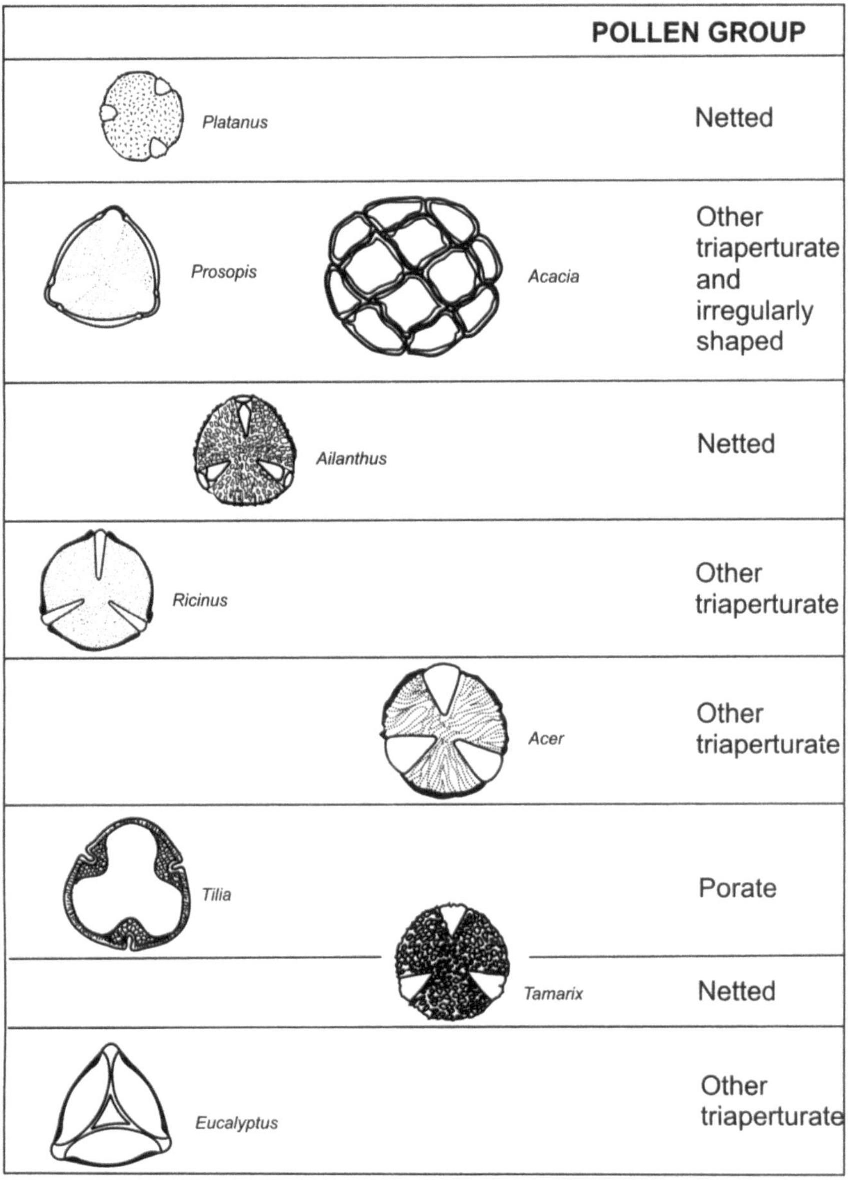

Fig. 5.6 Pollen Group: Netted: *Platanus, Ailanthus, Tamarix.* Other inaperturate and irregularly shaped: *Prosopis, Acacia.* Other Triaperturate: *Ricinus, Acer, Eucalyptus.* Porate: *Tilia.*

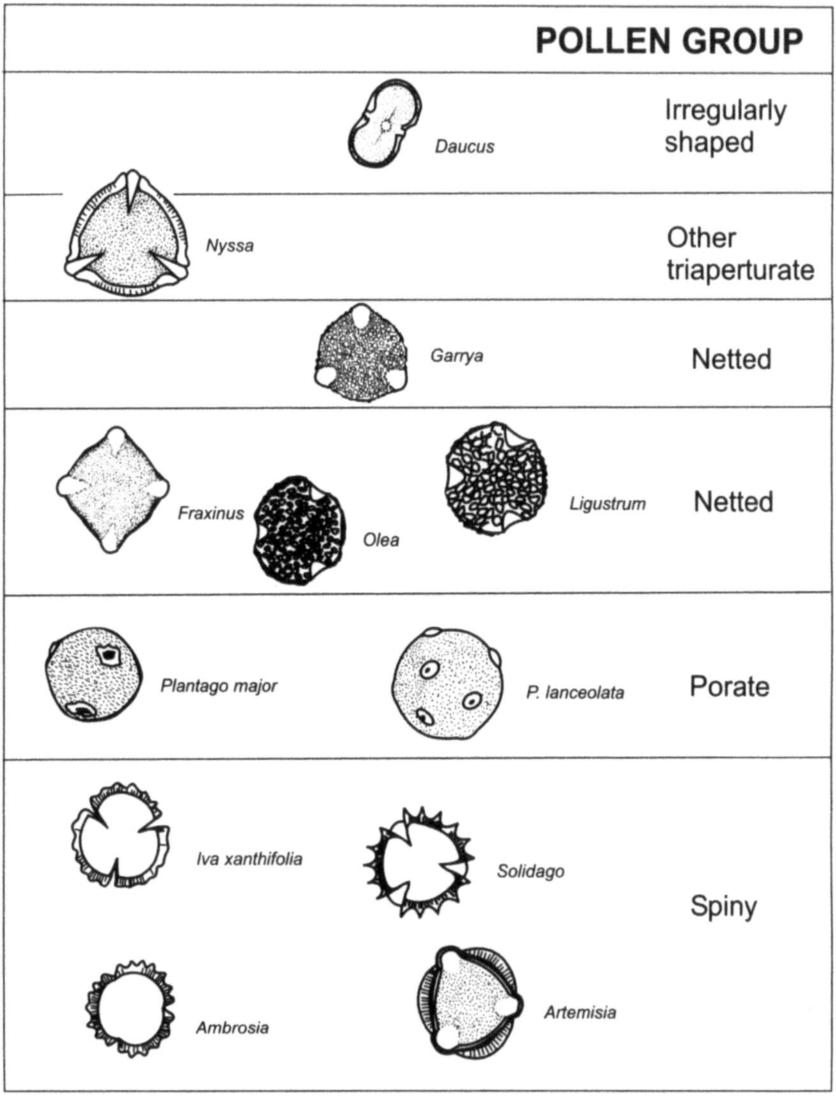

Fig. 5.7 Pollen Group: Irregularly shaped: *Daucus.* Other Triaperturate: *Nyssa* Netted: *Garrya, Fraxinus, Olea, Ligustrum* Porate: *Plantago major, P.lanceolata* Spiny: *Iva xanthifolia, Solidago, Ambrosia, Artemisia.*

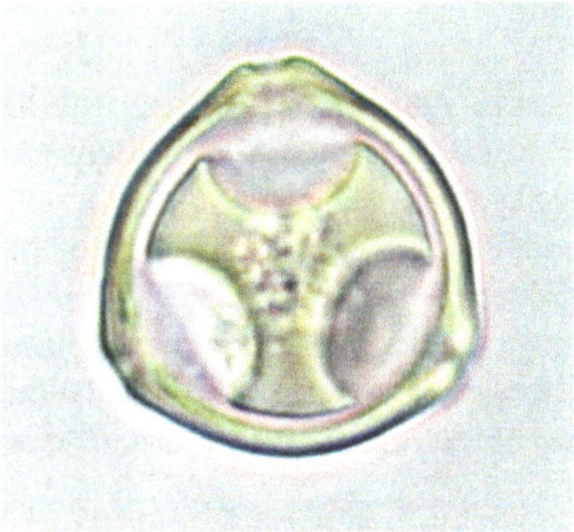

Fig. 5.8 Triporate pollen of *Corylus* with large onci (LM).

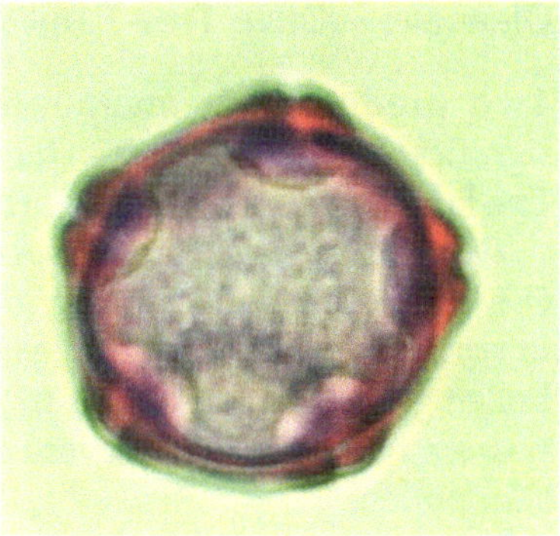

Fig. 5.9 *Alnus* (LM).

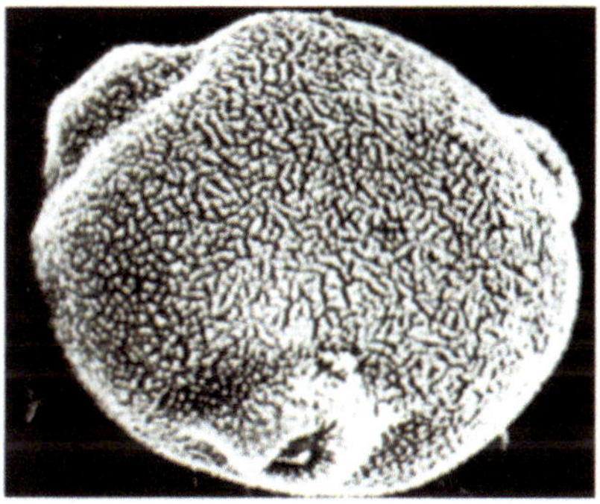

Fig. 5.10 *Betula* (SEM).

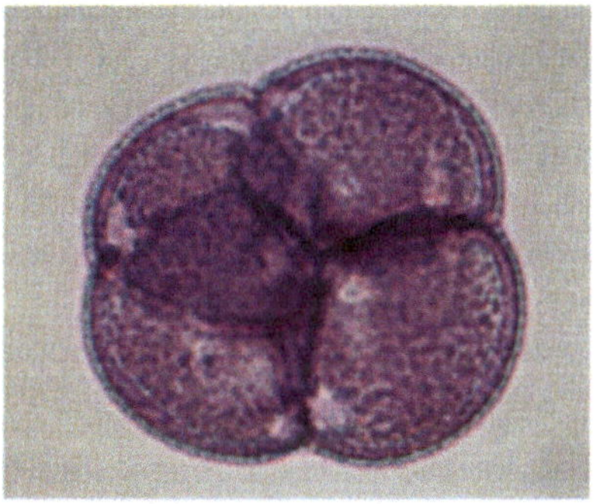

Fig. 5.12 *Erica* (LM).

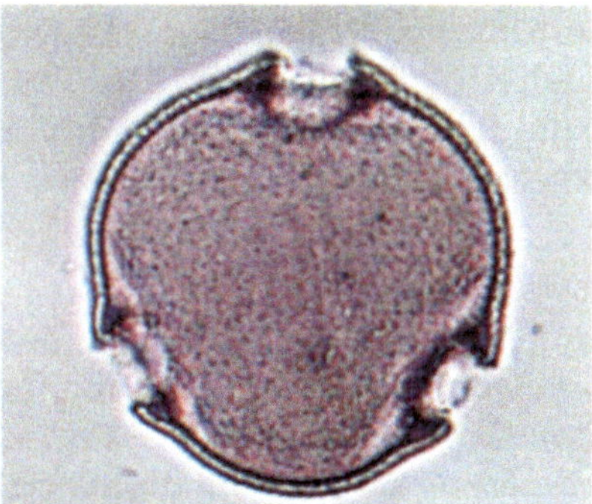

Fig. 5.13 *Tilia* (LM).

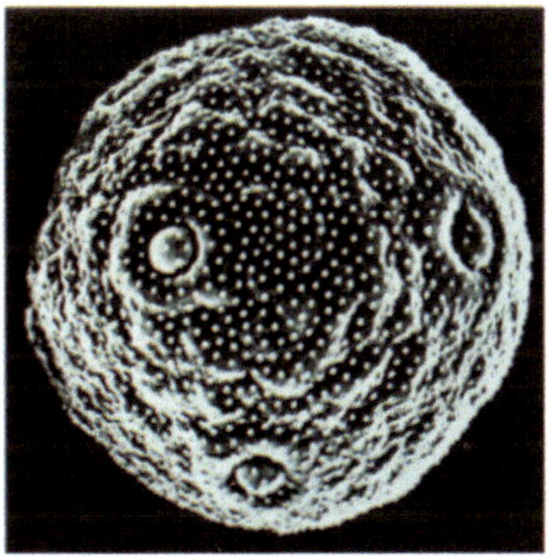

Fig. 5.14 *Plantago* (SEM).

Fig. 5.8 to 5.14 Photomicrographs of pollen of angiospermous plants from temperate region.

Oleaceae – Olive Tree Family

Two predominant anemophilous plants producing allergenically significant pollens belong to the genera *Fraxinus* (ash) and *Olea* (olive tree). The latter one is also an important source of olive oil.

Fraxinus excelsior is widespread in north western and central Europe. *Olea europea* trees are common in the Mediterranean. *Fraxinus* flowers mostly in spring, during April and May. *Olea europea* flowers in May and June. Pollen grains are sub-oblate, tricolpate, measuring about 20 μm in size and possess reticulate wall. Severe symptoms of pollinosis are observed from *Olea* pollen.

Betula – Birch

The trees of this genus are a favourite with landscape architects on account of their ornamental value, white stems and drooping branches. The trees are most dominant in Nordic countries forests, western and central Europe. The most common species of *Betula* are *B. verucosa* (*B. pendula*) and *B. pubescens*. The male catkins are pendulous, yellow green in colour. In contrast, female catkins are erect and green.

Normal flowering of this plant is from April to the first half of May. However, sometimes delayed flowering from May to June is also observed, particularly in Nordic countries.

Betula pollen grains are sub-oblate to spheroidal, triporate with small onci, measuring 20-25 μm **(Fig. 5.10)**.

Chenopodium, Amaranthus – Pigweed

Airborne pollen of Chenopodiaceae and Amaranthaceae cannot be distinguished from each other and hence they are always mentioned together as they are morphologically similar.

Amaranthus is widespread in southern Europe. It usually flowers in summer. The pollen grains are spheroidal, pantoporate, with as many as 60-80 pores and measure around 25-30 μm **(Fig. 5.11)**. The pollen exhibit allergic sensitivity, but the significance for symptoms of pollinosis is considered to be rather small.

Artemisia – Mugwort

Similar to *Ambrosia, Artemisia* also belongs to the family Asteraceae. It grows abundantly on disturbed soils. The most common spp. is *A. vulgaris*. The other two species *A. annua* and *A. verlotorum* are also common. In addition, the pollen of sage brush or *Artemisia tridentata* is known from the state of Missouri. The pollen of ragweed and sage brush cause severe allergic symptoms. However, both these show a very high degree of cross reactivity. In addition, the cross reactivity is also shown between giant and short ragweeds.

It flowers from the end of July to September. The plant produces allergenic pollen, which are oblate, spheroidal, about 20 μm in diameter, tricolporate.

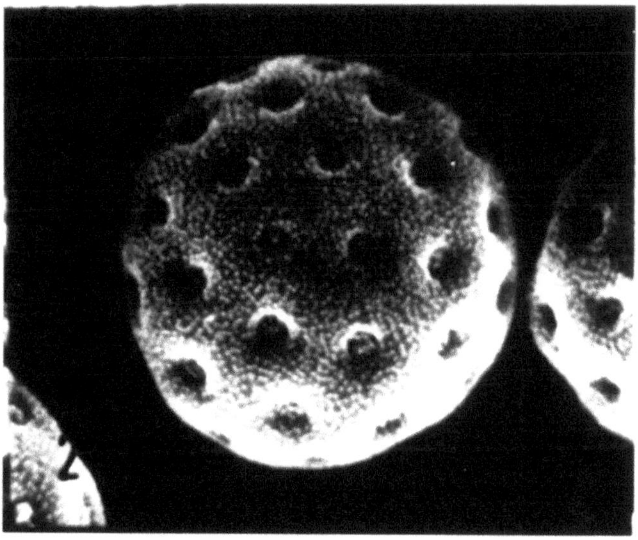

Fig. 5.11 Pantoporate pollen of Amaranth Chenopod (SEM).

Rumex – Sorrel

The plants of this genus are mostly weeds growing on wet soils found all over Europe. The plant bears red coloured, bisexual inflorescence often giving the meadow landscape a very attractive aspect. *Rumex crispus* is the most common species. Pollen grains are oblate, spheroidal, tricolpate with narrow furrows, measuring 20-30 μm. Pollen is mildly allergenic.

The other pollen which have been illustrated are *Erica* or Heather **(Fig. 5.12)**, *Tilia* or Lime **(Fig. 5.13)** and *Plantago* or Plantain **(Fig. 5.14)**.

Pollen morphological descriptions of selected european plants have been summarised in tabular form in Table 5.1.

Table 5.1 *Pollen morphological description of selected European pollen in tabular form.*

Name	*Grains*	*Exine*	*Furrows*	*Pores*	*Intine*	*Interior*
Artemisia vulgaris L. Mugwort Asteraceae	Tricolporate, Spheroidal, Small 16.9 × 17.9 µm	Thick up to 2.9 µm in middle of intercolpia ectexine	Three, 2.0 × 4.7 µm narrowing outside the pore to 3.4 µm	Three, one in the middle of each furrow, elliptical	Thick	
Carex caryophyllea L. Spring sedge Cyperaceae	Inaperturate, pyriform, medium sized to large, length 37.0 µm, width at greatest 27.3 µm	Thin			Upto 5 µm thick Poroids – an indistinct rounded aperture – usually one at thick end of grain and 3 on sides. Exine thinat these points	
Chenopodium album L. Fat hen Chenopodiaceae	Porate, spheroidal, small 25.8 µm	Medium ectexine thicker than endexine, thicker between pores forming a wavy outline in optical section		50-70 evenly scattered over whole surface 1.7-3.0 µm apart, circular 2.5 µm diameter, surface smooth except for small operculum	Thin	
Corylus avellana L. Betulaceae	Porate, suboblate or subtriangular, aspidate, small 20.2 x 24.8 µm	Thin, thickened a little round pores, raised to form aspis 10.2 µm diameter. Ectexine thicker than.		Three, evenly spaced around equator, circular to slightly elliptical 2.6 µm diameter	Thin, swelling beneath pores to more or less markedly convex onci, 10.1 × 5.6 µm	

Contd.

Table 5.1 *Contd.*

Name	*Grains*	*Exine*	*Furrows*	*Pores*	*Intine*	*Interior*
		endexine, surface smooth or slightly granular				
Galium mollugo L. Great Hedge Bedstraw Rubiaceae	Colpate, suboblate or oblate, spheroidal, small 16.6 × 18.9 µm	Medium, ectexine equalising endexine, surface faintly reticulate	Seven or eight 11.5 × 1.6 µm acute		Thin	
Heracleum sphondylium L. Cow Parsley,	Tricolporate, medium sized	Thick, ectexine thicker than endexine, comprising thick rods	Three, 2.0.0 × 2.3 µm narrowing, long, acuminate	Transverse, three, one in middle of each principal furrow	Thin	
Juncaceae common rush *Juncus acutiflorus* Hoffm. Sharp flowered Rush Juncaceae	Compound (tetrads) 40.7 µm Constituent grains: more or less spheroidal, inaperturate, united in tetrahedral tetrads, small 25.6 µm	Thin, large thin areas (poroids) of discrete			Thick below poroids, thin granules elsewhere	
Quercus robur L. English Oak Fagaceae	Tricolpate, oblate to spheroidal, small-medium sized 20.7 × 28.9 µm	Thin, ectexine thicker than endexine, surface warty-granular	Three, 16.0 × 7.6 µm tapering membrane protruding		Medium thickness	

Contd.

Table 5.1 *Contd.*

Name	*Grains*	*Exine*	*Furrows*	*Pores*	*Intine*	*Interior*
Rumex crispus L. Curled Dock Polygonaceae	Colporate, oblate to spheroidal, small to medium sized 28.0 × 29.2 µm	Medium thickness, ectexine thicker than endexine – thicker around pores. Surface finely reticulate.	Three, rarely four, longitudinal 22.2 × 1.3 µm, tapering	Three, rarely four, one in the middle of each furrow, narrowly elliptical	Thin	Packed with tiny starch granules
Salix caprea L. Goat Willow Salicaceae	Tricolpate, oblate, spheroidal, small 15.0 × 16.4 µm	Thin, ectexine thicker than endexine, surface reticulate, medium sized in middle of intercolpia be-coming smaller towards poles	Three, 12.5 × 4.0 µm, acute membrane protruding beyond outline of grain		Thin	
Senecia jacobea L. Ragwort Asteraceae	Tricolporate, oblate, spheroidal, echinate, small 20.4 × 21.5 µm	Medium thickness, surface granular, sharply pointed spines, 2.2 µm high, 1.7 µm broad at base	Three, 18.2×6.0 µm, abruptly narrowing	Three, one in middle of each furrow, Circular 6.0 µm mem-brane protruding	Thin, thickening over surface of pore	
Urtica dioica L. Nettle Urticaceae	Porate, spheroidal, small 13.4 × 14.5 µm	Thin, slightly raised round each pore forming aspis, surface smooth	Three or four, equally spaced around equator, circular 1.6 µm diameter	Thin, swelling beneath pores forming onci		

CHAPTER 6

Pollen Morphology of Gymnosperms

Gymnosperms along with angiosperms are classified as seed plants. Seed plants can be simply defined as plants producing seeds. The seed is an organ originated from an ovule. It has a resistant outer seed coat (testa) enclosing mainly a nourishing tissue called endosperm for the developing embryo.

Thus, ovule can be defined as a megasporangium (nucellus) containing a single functional megaspore (embryo sac) enveloped by coat (integumant). In other words, an ovule develops into a seed after fertilization in seed plants. In fact achievement of seed habit from heterospory is a very important step in the evolution of plant life. This is accompanied by evolution of pollen ultimately resulting in the diversity of plants. Conventionally microspores produced in the microsporangia of male cones in gymnosperms are termed as pollen grains. Paleobotanical evidence shows that fossil pollen of almost all groups of gymnosperms such as cycads and conifers in addition to extinct pteridosperms have been well preserved in the sediments dating back to the Devonian period.

As mentioned earlier, basically the terminology used for describing pollen morphological characters in gymnosperms and angiosperms is the same. In the present chapter only a brief account of pollen morphological variation observed in various groups of modern and fossil groups of gymnosperms will be given. Fossil gymnospermous pollen have been extensively used in palynostratigraphy, which are of great significance in the exploration of fossil fuels such as coal and oil.

It seems appropriate to discuss pre pollen and pollen of pteridosperms or the seed ferns before describing the pollen morphology of modern gymnosperms. The geologically oldest pollen has been reported from the Devonian rocks. This pollen is often referred as pre pollen. It appears to have retained the morphology of spores and germinated proximally as do spores of vascular plants. Chaloner (1970) has summarized the distinction between spores, pre pollen and true pollen and their significance in plant evolution. There appears to be evolutionary trend from proximal emergence of pollen tubes (proximal germination) in pre pollen to distal emergence of pollen tubes (distal germination) in true pollen occurring in advanced gymnosperms.

POLLEN MORPHOLOGY OF FOSSIL GYMNOSPERMS (THE PTERIDOSPERMS OR CYCADOFILICALES)

The most outstanding characters of the pollen grains of the cycadofilicales are their large size and pluricellular structure. There is a lot of variation in their size, ranging from 70-500 μm in diam. in the different species which have been described. There is also variation in their cellular structure; in some the cavity of the grain contains only two cells, while in others it contains as many as 30. Nevertheless, it is almost universally true that the entire cavity is filled with cellular tissue with well-developed cell walls.

Many of the pollen grains of this group are found in the pollen chamber, which probably increased in size after reaching there. Saporta and Marion (1935) have described pluricellular pollen grains in the pollen chamber of *Pachytesta,* which are 500 μm long. In the pollen chamber of *Aetheotesta,* Renault (1896) described similar grains about 400 μm long and without exines. The pollen grains are thus fairly large compared to that of angiosperms.

Cycadofilicinean pollen grains from the staminate inflorescenes of *Crossotheca* were one-celled and in all their characters which have been preserved, they are remarkably like many of the fern spores of today. In an exceptionally well-preserved fructification in coal ball, Rothwell (1972) has shown the occurrence of pollen tubes in pollen *Idanothekion* of Callistophytaceae. It seems likely that in the cycadofilicales the pollen grains left their anthers as single celled spores, and germination, or any extensive cellular proliferation, did not take place until after they arrived at their destination.

The next stage of the evolution of the pollen grains was accompanied by a reduction of the prothallial tissue until it became represented by two or three nuclear divisions without the formation of any cell walls, which took place prior to shedding, for example in the grains of cycads and *Ginkgo,* leaving the broad furrow, which had been provided to accommodate it a useless organ and very probably an encumbrance to the grain.

DESCRIPTION OF POLLEN GRAINS FOUND IN *CROSSOTHECA HÖNINGHAUSII*

Grains globular or slightly oval, 50-70 μm in diam. Their outer surface roughened by numerous closely placed, very minute, blunt points. Each grain is provided with a distinct triradiate crest, though this is often difficult to see on account of the crumpling of the pollen grain wall **(Fig. 6.1)**. It is believed that the presence of the triradiate crest indicates that these grains were developed as members of tetrahedral tetrads. Nucellular tissue is not observed in any of them, but this may be due to the fact that they were immature when fossilized, since they were dissected out of unopened anthers. *Crossotheca höeninghausii* is known to be the staminate inflorescence of *Lyginodendron oldhamium* Williamson.

Fig. 6.1 Pollen of *Crossotheca Honinghausii* showing triradiate mark.

Stephanospermum caryoides: Grains found in the pollen chamber of the seed. Grains ellipsoidal, flattened 91 x 72 μm provided with a wind like bladder, which completely encircles the grain (looks like grains of Podocarpinae) **(Fig. 6.2)**.

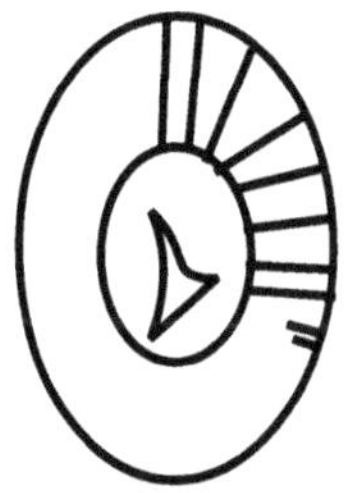

Fig. 6.2 Monosaccate pollen of *Stephanospermum caryoides.*

Pachytesta gigantea Renault. Grains ellipsoidal, 320-400 μm long and 270-310 μm broad. One sees in the interior a number of walls dividing the cavity into a certain number of cells. The grains do not have any exine **(Fig. 6.3)**.

Fig. 6.3 Pollen of *Pachytesta gigantea* showing pluricellular internal structure.

POLLEN OF CORDAITALES

The pollen grains of the Cordaitales are generally ellipsoidal in shape, rather large, measuring about 100 μm in length, and with a characteristically roughened exine. They are always provided with a single, deep, longitudinal furrow and exhibit a pluricellular internal structure.

As compared with the grains of Cycadofilicales they possess certain rather striking differences. There is generally a little less prothallial tissue, and its development takes place earlier. There is no doubt that the germination of cordaitalean pollen grains took place prior to their release from the anther.

These pollen grains possess a single longitudinal furrow, which appears to function as a proliferation chamber, enabling the prothallus to develop without rupturing the spore wall, as it did in the grains of the Cycadofilicales, which were without a furrow. In the grains of Cordaitales the floor of the furrow became pushed up by the development of the prothallus within, finally separating from the rest of the surface of the pollen grain along its rim and opening as a lid, permitting the escape of the antherozoids.

In these characteristics the pollen grains of Cordaitales show a distinct advance over those of the Cycadofilicales, and a step toward the form of the conifers and angiosperms. But they still differ from the grains of the conifers in their pluricellular internal structure with well-developed walls partitioning off the whole device for their wide-open furrow.

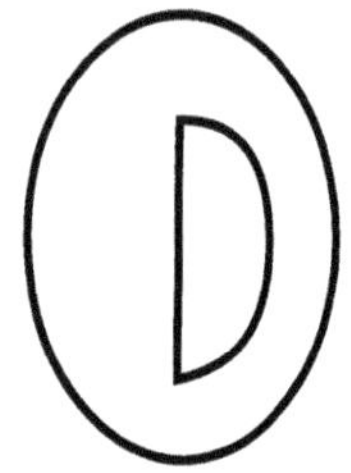

Fig. 6.4 Pollen of *Dolerotheca fertile* with longitudinal furrow.

Dolerotheca Renault.: Grains ellipsoidal, about 280 μm long. Exine finely roughened. On one side are two deep furrows joining together at one end and marking out an elliptical shaped operculum by which dehiscence may take place. The interior of the grain is divided into 8-10 cells with thin walls **(Fig. 6.4)**.

POLLEN MORPHOLOGY OF BENNETTITALES

The pollen grains of the Bennettitales are, with one or two exceptions, scarcely different from those of the cycads. They are boat shaped and provided with a single longitudinal furrow as in *Cycadeoidea etrusca* **(Fig. 6.5)** and *Cycadeoidea dacotensis* **(Fig. 6.6a)**, which appears to have been as ineffective in closing as that of the cycads and *Ginkgo.* Their prothallial tissue was much less extensive than in the grains of Cycadofilicales, or even of the Cordaitales but considerably more extensive than that of the Cycadales as in *Cycadeiodea dacotensis* **(Fig. 6.6b)**.

Fig. 6.5 Pollen of *Cycadeoidea etrusca* showing longitudinal furrow.

In size, they range from 20-67 μm in length. They are thus generally a great deal smaller than those of the Cycadofilicales or even the Cordaitales

(a) (b)

Fig. 6.6 *Cycadeoidea dacotensis* pollen: a. Showing two longitudinal furrows, b. Optical section of pollen showing multicellular internal structure.

but are larger than those of the Cycadales and Ginkgoales. In the development of their prothallial tissue and their size they thus occupy a position intermediate between the Cycadofilicales and the Cycadales. Though they are found in the Carboniferous, they become more abundant in the Mesozoic.

MORPHOLOGY OF POLLEN OF EXTANT GYMNOSPERMS

Living gymnosperms are broadly divided into four orders: Cycadales, Ginkgoales, Coniferales and Gnetales. Prominent pollen morphological features of certain members of these orders will be described here.

POLLEN GRAINS OF CYCADALES

In Cycadales the pollen grains are small, broadly ellipsoidal, boat shaped, with a single deep longitudinal furrow reaching from end to end, essentially as in the grains of *Ginkgo*. The pollen grain is bilaterally symmetrical in the sense that its two sides and two ends are exactly alike; but its remaining two sides are dissimilar, since one of them bears the furrow and other does not.

When the pollen grain is dry, the edges of the furrow arch inward toward each other, tending to close the opening, and may even touch in the middle. When the pollen grain is moistened the furrow gapes widely open as in *Cycas* **(Figs. 6.7 - a, b, c)**. In *Zamia* **(Fig. 6.8)** and *Ceratozamia* the spore coat is quite smooth. In *Cycas* it is minutely warty on the outside suggesting the flecked surface of the grains of *Juniperus*.

In the grains of the cycads the exine is thinner than in other grains. This one furrowed or monocolpate type of grain, besides occurring throughout the Cycadales, is characteristic of many Monocotyledons and primitive Dicotyledons for example the Palmaceae, Magnoliaceae and Nymphaeaceae.

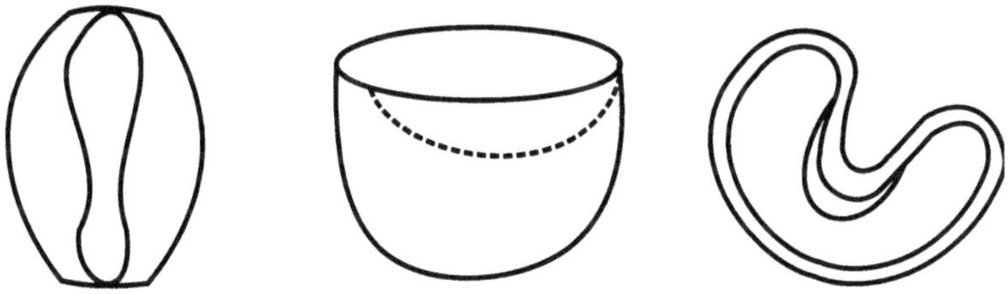

(a) (b) (c)

Fig. 6.7 Pollen of *Cycas*: a. Ventral view showing wide, longitudinal furrow, b. Longitudinal view, c. Optical section showing furrow.

Ginkgoales

The pollen grains of the Ginkgoales, like those of the cycads, are notable for their single deep and broad unprotected furrow and their lack of prothallial tissue **(Figs. 6.9 - a, b, c)**. At maturity they exhibit nothing of the pluricellular tissue, which characterized the grains of the extinct Cordaitales and Bennettitales. The male gametophyte is reduced to nominally three cells, that is one vegetative, one generative and one tube cell **(Fig. 6.10)**.

Fig. 6.8 Polllen of *Zamia* showing longitudinal furrow.

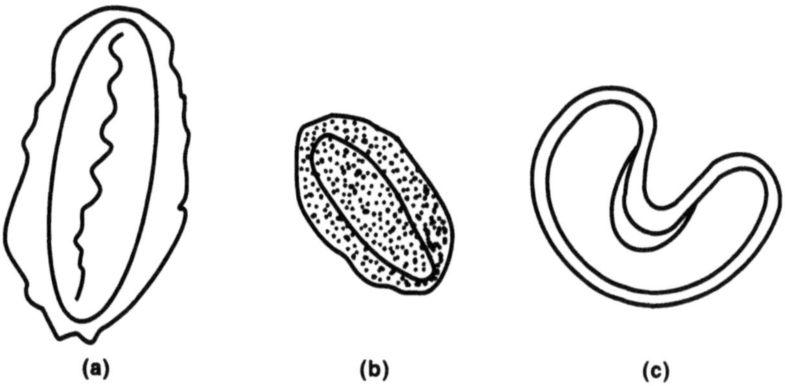

Fig. 6.9 Pollen of *Ginkgo biloba*: a. Lateral longitudinal view, b. Ellipsoidal monosulcate pollen, c. Median longitudinal section showing a single furrow.

The furrow, on the other hand, is of practically the same form as that of the grains of the Cordaitales, except that there being no internal tissue developed within, it remains deeply invaginated until the pollen tube begins to emerge.The pollen grains of *Ginkgo biloba* are similar in all their major features to that of *Cycas* but may be distinguished from the latter by their more elongated shape, smoother surface, and the slightly wavy margins of its furrow.

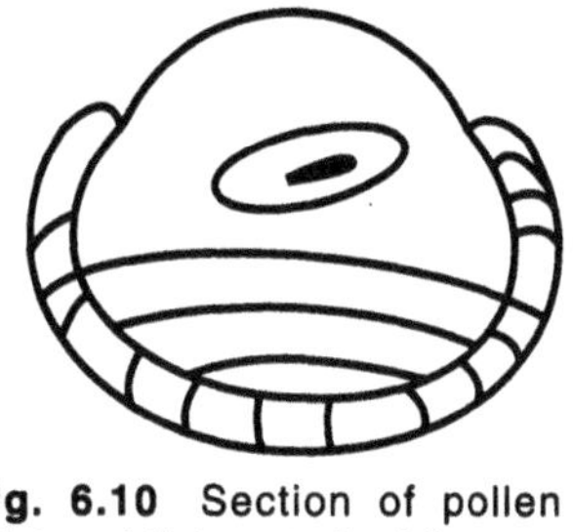

Fig. 6.10 Section of pollen of *Ginkgo biloba* at shedding stage. (Three-celled male gametophyte).

CONIFERALES

The pollen grains of the Coniferales are known for their extraordinary diversity of form. As compared with the grains of the more primitive gymnosperms, we find little trace of the pluricellular gametophytic tissue, which was very evident among the Cycadofilicales and Cordaitales.

POLLEN DEVELOPMENT

In majority of conifers the tetrad of microspores is produced by simultaneous division resulting in a tetrahedral arrangement of the developing pollen grain. In this tetrad the proximal pole (ventral) is toward the centre and the distal pole (dorsal) is away from the centre of the tetrad. However in nonsaccate pollen, polarity of mature pollen grains is not very clear. The tetrahedral type includes both saccate and nonsaccate pollen whereas the bilateral type produces only nonsaccate pollen. The pollen wall in conifers and other gymnosperms has an outer exine composed of sporopollenin and an inner intine consisting of hemicellulose.

SACCATE POLLEN

In living conifers, pollen that develops two or more, usually hollow, extensions of the infratectum (sacci, wings or bladders) occur only in *Phyllocladus, Pinus, Picea, Abies, Cedrus, Pseudolarix* and *Podocorpus*. The pollen lack sacci in *Larix, Tsuga,* and *Pseudotsuga* and *Saxegothaea*. The sacci are developed laterally from the body of the grain, that is their proximal limit determined by the limits of contact of the cells in the original tetrad as shown in **Fig. 6.11**.

In saccate pollen of gymnosperms normally there are two sacci. However three sacci occur in pollen of *Dacrycarpus, Microstrobus,* and *Microcachrys*. In the bisaccate pollen of *Pinus* the proximal part is described as the 'cap',

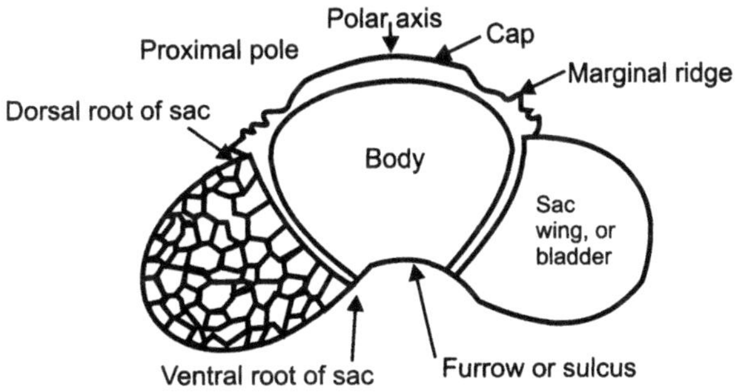

Fig. 6.11 A typical bisaccate pollen showing different features.

the distal part, delimited by sacci is referred as 'germinal furrow' **(Fig. 6.12 and 6.12a)**. The inflated portion of the saccus is supported internally by the alveolar or honeycomb-like structure of the ektexine. The outer surface of the saccus is smooth but minutely porous.

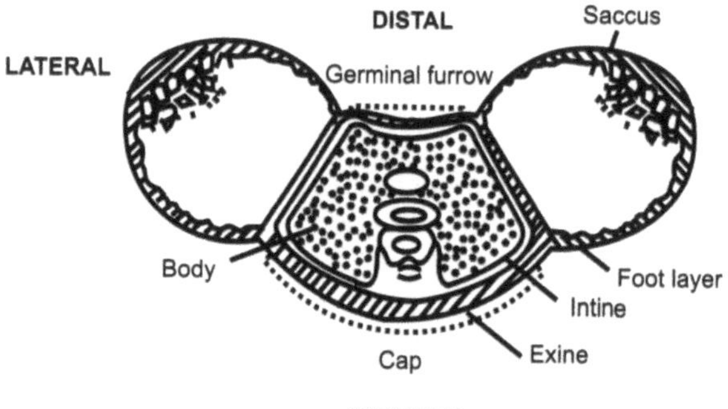

Fig. 6.12 Mature saccate pollen grain of *Pinus* in median section showing different features.

The non-saccate pollen is essentially spherical or biconvex and of simple organization. The intine in this pollen is much thicker than the granular exine. The pollen of Araucariaceae has a moderately thick intine and little differentiated sexine. The pollen grains are multicellular.

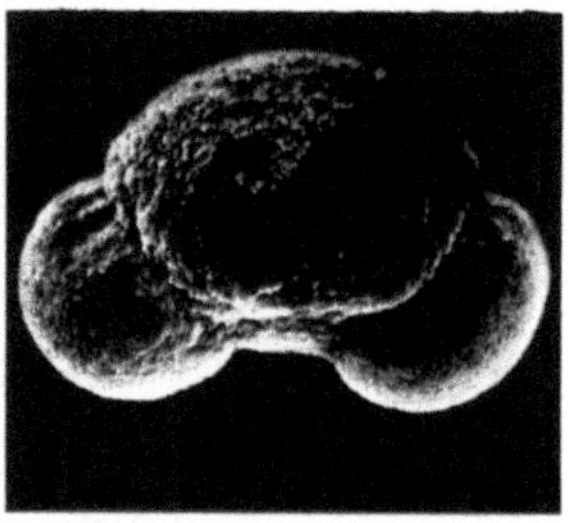

Fig. 6.12a Pinus Pollen (SEM).

In the Abietineae, the wings when present are nearly always two, one on each side of the furrow and forming for it a protective cover when the grain dries. While in the grains of the Podocarpineae, there may be 2, 3, 4, 5 or 6 bladders, and in the pollen of one species of *Podocarpus* some grains have a single bladder encircling them completely like a frill. But in both tribes there are genera with grains entirely lacking bladders. Similar wings are known to have occurred on the microspores of some Palaeozoic Cycadofilicales and Lycopodiales and still occur on the spores of some of the modern ferns and lycopods, and their presence among the Coniferales suggests the great antiquity of the group.

The capacity to develop wings has apparently been inherited from the remote past, but only in these few genera has it been called forth here, apparently, in response to a need of protection for the broad, open furrow with which such wings are here associated. Modern gymnosperm families characterized by inaperturate pollen are the Araucariaceae, Cephalotaxaceae, Cupressaceae, Taxodiaceae and some genera of the Pinaceae.

Araucariaceae

The Araucariaceae are undoubtedly among the most ancient living conifers. The pollen grains of both *Araucaria* and *Agathis* are without a true furrow or pore. In the grains of *Agathis* no vestige of a furrow can be found, but in those of *Araucaria* there is an annular thickening which corresponds in position to the furrow rim in the expanded Cycad grains and which appears to be the vestige of the cycadean furrow rim. The pollen grains are approximately spheroidal, non-saccate without any vestige of a germ pore.

Abietineae

Within the group are found three distinct types of pollen grains. The grains of five genera such as *Pinus, Cedrus, Picea, Abies* and *Pseudolarix* resemble each other in their common possession of bladders and in all the major features of their construction, such as the differentiation between the dorsal and ventral side and their possession of a single long furrow. In contrast to these, the pollen grains of *Tsuga* have no true furrow or wings but resemble those of the winged grained Abietineae in the character of their exine. They are also referred as monosaccate. The pollen grains of *Larix* and *Pseudotsuga* are also entirely without furrow or bladders having perfectly smooth exine and rather thick intine.

Taxodiaceae

The pollen grains of both the Taxodiaceae and Cupressaceae are entirely without prothallial cells, which decidedly has given them a modern aspect among the Coniferales. A unique feature of some of the pollen of Taxodiaceae is the presence of a conical projection (termed papilla or ligula) rising from the spherical body of the grain as in *Sequoia* (**Fig. 6.13**). The wall layers at the tip of the papilla are thinned where a germinal pore may be present.

Fig. 6.13 Non saccate pollen of *sequoia* showing papilla-like projection.

The pollen grains of Taxodiaceae are nearly spherical or sometimes angular with thin exine and thick intine. In the grains of *Taxodium* the furrow appears to be reduced to vanishing point, represented by only a small protuberance.

Podocarpaceae

The pollen grains of the Podocarpaceae are known for the possession of bladdery wings in most of their species, causing them to bear a superficial resemblance to the grains of the Abietineae. But in neither the Abietineae nor the Podocarpineae are the grains of all species winged.

Podocarpus sp. usually have two wings. *P. dacrydioides* normally have three wings, *Microcachrys* the bladders may be 3-4-5-6. *Phyllocladus* – two bladders. *Saxegothaea* has no wings. The bladders are associated with an ancestral type of furrow. The pollen grains of *Podocarpus* have two lateral sacs, which float upwards in a fluid as in *Pinus*. The pollen grains of *Saxegothaea* are more or less spherical and non-saccate **(Fig. 6.14)**.

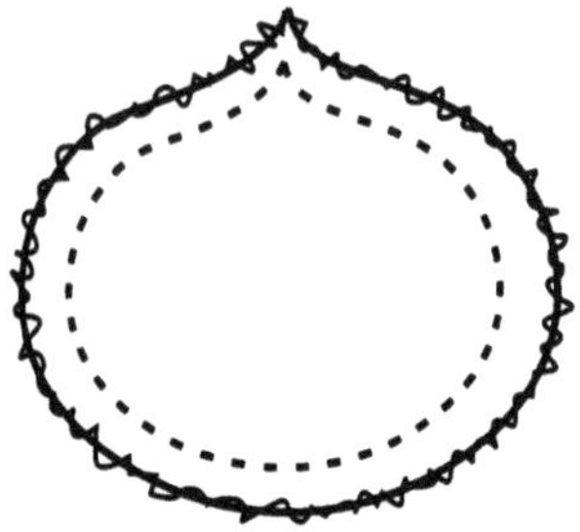

Fig. 6.14 *Saxegotheca* pollen which is spherical, non saccate

Cupressaceae

The pollen grains of cupressaceae are similar to those of Taxodiaceae. The intine is very thick.

Cephalotaxaceae

In *Cephalotaxus* the plant is monoecious. The pollen grains are round with thin exine and thick intine. When the grains are left in water or in a weak sucrose solution, the exine cracks and the intine and its contents are released.

Pinaceae

Pinus: It has characteristic needle-like leaves **(Fig. 6.19)**. Pollen grains are fairly large and possess two ventrally borne air-filled sacs with a single long furrow situated between them **(Fig. 6.20)**. The dorsal surface has a thick and rugged exine. The sacs diverge sharply in moist conditions, but in dry pollen they are pressed together. The sacs on their inner surface are thrown into a reticulum of ridges, which impart rigidity and prevent them from collapsing. The pollen grains of *Pinus* are very buoyant due to the presence of sacs. They rise upwards and float in the fluid. Thus winged pollen grains are characteristic of *Pinus* and other genera of Pinaceae. Whether or not wings are organs of flight is doubtful though they do impart a buoyancy to the grains.

Takeshi N. *et al.* (1996) thoroughly explored the pollen morphology of modern gymnosperms from the Himalayan region. *Pinus roxburghii* and *P. wallichiana* pollen grains were examined by them with Scanning Electron Microscope (SEM) in an attempt to identify fossil pollen grains of the Himalayan region. The exine sculpture of *P. roxburghii* is rugulate whereas that of *P. wallichiana* is smooth or slightly rugulate.

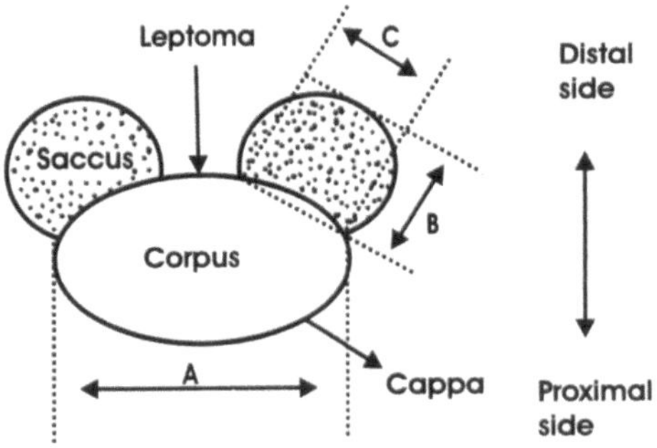

Fig. 6.15 *Pinus* pollen showing details of measurement of various dimensions.

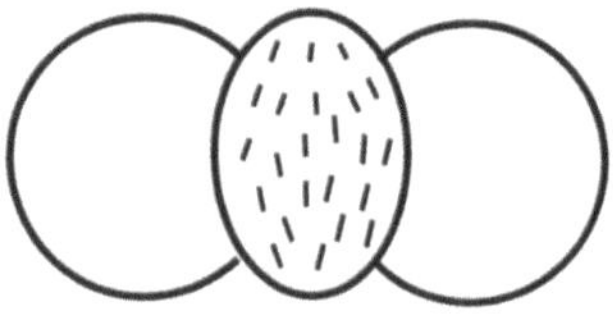

Fig. 6.16 Pollen of *Pinus excelsa*.

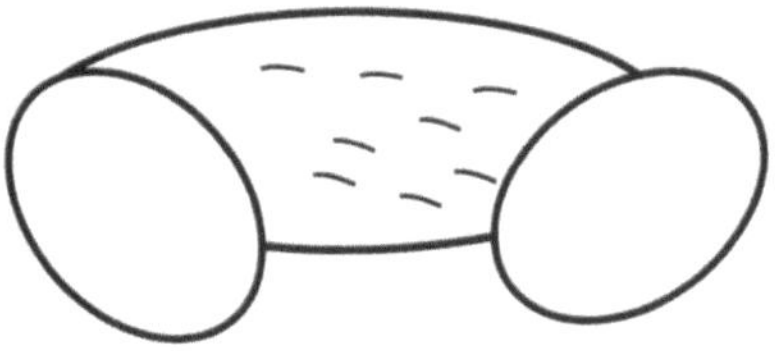

Fig. 6.17 Pollen of *Pinus orientalis*.

Fig. 6.18 Striate, ellipsoidal, ridged pollen of *Ephedra glauca*.

Fig. 6.19 A branch of *Pinus* with needles.

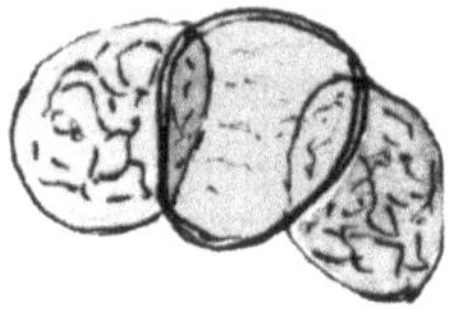

Fig. 6.20 Bisaccate pollen of *Pinus*.

Various dimensions of *Pinus* pollen grains, which are taken into account, were clearly described by them as shown in the **Fig. 6.15**. They stated that saccus height, saccus width at base, corpus breadth, and surface sculpture of 'Cappa' (proximal side of corpus) were the important characteristics for the description of the Himalayan *Pinus* pollen grains. It is also known that the size and orientation of body and sacci differ in different species of *Pinus* **(Figs. 6.16 and 6.17** and **Table 6.2)**.

In *Larix* and *Pseudotsuga* the pollen grains are non-saccate and larger but they too, are borne by the wind as effectively as the saccate ones. In gymnosperms the non-saccate condition is common but this in no way hinders the pollen flight. These sacs fold under dry conditions during pollination. Wodehouse (1935) remarks, "If the bladders are organs of flight, pollen grains are possibly the only flying organisms of which it can be said that they fold up their wings and fly away". According to Doyle and O'beary (1935) and Doyle (1945) the sacs have a definite floation function, which is adapted to the inverted position of the ovules. The distended sacs of the pollen grain make it float up in the fluid – filled inverted micropyle. In the process the sacs point upwards so that the germ pore comes directly in contact with the nucellus.

MORPHOLOGY OF SACCATE POLLEN AND POLLINATION IN CONIFERS

Saccate pollen are restricted to, but not ubiquitous in, Phyllocladaceae, and are associated with the development of an inverted pollination drop except in *Phyllocladus*.

Tomlinson (1994) studied these pollen grains with greater emphasis on the pollination mechanism. He found that the saccate pollen is also non wettable, in the sense that it floats on water and it functions as a distinctive antigravity device. Pollen entry into the micropyle is dependent on resorption of the pollination drop. In contrast, non-saccate pollen is wettable and sinks in water. In coniferous families such as Cephalotaxaceae, Cupressaceae, Sciadopityaceae, Taxaceae and Taxodiaceae, there is a pollination drop without preferred orientation. Pollen swells and ruptures in water, shedding the exine by a mechanism determined by the apparently simple wall structure. In nonsaccate pollen of other conifers such as *Agathis, Araucaria, Larix, Pseudotsuga, Tsuga* and *Saxegothaea* there is no pollination drop, though pollen is wettable. They do not rupture in water and there is either extended siphonogamy or micropylar invagination **(Table 6.1)**.

Pollen reception in most of the living gymnosperms is mediated by pollination drop, exuded by the micropyle of the ovule. Therefore, the site of pollen reception is wet, and in a way pollination is hydrophilous. Doyle (1945) had emphasized the significance of the pollination drop, showing the diversity of mechanisms that exist for the engulfment of the pollen by the micropyle. An overview of gymnospermous pollen with special emphasis on conifers was summarized by Ueno (1960).

Picea

The pollen grains are larger but similar to those of *Pinus*. The sacs are small in proportion to the size of the grain. The furrow forms a shallow groove between the sacs (Wodehouse 1935).

Table 6.1 *A summary of information on coniferous pollen is incorporated below.*

Pollen Structure	*Family*
Grains nonsaccate; exine granular:	
Intine considerably thicker than exine; numerous orbicules adhering to the exine	Cephalotaxaceae, Cupressaceae, Sciadopityaceae, Taxaceae, Taxodiaceae
Intine equal to or somewhat thicker than exine; orbicules usually absent	Araucariaceae
Grains usually saccate; exine tectate, alveolar	Phyllocladaceae, Pinaceae, Podocarpaceae (sacci absent: *Larix, Pseudotsuga, Saxegothaea, Tsuga*)

Larix

The pollen grains of *Larix* are nonsaccate, spherical and comparatively large. The exine is thin, smooth, generally rupturing, and is cast off when moistened, intine is thick and hyaline. The pollen grains of *Larix* resemble those of *Pseudotsuga.*

Abies

The pollen grains of *Abies* are rather large with two bladders, which are variable in size, generally small in proportion to the grain size.

Cedrus

The pollen grains are bi-saccate, the sacs being smaller and more laterally placed than those of *Pinus*. In *Cedrus atlantica* the pollen grains show variable behaviour, some float, others sink or remain suspended in water for some time.

Gnetales

The pollen grains of *Welwitschia* are ellipsoidal, monocolpate with smooth exine marked with longitudinal ridges and grooves. *Ephedra* has ellipsoidal-ridged grains. *Welwitschia* and *Ephedra* have ellipsoidal, striate **(Fig. 6.18)** or polyplicate pollen, which are unique among extant gymnosperms. The pollen grains of *Gnetum* are spheroidal in shape with an echinate exine that is with conical spines.

Table 6.2 *Pollen morphological description in tabular form.*

Name	*Pollen Grains*	*Exine*	*Furrows*	*Intine*
Pinus nigra var. *poiretiana* (Ant.) Aschers & Graebn. Corsican Pine	In 3 parts, body and 2 air-filled bladders, the body an oblate ellipsoid with a roughened dorsal surface (cf. Wodehouse) surrounding the proximal pole and a ventral surface occupied mainly by the attachment of three-parts spherical bladders leaving a smooth area (the so-called 'furrow') in between, the whole usually presented either in long lateral view, symmetrical about the polar axis only or in one or other of the two possible polar views (a) proximal pole uppermost and bladders below or (b) distal pole, 'furrow' and bladders uppermost and proximal pole below. Body of grain rather large or large with polar axis 41.6 (37.4-50.0) µm and axes of the ellipse seen in polar view 49.5 (45.1-56.1) µm and 46 (38.3-51.0) µm respectively, body and bladders together in either view 78.0 (71.4-89.3) µm	Exine of body thickest over dorsal cap, where ectexine c.1.7 µm and endexine 0.8 µm, and the surface patterned with short bent worm-like grooves and interspersed ridges of similar width (vermiculate), about 10-12 grooves to every 10 µm, external walls of bladders consisting only of locally inflated ectexine, which here forms an open 3-dimensional network c.5 µm deep built up of rod-like bodies, the meshes nearest the surface varying from 1 to 3.5 µm across and not all closed, those below the surface much larger, up to 9 µm across, the bladders bounded internally by the endexine of the body.	One on the distal (ventral) surface, crossing the distal pole between the two bladders, rather ill-defined, width between roots of bladders 8.3 (5.2-11.2) µm, membrane thin, its surface smooth.	Intine of body is variable in thickness, 1.7-9.5 µm.

CHAPTER 7

Spore Morphology

A: SPORE MORPHOLOGY OF ALGAE, FUNGI AND BRYOPHYTES

Algae produce different types of spores known as akinetes, endospores, exospores, zoospores, aplanospores, oxospores, zygospores, oospores and cysts. Most of these spores are produced asexually, where as few of them are produced during sexual reproduction.

The morphology of some of these spores is helpful in algal taxonomy. However, as the algal spores do not find any applications, the algal spore morphology has been the most neglected.

In the family Zygnemaceae, the genera can be identified on the basis of the colour of zygospores. For example, *Sirogonium* produces black zygospores where as the zygospores with various shades of blue are produced by the species of *Zygnema, Zygnemospsis* and *Zygonium*. In addition to colour, the wall of the zygospores is known to be variously sculptured. Some of these sculptured patterns include granulate, verrucose, wrinkled, reticulate, foveolate, denticulate, punctate, scrobiculate, grooved and ridged.

The spore wall in *Bulbochate elatior* is scrobiculate whereas in *B. intermedia*, it is longitudinally ribbed. The spores of *Oedogonium echinospernum* have a typically echinate spore wall.

Charophytes produce oospores, which have been worked out in detail on account of their occurrence in the fossil sediments also. The oospores, which are either colourless or pale yellow to black in colour are characterized by a wall-possessing curved ridges. The number of these ridges is an important character in the classification of the living and fossil Charophytes. The oospores are further characterized by a spore wall which may be finely granulated, reticulated or tuberculated.

The fossil charophyte fruits are represented by a calcarious hollow body showing clearly the spiral arrangement of the envelope. Such fossil charophyte oogonia have been described as *Gyrogonites* from the late tertiary limestones in the vicinity of Gokak falls in Karnataka, India.

Akinetes, which are known to be thick-walled resting spores among the blue green algae, are very characteristic. These akinetes are either smooth, sculptured or papillate or without papillae. The akinetes vary considerably in their shape such as spherical, cylindrial or ellipsoidal.

Thus, it can be concluded that algal spore morphology is useful only in the interpretation of phylogeny and taxonomy of algae.

Palynology of Fungi

The variety of fungi represented by a large number of genera and species perhaps outnumber all other types of plants.They produce different types of reproductive bodies, which are collectively called 'spores'. The analogy between a spore and a seed can be explained by the fact that both of them germinate and produce the mother plant. The spores of fungi are unicellular or multicellular. The morphological features of fungal spores are studied in the same way as pollen morphology. Thus, the study of spore morphology includes the characters such as spore size, shape, wall components, and ornamentation.

Historically, the fungal spores were observed for the first time during the latter half of the 16th century by the Italian botanist J. P. Porta. The role of fungi in causing diseases by means of dissemination of spores was responsible for drawing the attention of botanists and medical men. A detailed account of certain airborne fungal spores and their significance in allergy and immunology has been dealt with separately in Chapter 16 'Significance of fungi as aeroallergens'.

As early as 1899, Saccardo attempted to classify fungal spores based on septation and colour. The germinal aperture, which is so significant in pollen germination is also noticed in a rudimentary form in certain fungal spores such as uredospores of rust fungi (Uredinales). Ainsworth and Bisby had recognized various types of fungal spore surface ornamentations such as punctate, verrucose, echinulate, aculeate, foveoate, reticulate and striate.

Similar to that of pollen grains, fungal spore morphology plays a significant role in the taxonomic considerations of fungi.

Palynology of Bryophytes

Bryophytes have a very long geological history and are known from Carboniferous to the present. They have played a very important role in forming ground vegetation since the remote past.

The spore morphology of Bryophytes has not received adequate attention though it has been proved that bryophytic spore morphology is very useful in taxonomy. Roth, at the beginning of the 20th century described spore morphology of mosses paying attention to their colour, size and sculpturing of the wall. Knox (1939) emphasized the exine ornamentation of spores of liverworts. Actually, moss spore morphology has been studied in greater detail than in liverworts. The spore wall was distinguished into three major layers from inside to outside, intine, exine and perine.

Among the liverworts spore morphology of different species of *Anthoceros* has been worked out extensively. Different species of *Anthoceros* can be identified on the basis of the characters of a triradiate mark and wall sculpturing pattern.

In the species of *Cyathodium* four evolutionary lines have been recognized based on spore morphology.

1. *Cyathodium acrostichum, C. africanum, C. tubrosum* have spiny exine in the spores.
2. Spores of *C. smaragdimum* have granulose exine.
3. Varrucose exine is seen in the spores of *C. foetidissimum*.
4. Spores with reticulate exine occur in *C. spruceanum*.

The noteworthy feature of bryophytic spores is the absence of aperture in the spore wall for the liberation of germ cells. It can be concluded that bryophytic spores have a limited value in the taxonomy and do not have any practical applications.

B: SPORE MORPHOLOGY OF PTERIDOPHYTES

Pteridophytes have been classified in many different ways by different pteridologists. However, according to the presently acceptable view, pteridophytes are classified into four major groups such as Pteropsida, Psilopsida, Lycopsida and Sphenopsida.

All these groups have well preserved fossil ancestors and representative genera and species. Isospores, miospores, microspores, megaspores and sporangial structures producing them have been well preserved in the form of fossils. Some of them find applications in palynostratigraphic correlation, which is widely used in the exploration of fossil fuels such as coal and oil.

Of all the groups of pteridophyta, spore morphology of ferns has been well worked out. Hence in this chapter spore morphological characters of ferns will be briefly described. The main objective of this account is to facilitate identification of fossil fern spores. Although the fern spore is in many ways similar to the pollen grains of seed plants, it occupies a position equivalent to the seeds of higher plants, being the chief instrument of dispersal. In fact, the spore is often loosely termed as the 'fern seed', and recognized as such in fern propagation.

The spores of pteridophyta, like the pollen grains of seed plants, have a highly resistant outer protective coat. Consequently they are well preserved in most of the fossil-bearing horizons, ranging from the Palaeozoic to the present, making them favourable objects for microfossil analyses. A detailed study of spores of living pteridophytes is essential for properly indentifying fossil pteridosphytic spores.

Historical Account

A general comparative study of the spores of living pteridophyta has been attempted by Knox (1938), to provide "students of fossil spores with some basis for comparison with recent types". She described spores of 39 genera of ferns and 7 of fern allies.

This was followed by a detailed and well-illustrated account of the spore morphology of the Lycopsida by Knox (1950) in which spores of over 480 sp. of *Lycopodium, Phylloglossum drummondii, Isoetes* and *Selaginella* are described. Erdtman (1957) published detailed illustrations of spores of over 100 genera of pteridophytes.

Over 300 sp. of Indian ferns belonging to Aspidiaceae, Aspleniaceae, Blechnaceae, Polypodiaceae Pteridaceae and Grammitidaceae have been palynologically investigated by Nayar and Santha Devi (1963).

Over 75 % of the pteridophytes are tropical plants, many of them with a restricted geographical range and often growing in inaccessible tropical jungles. The difficulties of procuring material, and of proper identification of the sample, may well discourage an amateur worker. The external morphology of the pteridophyte spores, unlike that of pollen grains, often undergoes marked changes immediately before the time of natural shedding.

In addition to acetolyzed preparations, it is advantageous to study fresh spores as well, as acetolysis sometimes alters spore-morphology (mainly the spore size) affecting differently the various species (Nayar and Santha Devi 1964).

There are some examples of acetolysis altering the general morphology of the spore coat. Thus, in *Plagiogyria,* the verrucae on the exine in many species are deciduous and are nearly always shed on acetolysis, leaving a smooth surface. Majority of the verrucae on the exine of some sp. of *Pyrrosia* and the Grammitid ferns are often shed on acetolysis. In *Pyrrosia nummularifolia* a thick ornamental layer of the exine is completely dissolved, and removed without a trace during acetolysis **(Figs. 7.1 a, b)**.

Spore Morphology

The following details are considered essential for the description of spores. Symmetry, shape, size and the nature of the spore wall.

Symmetry

Pteridophyte spores, similar to pollen grains are formed in tetrads. Depending on the arrangement of the individual spores in the tetrad they

a. b.

Fig. 7.1 Bilateral spore of *Pyrossia nummularifolia*: a. Spore before acetolysis showing perine b. spore after acetolysis.

are either bilaterally symmetrical **(Fig. 7.2)** or radially symmetrical (tetrahedral) **(Fig. 7.3)**.

The surface of the spore facing the centre of the tetrad is the 'proximal surface', and diametrically opposite to it, facing the periphery of the tetrad, is the 'distal surface'. It is often easy to locate the proximal surface in isolated spores, because it is on this side that the laesura is found.

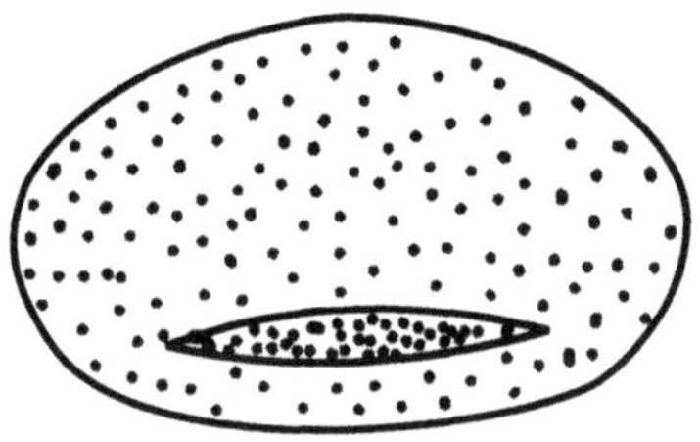

Fig. 7.2 Monolete bilateral spore with a single laesura as in *Dryopteris*.

Shape

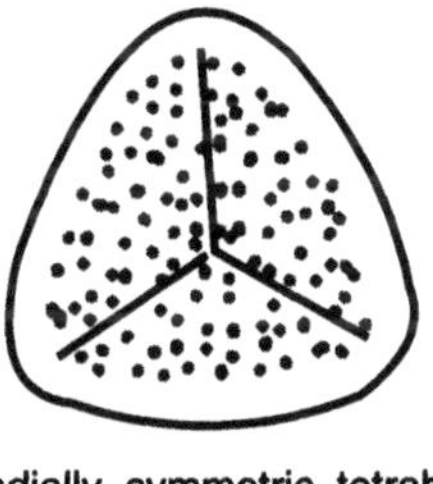

Fig. 7.3 Radially symmetric tetrahedral spore.

Shape in the polar as well as in the lateral view of the spore is significant. Pteridophye spores are rarely globose, but the tendency to attain this form is met with both among bilateral and tetrahedral types. The outline of the tetrahedral spore seen in polar view ('amb') is often triangular with either straight sides or variously concave or convex sides.

The amb of the bilateral spore may be 'ovate', 'oblong', elliptical or 'circular'. In the lateral view, the bilateral spore can be most commonly 'plano – convex' (proximal side flat, distal convex) or 'biconvex', the latter being uncommon.

Typically, a tetrahedral spore has a pyramidal proximal side and hemispherical distal side. The proximal side is sometimes flattened to various degrees or may even be concave.

Size

In the case of tetrahedral spores two measurements are often given which most probably represent the length of the longer side of the triangular amb × (MN × OQ) that of the short side, or the longest diameter in polar view × the shortest diameter in the same view if the amb is triangular with sides of uneven length as it often is and, or alternatively the length of one side of the amb × diameter in polar view, or rarely the palynological concept, polar diameter × longest equatorial diameter.

Currently it is accepted that spore size is to be expressed as the length of an imaginary line connecting the proximal and distal ends of the spore and passing through the centre (polar diameter p) × the diameter of the spore at the broadest region (equatorial diameter, E. For bilateral spores the equatorial diameter along the plane in which the spore is elongated (E_1, the length of the spore AB) as well as the diameter in a plane perpendicular to it (E_2, the thickness of the spore) are recorded **(Figs. 7.4 a, b)**.

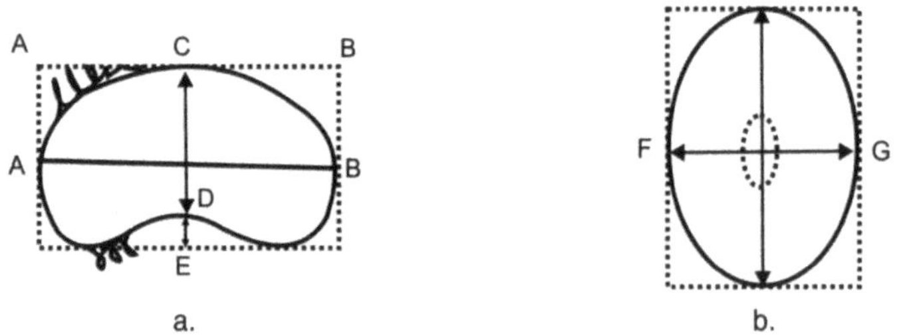

Fig. 7.4 Bilateral spore: a. Lateral view of spinose bilateral spore showing AB – equatorial diameter and CD=CE-DE=Polar diameter, b. Polar view of bilateral spore showing E1=AB-Longest equatorial diameter E2=FG – Shortest equatorial diameter (thickness of the spore).

In the case of concavo – convex bilateral spores, P (CD) is expressed as the distance from the distal pole to an imaginary line on the proximal side connecting two curved ends of the spore. CE minus the distance from this line to the proximal pole (DE), so that the extent of concavity of the proximal surface can be assessed.

For tetrahedral spores with amb having curved sides (convcave or convex), E is best expressed as the distance from one of the corners to the middle of a line connecting the two opposite corners, plus (in case of convex sides) or minus (in case of concave sides) the shortest distance from this point to the periphery of the amb (MN or OQ=OP + PQ or RS=RT-ST) **(Fig. 7.5)**.

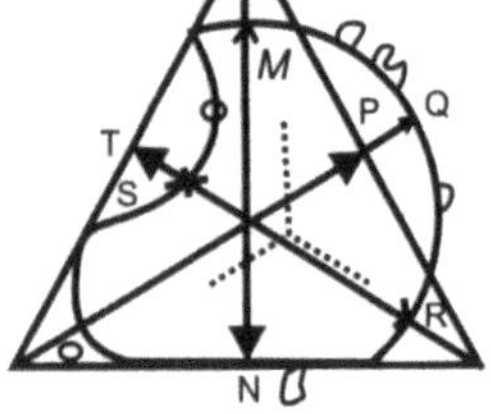

Fig. 7.5 Polar view of a verrucate tetrahedral spore with a triangular amb.

Nature of Laesura

The laesura is the mark at the proximal pole of the spore, and is an 'aperture' in the exine. In tetrahedral spores it is triradiate (trilete laesura), and in bilateral spores appearing as a straight line ('monolete' laesura). Only these two types commonly occur among modern pteridophyta, and their spores are either trilete or monolete, ('alete', spores, that is devoid of a laesura, are reported infrequently, for example *Equisetum*).

The average length of the laesura (one of the arms alone of the trilete aperture) is a useful character. The margins of the laesural aperture are sometimes thickened on the outer and or inner surface (crassimarginate and incrassate, Erdtman 1952) forming lip-like ridges, which in their turn may be sculptured.

The three arms of the laesura in a trilete spore occupy the margins of the three contact surfaces (the areas where each spore of a tetrad is in contact with the three other spores of the same tetrad during development), so that the laesural arms separate one contact surface from the other on the proximal surface of the spore.

Possibly depending upon the early and late separation of the individual spores from the tetrad, the contact surfaces are either nearly flat and thin, conspicuously separated from each other, or more or less rounded and then less conspicuously so, the laesura is more prominent with comparatively long arms in the former case and tending to be inconspicuous with short arms in the latter (some of the typically globose spores like those of some sp. of *Cheilanthes, Pellea,* and *Osmunda,* have faint short laesura).

Outline drawings of spores of *Loxogramme lanceolata* **(Figs. 7.6 a to f)** showing trilete, monolete and intermediary forms are given below.

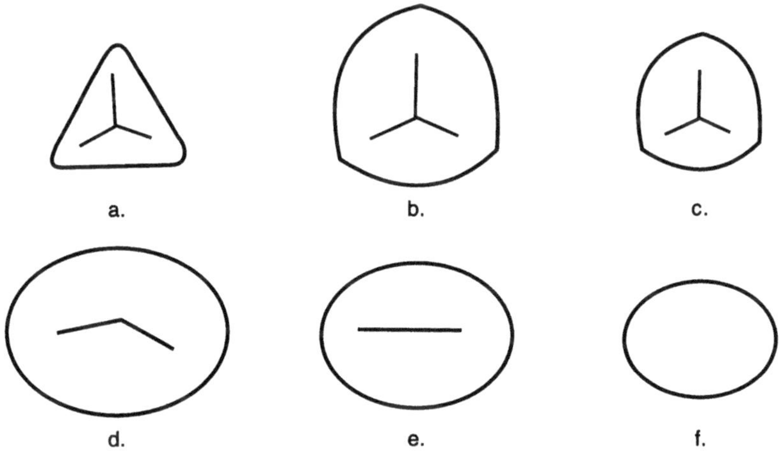

Fig. 7.6a-f Outline of spores of *Loxogramme lanceolata* showing trilete, monolete and intermediate forms.

In the case of bilateral spores, there are only two contact surfaces, and these are generally rounded. The laesura is a single, thin, elongated aperture on the proximal side between the two contact surfaces of the spore, resulting in the 'monolete' condition. The term 'monolete' denotes that there is only one letus constituting the laesura of a bilateral spore, but in all probability there is a pair of letae placed end to end.

The proof for this contention is found in the genus *Loxogramme*, in which some sp. possess both trilete and monolete spores as well as many intermediate forms. The laesura of the bilateral spore in this case is double the length of each letus of the typical trilete spore of the same sp. (Nayar and Devi 1964). In the majority of the spores of *Loxogramme lanceolata* and *L. forbesii* (in both of which trilete, monolete and intermediary forms occur mixed in the same sporangium) one of the arms of the trilete laesura is shorter than the two to a varying extent.

Accompanying this reduction in length is the gradual reorientation of the two longer arms. The angle of divergence between the long arms increases from the typical 120° of the normal trilete spore, till it is 180° when the third arm is completely suppressed (typical 'monolete' condition) and the spore is bitateral.

Nature of Spore Wall

The total thickness, colour and stratifiation of the exine are important morphological characters. The outer surface of the exine often bears distinctive markings ('ornamentations', or 'sculpturing'), which are of diagnostic value. The same descriptive terms as are employed in pollen descriptions, are used to describe the ornamentation pattern (granulose, verrucate, tuberculate, striate, spinose, etc.).

In some spores there is a covering outside the exine ('perispore' or 'perine'), which in many cases is loose, and separated by a cavity (or cavities) from the exine. The perine is often a thin membraneous structure, which is apparently fragile, but quite resistant and firm; it maintains its shape, characteristic markings and foldings even when acetolyzed and repeatedly centrifuged (**Fig. 7.7**). The perine is of as much diagnostic value as the exine. Important phyletic conclusions are often drawn from the nature of the perine in many groups. The same terms, as are used in the case of the exine, are employed to describe the ornamentation on the perine.

Fig. 7.7 Radial spore of *Pteridium* showing sporoderm with perine.

Spore morphology is an important character (mainly that relating to perine and the exine ornamentation) in taxonomic considerations of

Aspleniaceae, Lomariopsidaceae, Polypodiaeceae Pteridaceae and also a few other taxa of Indian ferns. Between the two species of *Diplazium, D. asperum* spores are non-perinous, whereas *D. polypodidoes* is perinous.

Morphology of some Cheilanthoid ferns has been described below.

Spore ornamentation varies in different species; sporoderm of the spore is laminated with distinct inner strata overlying the spore exine in Cheilanthoid ferns. Species of *Cheilanlhes* and *Aleuritopteris* can be distinguished on the basis of spore wall ornamentation. *Cheilanlhes albomarginata* **(Fig. 7.8)** has papillate spores whereas reticulate spores occur in *A. anceps* and rugulate spores are found in *A. dealbeta*. According to Nayar, (1963) spores in *C. bullosa* are verrucate, faintly granulose spores in *C. mysurensis* whereas psilate spores occur in *C. tenuifolia* and flap- like protuberances occur in *C. farinosa* **(Fig. 7.9)**.

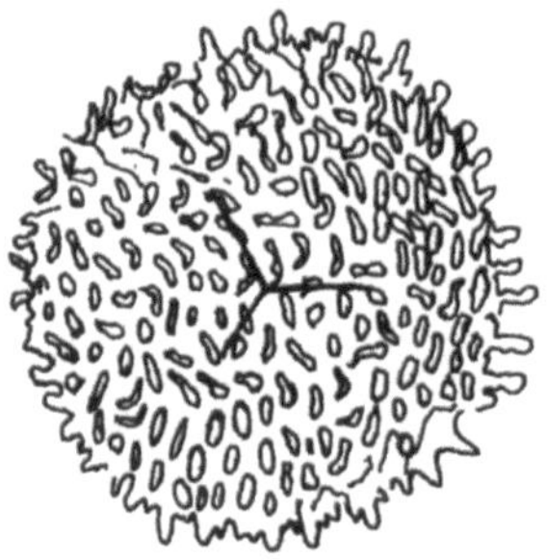

Fig. 7.8 Radial triradiate spore of *Cheilanthes albomarginata.*

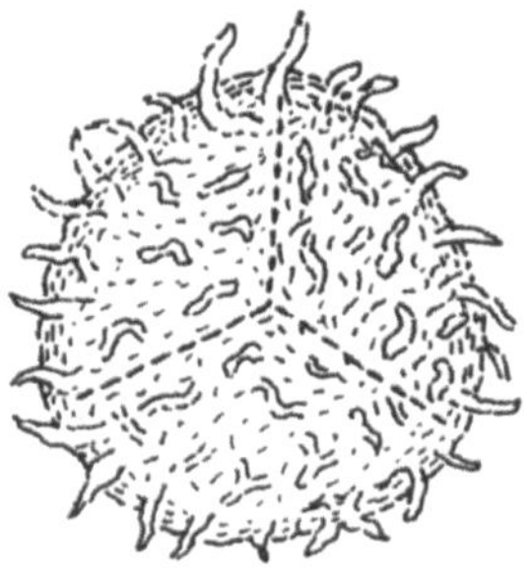

Fig. 7.9 Spore of *Cheilanthes farinosa* showing wall with flap like protuberances.

Spores of Cheilenthiod ferns are usually trilete and tetrahedral. Spore dimorphism in size and shape is observed in *Hemionitis arifolia* having both trielete and tetrahedral **(Fig. 7.10)** and monolete bilateral spores **(Fig. 7.11)** (Madhusoodanan and Jyothi 1993). It is also observed that within a sporangium, the bilateral and tetrahedral spores vary in size.The exine is smooth and the perine is highly ornamented with flap-like protuberances, which unite to form a reticulate pattern.

Fig. 7.10 Trilete spore of *Hemionitis arifolia.*

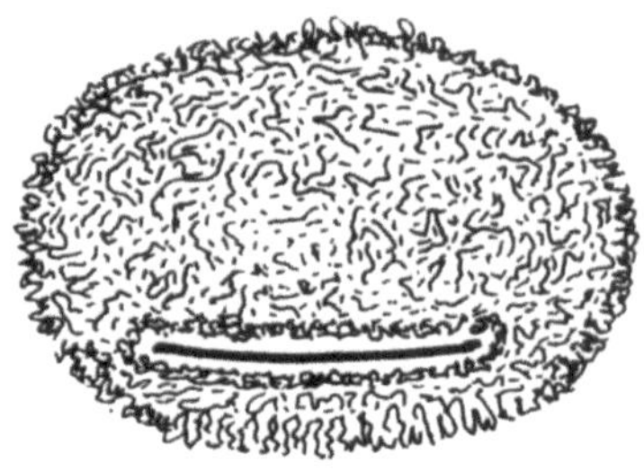

Fig. 7.11 Bilateral monolete spore of *Hemionitis arifolia.*

Cheilanthes albomarginata Clarke

Spores trilete, globose-tetrahedral, 45 x 51 µm, with little variation in size. Amb nearly circular. Proximal face convex to flatly conical, with straight or convex sides; Laesura arms 20 µm long, with faintly thickened outer margins.

The exine is dark brown, 3 µm thick; slightly thicker than the nexine and smooth. The perine is deep brown, closely adherent to the exine (skin-like), densely granulose and bearing crowded, thin, flap-like, irregular folds; the folds are apparently solid protrusions **(Fig. 7.8)**.

Tectaria polymorpha

Spores plano-convex (lat), 28 x 44 x 30 µm. Laesura 31 µm, its margins faintly thickened. The exine is 2.2 µm thick, brown. The sexine is considerably thicker than the nexine, densely spinulose, the spinules having sharp apices. The perine smooth, folded into a few elongated thin ridges protruding up to 8 µm from exine surface **(Figs. 7.12 a, b)**.

a. b.

Fig. 7.12 Bilateral spore of *Tectaria polymorpha*: a. Equatorial view of spore, b. Spore wall showing exine ornamentation.

Asplenium lunulatum Sw.

Spores 22 x 32 x 22 µm. Laesura margins faintly thickened. The exine is light brown, uniformly granulose. The perine is pale brown, closely adherent to the exine, minutely spinulose (spinules hyaline). The folds of perine are broadly conical in the optical section **(Figs. 7.13 a, b)**.

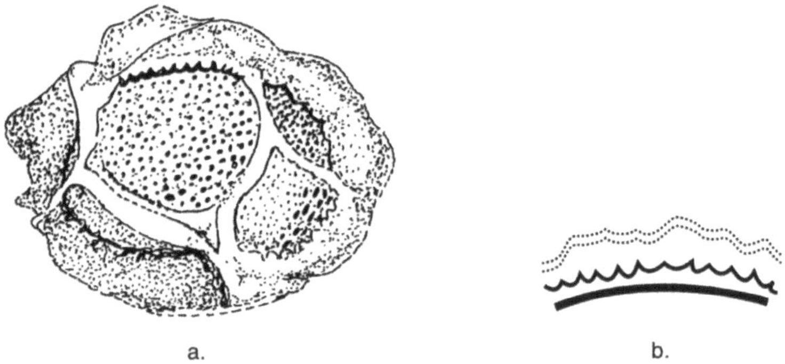

a. b.

Fig. 7.13 Bilateral spore of *Asplenium lunulatum*: a. Lateral view of spore, b. Exine stratification showing broadly conical perine fold.

Morphology of selected European Pteridophytes has been illustrated in **Figs 7.14a-g**.

Spore productivity in ferns resembles in its abundance some of the anemophilous gymnosperms and angiosperms. Calculations, for example, undertaken by R. C. Moran on the fern *Dryopteris carthusiana* revealed a spore count of over 7.3 million spores on a frond 60 cm (25 inches) long (Moran, 2004).

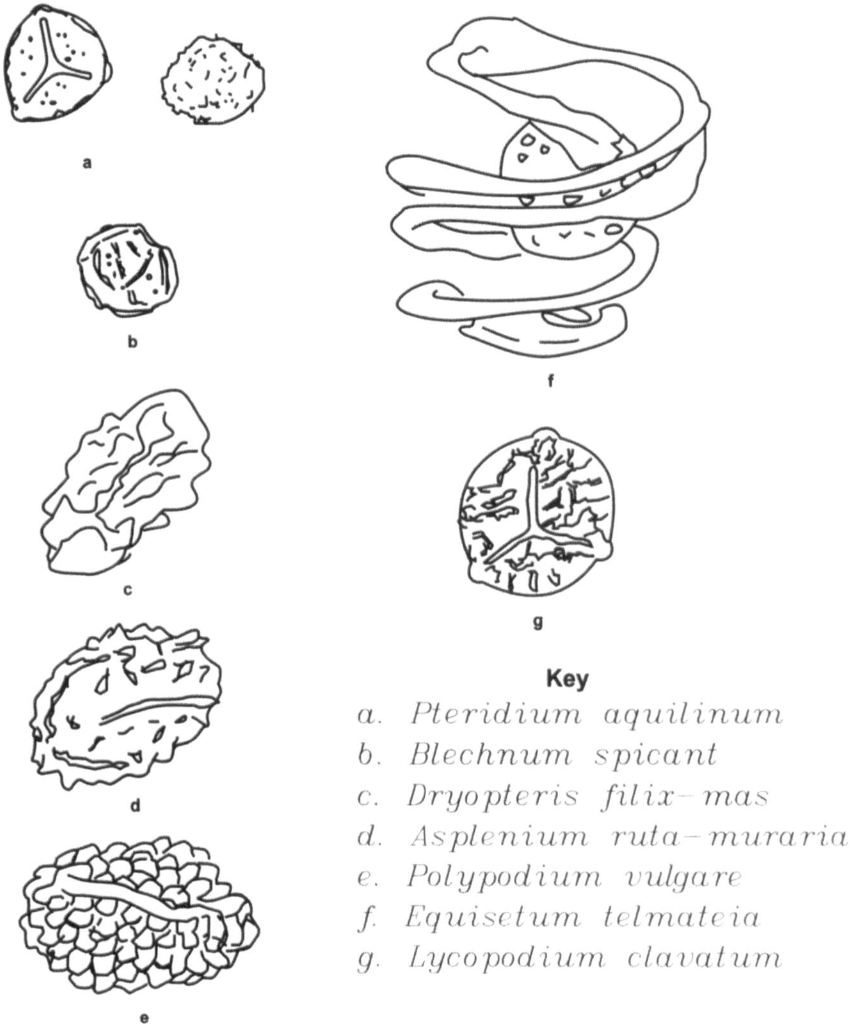

Fig. 7.14a-g Spores of selected European Pteridophytes.

Table 7.1 *Morphology of selected European Pteridophytes.*

Size	*Shape*	*Aperture*	*Surface*	*Structure*	*Exospore*
Pteridium aquilinum (L.) Kuhn. Spores (in genus)	23-29 µm	Tetrahedral to globose or bean shaped	Trilete – arms 2/3 of radius of spore	Irregularly granulate	Exospore thin, surface slightly undulate, perispore a thin continuous layer
Blechnum spicant (L.) Roth. Spores (in genus)	27-90 µm	Ellipsoidal to more or less spherical	Monolete ½ - ¾ length of spore	Plain to irregularly granulate, papillate or rugulate often foliose or with projecting folds	Exospore plain or gemmulate, perispore laminate, pillared with small rods, compact outer space
Dryopteris filix-mas (L.) Schott Spores (in genus)	20-70 µm	Ellipsoidal	Monolete ½ - ¾ length of spore	Regulate often tuberculately contoured by inflated folds, sometimes cristate or echinulate	Exospore plain, perispore single layer in short scales below
Asplenium ruta-muraria (L.) Spores (in genus)	23-60 µm	Ellipsoidal, rarely spherical	Monolete ⅔ - ¾ length of spore	Prominent folds to long wings, alete or cristate or reticulate or echinate	Exospore plain, perispore cavate with pillars, long laminate in inner part
Polypodium vulgare (L.) Spores (in genus)	33-80 µm	Ellipsoidal to fusiform	Monolete ½ - ¾ length of spore	Coarsley verrucate or tuberculate or papillate or echinate or winged	Exospore 2 layers forming surface contours, perispore thin sometimes folded or echinate
Equisetum telmateia Ehrh. Spores (in genus)	34.7-67.4 µm mean diameter of fresh spores	Spheroidal with spathulate elators	Very small, circular to oval	Granulate with scattered spherules	Exospore plain, epispore of one layer forming outer wall fused with exospore, elaters near aperture
Lycopodium clavatum (L.) Spores (in genus)	30-45 µm	Spheroidal, may be prolonged at triradiate arms, subtriangular in outline	Trilete, arms ¾ to equaling radius of spore	Often reticulate	Exospore reticulate or baccate surface, apertural fold clearly projecting, perispore thin, usually one layer

CHAPTER **8**

Morphology of Microfossils

During the maceration of sediments along with the pollen grains and spores, a large quantity of microfossils are also recovered, though their affinity with plants and animals is uncertain. These microfossils have abundant applications in interpreting stratigraphical and paleoenvironmental conditions, which have applied scope in the fossil fuel exploration. Technically, the study of these microfossils also forms part of the palynological investigation.

The overall composition of organic remains of these microfossils furnish very useful information concerning the environment in which the sediment accumulated and can aid in the evaluation of petroleum resources (Batten 1981).

As the name indicates the microfossils comprise minute remains of fossil organisms. These organisms lived in the past in a variety of habitats. The detailed study of microfossils got a real boost after realization of their applied importance, particularly in the exploration of fossil fuels.

A variety of microfossils belonging to different groups of plants, animals and intermediate types occurs abundantly in the oil-bearing sediments of different geologic ages. The microfossils of plant origin include pollen grains, spores, fungal filaments, bacteria, dermal appendages, cuticles, vascular elements and remains of algal groups such as Chlorophyceae, Charophyceae, desmids and diatoms. Among the animal microfossils the most prominent are Foraminifera, microforaminifera, ostracods, discoasters, Chitinozoa, Radiolaria, scolecodonts, and otoliths, whereas the intermediate types, i.e. types without definite ancestry, include members of the coccolithophorids, dinoflagellates, silicoflagellates and hystricosphaerids. These forms are considered to be an intermediate between plants and animals by virtue of their mixed characters.

DINOFLAGELLATES

Dinoflagellates comprise unicellular planktomic organisms ranging in size from 2-2000 µm, but most of them fall in the size range of 60-150 µm. The main body of these organisms is globular in shape and consists of cellulosic or proteinaceous cell walls. These organisms are further characterized by the distinctive character of the nucleus, position of vacuole-like structures called pustules, their biochemical constitution (with unique sterols and xanthophylls) and, most conspicuously their possession of flagella.

The Dinoflagellates reproduce by both asexual and sexual means. The sexual reproduction is known to take place during optimal nutritional conditions, whereas during less favourable environmental conditions asexual reproduction takes place, in many genera involving the formation a cyst. The cysts are formed primarily inside the shell of the motile Dinoflagellates.

The ornamentation of the cell wall of these organisms consists of polygonal plates and spines of various sizes. The Dinoflagellates commonly occur in the rocks ranging from Triassic to the present. They are known to occur mostly in marine environments but also found in sediments deposited in fresh and brackish water, living in lakes, estuaries. **(Refer Fig. 18.1a, b in Chapter 18)**.

ACRITARCHS

The term 'acritarchs' is derived from the Greek words *Akritous* meaning uncertain; *Arche* meaning origin indicating the organisms of doubtful origin. Earlier these organisms were classified as hystrichospheres.

The term was introduced as early as in 1963 by Evitt. It includes a large range of presumed algal bodies, mostly marine but also include brackish water or fresh water forms. Acritarchs are among the oldest of the fossil groups occurring abundantly in Proterozoic (Late Pre-Cambrian to the present). The principle character of an acritarch is its possession of a single central cavity partly or fully enclosed with in an organic vesicle. They range up to 100 μm in size. The wall of these forms is composed of sporapollenin. The wall pattern of these organisms is scabrae, spiny, reticulate, and an array of other sculpturing types. The study of acritarchs has helped considerably in the palaeoecological and stratigrafical correlations.

COCCOLITHOPHORIDS

These are also unicellular, spherical flagellates covered with numerous plates. These plates are either perforated or non-perforated and readily become separated from the main body of the organism. Though the coccolithophorids occur mostly in the marine sediments from the Jurassic to recent, some members have been reported from rocks as old as the Permian or Silurian. Given their restricted habitats, coccolithophorids are ideal indicators of marine environments.

CHITINOZOANS

Their geological range is from Cambrian to Permian. They are always found in marine rocks. They were partially or wholly planktonic. They may usually occur as chains but also as single individuals. They are characterized by thick, more or less opaque walls. Their origin and affinity are uncertain, though they have been consistently placed in the animal kingdom.

FORAMINIFERA

They have been classified in the group Protozoa. They have easily deformable shells composed of an organic substance variously referred to at different times as chitin, pseudochitin, keratin or tectin. They occur from Upper Cambrian to the present, though they are most abundant in Palaeozoic rocks. The walls of the Foraminifera are mostly calcarious, hyaline shells. These shells have a system of perforations, which permits the passage of pseudopodia. The Tertiary Foraminifera have been used for the correlation of oil-bearing beds.

MICROFORAMINIFERA

The forms frequently occur in marine rocks ranging from lower Createaceous to the present. They present chitinous inner shells of Foraminifera. As the name indicates the forms are much smaller than Foraminifera.

SCOLECODONTS

The geological range of these microfossils is from Cambrian to the present. They are believed to be chitinous mouthparts of marine annelid worms. Some of the scolecodonts can be easily detected in the macerated sediments, as they appear usually amber coloured and look like broken combs (hair brush type).

Microscopic Colonial Algal Forms

They range from Ordovician to the present. *Botryococcus* is a typical example of this group of colonial algae. It occurs in a wide range of fresh water to brackish aquatic environments. The walls of these algae are composed of hydrocarbon. *Pediastrum* is another common colonial alga, which occurs from early Cretaceous to the present. On account of rich hydrocarbon contents *Botryococcus* is referred as very important algal form used in the exploration of oil.

CHAPTER 9

Current Techniques in Palynology

Understanding and mastering the techniques for studying various aspects of pollen, both modern and fossil, is very important (Faegri and Iversen 1964). The primary purpose of using various palynological techniques is to undertake microscopic examination of pollen and spores. Essential pollen morphological characters can be studied only if proper techniques are employed by the investigators.

In case of fossil pollen and spores, maximum caution has to be taken to recover them in an undamaged condition from sediments in which they are embedded. Specialized techniques used in the study of pollen trapped in honey samples and airborne pollen are described in Chapter 12 on Melissopalynology and Chapter 13 on Aerobiology respectively.

In this chapter, first the techniques used for modern pollen are described, followed by techniques used for microfossils including fossil pollen and spores.

A: TECHNIQUES OF PREPARING MODERN POLLEN AND SPORES FOR MICROSCOPIC EXAMINATION

Fresh pollen grains when observed under a light microscope appear as dense objects and reveal only the colour, size, shape and faint outline of sculpture patterns and pollen wall excrescences. In order to study them more closely under a light microscope to reveal other important diagnostic pollen morphological characters, it is necessary to make some special pollen preparations.

Fresh or preserved pollen/spore materials are used for preparing type slides which may be used for detailed pollen morphological studies or for comparisons with slide material of airborne pollen/spores to confirm or otherwise identification. Such collection of slides over time enables 'back tracking' of taxa of pollen and spores for research purposes. Appropriate stains acetocarmine or basic fuchsin for pollen, cotton blue/methylene blue for fungal spores are used.

Wodehouse (1935) had suggested treatment of fresh pollen grains with absolute alcohol to dissolve the waxy coating of the outer pollen wall and staining them with methyl green. This way only exine is stained while the intine and cell contents remain unstained. In order to get a better contrast, use of weak aqueous cosine or fuchsin was proposed. This method is in use for preparation of index (reference) fresh pollen slides useful for comparison and identification of airborne pollen, which are essentially untreated. A revolutionary method of pollen preparation namely acetolysis was developed by Gunnar Erdtman and his chemist brother in 1933.

Acetolysis involves acetylation of pollen grains during which the carbohydrate fraction of grains (protoplasmic contents) is removed or dissolved, thus leaving only the exine with its diagnostic features such as ornamentation, apertures and exine stratification. The process of acetolysis can be referred as acid hydrolysis wherein cellulose; a polysaccharide is removed effectively. The function of acetolysis is to dissolve cellulose, hemicellulose, and chitin. A secondary effect of the procedure is the darkening of the pollen grains, which allows one to more easily examine aspects of pollen wall layers, apertures and surface patterns. The assets of acetolysis of pollen grains include bringing transparency and expansion of pollen to facilitate examination of pollen morphological characters.

Acetolysis really consists of two processes: The initial acid hydrolysis reaction is carried out by concentrated sulphuric acid. The secondary esterification process is carried out by the acetic anhydride. Acetylation, on the other hand refers only to the second process of esterification, yet in today's literature it has become synonymous with acetolysis. In general, the effectiveness of the acetylation process is increased directly as the temperature and time is increased. Thus far, there is no evidence to suggest that it has any harmful effect upon sporopollenin even after long periods of duration.

Some of the sulphur ions attach to the pollen grains and thus tend to stain them brown. An oxidant will effectively remove excessive sulphur ions from the grains. The sulphuric acid acts first as the components for the acid hydrolysis step and then secondly to hold the ester components in solution. The acetic anhydride carries out the actual acetylation step by breaking the cellulose into esters and ester derivatives

Acetylation works much more rapidly since it attacks the OH (hydroxyl) group along the periphery of the cellulose molecule thereby breaking it down. Also, acetylation does not produce the water-soluble compounds found in the acid hydrolysis method but does produce compounds, which must be removed with glacial acetic acid. This is one reason why after acetylation is completed; the material is washed in glacial acetic acid.

Step by Step Acetolysis Procedure

Erdtman's (1960) acetolysis technique is the most popular method, which is followed almost throughout the world. The different steps involved in this technique are given below:

1) Collection and preservation of pollen material: It involves collection of anthers from mature flower buds just before opening and anthesis of flowers by using clean forceps and preserving the anthers or sometimes anthers with a portion of filament of stamens in clean glass vials with 70% glacial acetic acid. Some palynologists recommend 70% ethyl alcohol for preserving the anthers in place of glacial acetic acid. In this condition, the anthers packed with pollen can be stored indefinitely.
2) Anthers are transferred to a centrifuge tube and washed thoroughly with distilled water. The anthers are crushed with a glass rod and the mixture is filtered through cheesecloth or a sieve (c. 180 μm mesh). The lower fraction is recollected in the centrifuge tube.
3) Freshly prepared acetolysis mixture consisting of acetic anhydride and concentrated H_2SO_4 (sulphuric acid) in the ratio of 9:1 was poured in the centrifuge tube containing washed pollen. Care has to be taken to add conc. H_2SO_4 to acetic anhydride drop by drop to avoid excess heating. The acetolysis solution reacts violently with water. Glacial acetic acid washes are necessary to replace water with acetic acid, which does not react violently with the acetolysis solution. Although acetolysis procedure varies from palynologist to palynologist, the following procedure seems to be most suitable. From the first to last acetic acid wash, caution has to be taken so that no water comes in contact with the solution. Only distilled water has to be used for washes to avoid any contamination.
4) The centrifuge tube containing pollen in suspension of an acetolysis mixture is placed in a water bath and heated at 70°C for 5-10 minutes, stirring the contents with a glass rod intermittently. The time required for acetylation varies from material to material. Pollen after completion of acetolysis process turns light to golden brown in colour.
5) Acetolyzed pollen suspension is centrifuged and supernatant liquid is decanted.
6) Pollen sediment in a centrifuge tube is immersed in glacial acetic acid for a few minutes and the superfluous liquid is decanted after 5 minutes centrifugation (c. x 2000 rpm). This is washed with distilled water 2-3 times.
7) The pollen material is ready for mounting on the slides for microscopic examination. However, it can be stored in this condition in vials containing 50% glycerine.

8) Acetolyzed pollen material is mounted in either glycerine jelly, polyvinyl alcohol, lactophenol, Canada balsam, corn syrup, or synthetic resin: piccolyte. The choice of the mounting medium depends on the individual palynologist. It is proved beyond doubt that molten glycerine jelly prepared by using the following method is the most ideal one. Pollen mounted in glycerine jelly remains unaffected even after 4-5 decades. Water solubility and viscosity of glycerine jelly makes it an ideal mounting material for pollen.
9) After mounting the cover glass on acetolyzed pollen in glycerine jelly, edges of the cover glass are sealed with paraffin wax or nail polish to avoid desiccation.

As mentioned earlier, the assets of acetolysis of pollen grains include bringing transparency to and expansion of pollen to facilitate examination of pollen morphological characters. However, there are some disadvantages of this technique. The pollen grains sometimes get over expanded and become unfit for original size measurements. It has been reported that fossil pollen and modern pollen when subjected to acetolysis expand to different degrees. Hence, it is advisable to use unacetolyzed pollen or spores when size determinations are critical.

Sometimes discolouration of pollen is required, which is achieved by chlorination of sample as follows:

Add 5 cc of glacial acetic acid to a centrifuge tube containing pollen material and stir the contents. Then add 1-3 drops of saturated solution of sodium chlorate with a 0.5 cc pipette. Follow this with 1-2 drops of conc. HCl. The solution changes colour in 1-2 minutes as a result of liberation of chlorine gas. Centrifuge sample, decant chlorination mixture, add water, transfer to the acetolysed fraction, stir and follow the other regular actolysis procedure.

PREPARATION OF GLYCERINE JELLY

Kisser's method of preparation of glycerine jelly is usually followed by palynologists. The following ingredients are required for glycerine jelly preparation:

Gelatin	50 gms
Distilled water	175 ml
Glycerine	150 ml
Phenol crystals	7 gms

Gelatin is first soaked in 175 ml cold water for 2-3 hours. Superfluous water is discarded and gelatine is heated till it melts. 150 ml of glycerine is added to molten gelatin and while still hot it is filtered through spun glass pressed into the lower part of the heated funnel. Glycerine jelly has its natural light golden yellow to brown colour. It is stored in Petri dishes or a suitable glass container in a refrigerator.

The acetoloysis procedure described above is good for fresh pollen or spores. However, when fresh pollen material is not available, palynologists have to rely on herbarium material. Several methods have been proposed for pretreating pollen and spore material extracted from herbarium specimens and then subjecting it to Erdtman's acetolysis method.

Dahl (1952) suggested softening the anthers extracted from herbarium material in 10% NaOH or KOH and later subjecting them to lactic acid. Canright (1953) preferred to mount dry pollen from herbarium specimens in lactic acid, which expands the pollen.

Normally the acetolysis procedure for preparing pollen slides takes about 30 minutes. Several palynologists have suggested quicker methods for pollen preparation. Ikuse (1956) preferred to dissect the pollen from the anther in a drop of 95% alcohol; followed by heating to evaporate the alcohol, and then staining with either gentian violet or fuchsin in 95% alcohol. The material is further washed and dehydrated in absolute alcohol followed with a drop of xylene before mounting with Canada balsam.

Avetissian (1950) recommended a simplified method of acetolysis for preparing modern pollen on the slide. It is a quicker method, which takes 3-4 minutes compared to 25 to 30 minutes for Erdtman's acetolysis procedure. It is especially valuable for use with small quantities of pollen as there is very little loss in the procedure. The whole process can be watched under the microscope. The following are the various steps involved in this time-saving procedure.

1. Place the anther on a slide and cover with a drop of the acetolysis mixture (acetic anhydride and conc. H_2SO_4 in the proportion of 9:1).
2. Dissect pollen from anther with a needle. Scatter the liberated pollen in a thin layer over the slide in a spot about the size of the cover glass.
3. Heat the slide over a spirit lamp and observe the progressive colour change under the microscope. Stop heating when the protoplasmic contents are removed and exine is clear to dark brown.
4. Cool and wash the preparation with a drop or two of alcohol.
5. The pollen can be stained with fuchsin, gentian violet or methyl green.
6. Mount in glycerine jelly.

TEM AND SEM OF POLLEN FOR MORPHOLOGICAL STUDIES

Transmission Electron Microscopy (TEM)

The exine stratification is best studied by making ultra-thin sections and examining them in transmission electron microscope. The sections are post-stained in, for example, lead citrate and uranylacetate to achieve good contrast.

Scanning Electron Microscopy (SEM)

In order to see exine surface details better, scanning electron microscope is a useful instrument. The exine architecture and the inside of exine and apertures may be studied with this microscope. Due to the depth of focus in SEM, three-dimensional views of the objects are observed.

DETAILED PROCEDURE OF PREPARATION OF POLLEN GRAINS FOR SEM STUDIES

The method proposed by Falk (1980) is found to be most suitable for preparation of pollen grains for SEM studies.

The pollen grains are first subjected to acetolysis and they are transferred to 20% alcohol. Gradual step-by-step dehydration of pollen grains is carried out by transferring acetolyzed pollen through series of alcohol concentrations like 30%, 50%, 70%, 95% and finally in 100% acetone for a period of 30 minutes. The dehydrated pollen grains are then dusted on the smooth surface of an adhesive tape pasted onto the stub. These stubs with pollen grains mounted on them are further dehydrated by placing them in an oven at 60° C for a period of 30 minutes.

The stubs are then coated with gold in a sputter coat (JEOL JSM 1100). The coated pollen grains are observed in the SEM (JEOL JSM 5600 LV) and photographed.

B: TECHNIQUES USED FOR RECOVERING MICROFOSSILS INCLUDING FOSSIL POLLEN AND SPORES FROM SEDIMENTS AND PREPARING THEM FOR MICROSCOPIC EXAMINATION

Fossil pollen was observed as early as 1836 by Goeppert and in 1839 by Ehrenberg. Since then many techniques have been invented and used by palynologists. As mentioned early it is difficult to have a standard uniform method as many times the method depends on the choice of the palynologist, type of material and objectives. However, common goals for their techniques are outlined below. It should be remembered that the main purpose of these techniques is to extract or recover pollen and spores from the variety of sediments such as peat, lignite, coat, shale, clays, soil in which they have been entombed in the past.

The techniques used for recovery of microfossils differ with various samples depending on the geological age, lithology, chemical composition of the cementing material, the presence and amount of organic matter and the degree of their carbonization. However, successful results in this regard can be obtained if the following guidelines and objectives are taken into consideration.

1) To get rid of the matrix material as far as possible without loss of organic matter.
2) Concentration of the organic material as much as it may be representative for different types of samples and the forms recovered are not damaged.
3) Processing of samples should take the least time to achieve best results.

Irrespective of type of sample, certain essential steps are involved in the recovery of microfossils, which have been shown in the following flow chart.

Pre-acid preparation
↓
Mineral removal
↓
Concentration/Preparation for microscopy

Precaution has to be taken to see that the rock material to be analyzed for pollen contents is crushed carefully, neither is it ground nor are the pieces left so large, as chemicals may not to be able to penetrate and react. Crushing of the sample should not be so violent so as to destroy microfossils preserved in it. Carbonates are broken by treating the samples with dilute HCI, whereas silica and other minerals are dissolved in HF (Hydrofluoric acid). The humus and other organic matter is dissolved by weak alkali treatment.

The following procedures are to be used for recovering microfossils from specific types of sediments. These techniques essentially involve maceration of samples.

Peat

It is a type of loosely bonded sedimentary rock, which is loaded with organic matter including pollen and spores with only a minimum of sand or clay. The most common method for collecting peat samples is to take cores from subsurface area of peat deposits by using Hiller type of core borer as illustrated in **Fig. 9.1**. Several palynologists employ different techniques but the one suggested by G. Erdtman and followed by several palynologists including Lagerheim, Wodehouse and Faegri and Iversen, appears to be simple and the most efficient.

The technique involves boiling a piece of peat sample in 10% NaOH or KOH on a slide until the water evaporates. Add a small amount of glycerine, mix thoroughly and transfer the material to a clean slide, mount a cover glass over it. The procedure can be performed in a small beaker. Erdtman's acetolysis technique including bleaching or chlorination can be used for

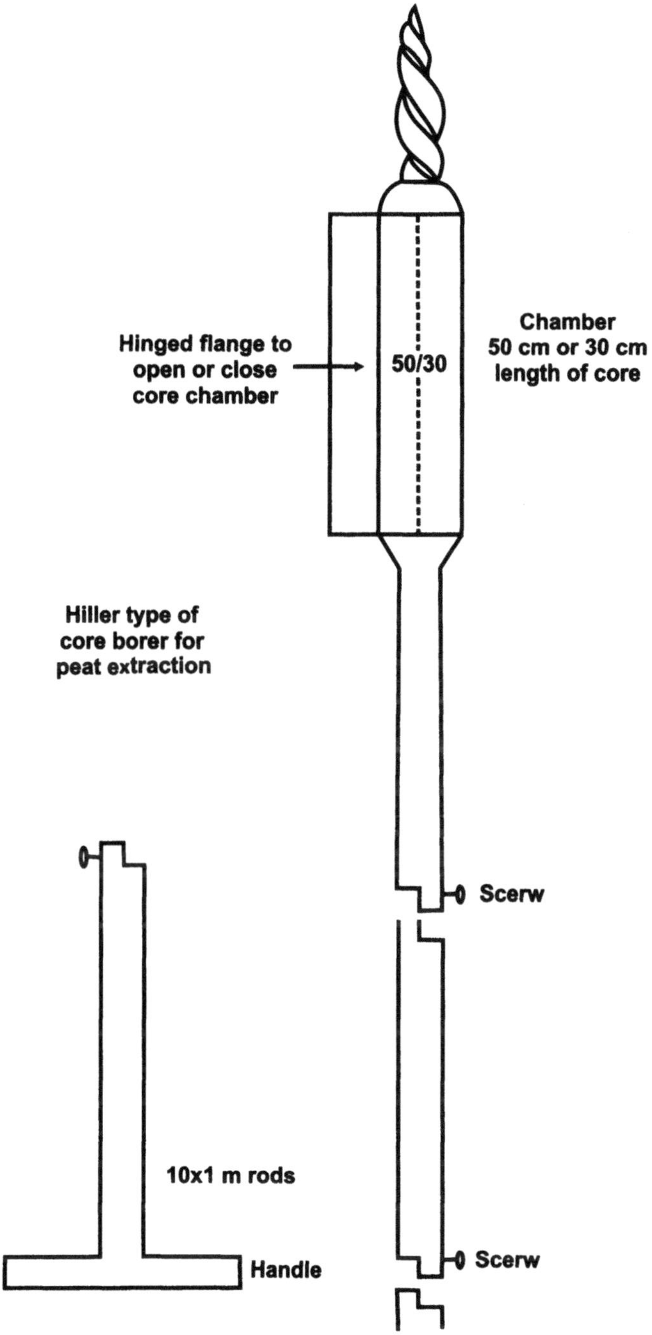

Fig. 9.1 Diagrammatic sketch of Hiller type of core borer used for collecting peat samples from subsurface area of peat deposits.

recovering well preserved pollen and spores from peat samples. If necessary, demineralization of peat sample can be done prior to alkali treatment or acetolysis procedure.

Lignite

It is geologically older than peat and occurs in early Tertiary sediments. It is also referred to as brown coal. The richest lignite deposits in India occur in Neyvile lignite mines in Cuddalore sandstone series of the Miocene age.

Most lignite samples can be macerated by soaking in dilute KOH (10%) for few hours to several days depending on the hardness of the material. The lignite sample can be boiled for 5-10 minutes to dissolve the humic materials. According to Clair A. Brown, 10% KOH is corrosive to some pollen and spores and therefore use of 5% KOH is recommended.

The following step-by-step procedure can be followed for maceration of lignite.

1. Boil 1 gm of powdered lignite in a beaker or small Erhlenmeyer flask in 20 cc of 5% KOH for 5-10 minutes.
2. Pour into centrifuge tubes, centrifuge and decant. Add remainder and repeat.
3. Resuspend in distilled water, centrifuge and decant. Repeat once or twice or until liquid is nearly clear, or until alkali is washed out.
4. Residue can be mounted with or without staining.
5. Alternately, as per Iversen's suggestion the residue can be dehydrated in glaciad acetic acid and subjected to the acetolysis process.

Traverse (1955) modified the above procedure by prior demineralizing lignite sample in HF and further bleaching in 5-7% sodium chlorite in dilute HCL, followed by acetolysis and mounting macerated sediment in glycerine jelly.

The following procedure for maceration of lignite for the recovery of microfossils adopted by Rao and Vimal (1952) was used for miocene lignites from the Warkali formation in Kerala, South India. This procedure appears to be most appropriate for Indian lignite samples.

1. Soak lignite in conc. HNO_3 for at least 12 hours.
2. Wash with distilled water several times.
3. Treat with 10% KOH for 2 hours.
4. Remove silica by heating residue in HF for 1 hour.
5. Rewash and mount residue in pure glycerine, seal with Canada balsam.

Coal

Coal is a very compact stratified organic sedimentary rock also classified as a perfect example of compression fossil. In addition to carbonized plant matter it contains several minerals, oils, etc. Coal samples when macerated show that they are loaded with microfossils such as gymnospermous pollen grains, pteridophytic spores, leaf cuticles, vascular elements, etc.

There are three major maceration techniques for liberation of pollen and spores from coal, namely the Schultze method, nitric acid combinations and bromine-nitric acid process. Each of these methods has undergone minor modifications by various palynologists. Schultze was primarily interested in the chemistry of wood and coal, treated some coal with a mixture of nitric acid and potassium chlorate, and followed this maceration with a bath of ammonia. This mixture of nitric acid and potassium chlorate used for maceration and oxidation of coal samples came to be known as Schultze's solution.

The macerated sediment revealed a good quantity of vascular elements, pollen and spores. This fortuitous discovery by Schultze (1855) has been a material aid to coal palynology. The use of potassium chlorate (used in explosives) which accelerates the oxidation process is not always necessary.

Kosanke (1950) had successfully used Schultze's maceration technique for coal samples. This method can be also used for other sedimentary rocks such as shales or sandstones. However, microfossil recovery from carbonaceous shales and sandstones requires treatment with HF.

The following steps are involved in Kosanke's maceration procedure:

1. Place coal fragments or powdered coal in Schultze's solution. (One part of saturated, aqueous solution of potassium chlorate and two or three parts of cold, concentrated nitric acid. This solution oxidizes the coal and different coals require different lengths of time.
2. After oxidation is complete, wash out the acid and add 10% KOH. The potassium hydroxide dissolves humic materials and liberates pollen and spores. The time varies from 15 minutes to 12 hours.
3. Wash the residue through a 65-mesh sieve.
4. Stain spores with Safranin Y for 10-12 hours.
5. Wash the sediment, decant, centrifuge, and run into solvents for mounting in diaphane.

Tschudy (1958) modified the maceration procedure as mentioned below. This procedure seems to be very successful in recovery of higher percentage of fungal spores, pollen and spores from the coal samples.

The following steps are involved in this procedure:

1. Take 1 gm of pulverized sample.
2. Soak over night in 52% HF to remove non-organic materials.

3. Wash out acid.
4. Treat residue with Schultze's solution (75% nitric acid and crystalline sodium chlorate.) Oxidation is rapid. Stop at end of 15 minutes. Centrifuge and decant.
5. Wash with water, centrifuge, and decant.
6. Treat with a mixture of 80% acetone and water for 15 minutes. Centrifuge and decant.
7. Wash through several changes of water until supernatant liquid is clear.
8. Transfer to sample vials with equal volume of water.
9. Mount for microscopic study.

Use of bromine for maceration of coal was recommended by Bharadwaj (1957). The steps involved are:

1. Crush 15 to 20 gm coal into pieces 2 to 5 mm sizes.
2. Place them in wide-mouthed, glass-stoppered one-litre bottles. (Set up 6 numbered bottles at one time.)
3. Place in acid-fast sink under hood. Equip sink with frame arranged so that horizontal bar with setscrews centre on top of each glass stopper.
4. Add 5 to 8 cc of bromine to each jar, shake thoroughly, place in sink and allow to act over night.
5. Next morning fill sink with running water to a depth of 6 cm. If temperature of water is above 5 to 10°C, add ice.
6. Open bottles and add 2 to 4 cc of fuming nitric acid to each jar. Adjust setscrews so that stoppers are not too tight. (The worker should wear rubber gloves, a rubber apron, and gas mask.) Stoppers should be tight if treating rich coals as higher pressure is necessary.
7. After every 20 to 30 minutes, add 15 to 25 cc of nitric acid until the jar is 1/3 full. Shake jars frequently to allow uniform maceration. Maceration takes place in 2 to 8 hours, depending upon the kind of coal.
8. Add small quantities of water at 5 to 10 minute intervals, when maceration is complete, until jar is full.
9. Pour contents of jar on Müller gauze (silk screen) and wash simultaneously with a thick spray of water. Add water to jar and pour until all contents are removed. Wash until all acid foam is gone.
10. Invert sieve over a large porcelain dish and work residue into the dish. Decant excess water.
11. Divide residue into two parts A and B. A. Boil 2 to 4 gm in 10% KOH. B. Soak remainder in cold 10% KOH for 5 to 10 minutes.
12. Centrifuge, wash, decant boiled fraction until clear. It contains microspores.

13. Resieve and wash residue B until water is clear. Dry on sieve at 50°C. This fraction contains megaspores and cuticles.
14. Mount these microfossils on a microscope slide with suitable mounting medium.

Surange *et al.* (1953) recommended the use of Harris' bulk maceration process to recover megaspores from coals of India as per the following procedural steps.

1. Wash coal in running water then place in commercial HNO_3.
2. Wash off acid after the coal has macerated.
3. Heat residue in 10% KOH and allow to stand for at least an hour.
4. Remove alkali by rapid decantation.
5. Transfer residue to screen and place under running water. Care is needed as the megaspores swell when heated in alkali and become very delicate.
6. Pick dry megaspores from screen or paper under screen; or pick wet megaspores directly from the water.
7. Pick dry megaspores under binocular microscope.
8. Treat individual megaspores in HNO_3 to reduce black colour.
9. Remove silica fragments by placing megaspores in 20% HF over night. Wash off acid.
10. Mount spores in glycerine jelly.

Gravity Separation Technique

When large number of sedimentary samples are to be investigated for microfossil contents in a short time as done by palynologists of oil and coal exploration laboratories, the gravity separation technique is most suitable.

As the matrix, organic matter, spores and pollen grains have different specific gravities, different portions can be separated by floating the processed samples in heavy liquids of about 2 specific gravity. Generally, spores and pollen grains range between the specific gravity 1.3 to 1.7. Funkhouser and Evitt (1959) have suggested the use of following:

1. Saturated water solution of zinc chloride whose specific gravity could be taken to 1.96. The basic ingredient here is cheap and need not be reclaimed.
2. Where the heavy liquid of higher specific gravity is required the mixture of zinc iodide, potassium iodide and cadmium iodide may be used mostly, retaining its specific gravity to 2.3. The ingredients are costly and can be reclaimed without any apparent loss and could be used again and again.

POST ACIDIZATION REMOVAL OF UNWANTED DEBRIS

Inorganic Remains

Heavy minerals

Undigested mineral matter is removed primarily by differential density centrifugation (heavy liquid separation). The commonly used heavy liquids are solutions of zinc chloride or zinc bromide, or a bromoform/alcohol mixture. Variations of density separation technique are outlined by Gray (1965a) and Traverse (1988). Some eastern European laboratories reportedly now use a 'potassium cadmium mixture' If opaque sulphide minerals are present, samples should be cleaned with an oxidizing agent before heavy liquid treatment. White (1988) indicated that the specific gravity of palynomorphs increases with the degree of carbonization, which may impact density separation. A survey of the literature indicates that specific gravities of heavy liquids used by palynologists range from 1.65 to 2.0. The specific gravity must be appropriate for the research goal. For example, most forensic studies require a 2.0 s.g. because it is less likely to be challenged in court. Many investigators of recent pollen advocate that lower specific gravity should be used. Litwin and Traverse (1989) recommended that pyrite be removed prior to density separation.

Fine minerals (clays)

Fine inorganic material can be removed either chemically or mechanically before or after heavy mineral separation. Sieving and swirling may also be helpful. A surfactant detergent will deflocculate and suspend clays during short or differential centrifugation (Gray 1965 a, b; Barss and Williams 1973; Litwin and Traverse 1989). The clay then can be decanted with the supernatant. The most widely used surfactants are Quaternary-O, Darvan No. 4 (Hills and Sweet 1972) and Brij 35. Algonox, and Alcojet (Peterson *et al.*, 1983). Surfactants may damage some microfossils (e.g. may etch Foraminifera), but there is no evidence that they harm organic-walled microfossils (Hodgkinson 1991).

Finally the standard extraction techniques of palynomorphs from rock samples are summarized in the following flow chart.

Raw sample
↓
Log in and wash sample
↓
Hydrochloric acid
↓
Siphon and flood with distilled water

↓

Hydrofluoric acid

↓

Siphon/pour and flood with distilled water
(acidified with hydrochloric acid)

↓

Siphon and transfer sample to
plastic centrifuge tube and pack

↓

Darvan and distilled water wash
(to suspend clays) 5 minutes @ 1,500 rpm-decant

↓

Wash in distilled water 5 minutes
@ 1,500 rpm-decant

↓

Wash in distilled water 5 minutes
@ 1,500 rpm-decant

↓

1:1 mixture Hydrochloric acid:
distilled water (hot bath) 30 minutes
(to remove possible Ca^{+} ions)

↓

Allow sample to cool, Pack 10 minutes
@ 2,000 rpm-decant

↓

Darvan and distilled water wash
5 minutes @ 1,500 rpm-decant

↓

Distilled water wash
5 minutes @ 1,500 rpm-decant

↓

Distilled water wash
5 minutes @ 1,500 rpm-decant

↓

Distilled water wash
5 minutes. @ 1,500 rpm-decant

↓

$ZnBr_2$ separation
5 minutes @ 1,500 rpm-25 minutes
@ 2,000 (30 minutes total)
(Note: some individuals prefer to

oxidize prior to this step)

↓

Pour float into 400 ml
Beaker with Darvan and distilled water

↓

Heavy fraction

check sink-
discard if barren

Light fraction

↓

Siphon-transfer or pour to
Centrifuge tube and pack
5 minutes @ 200 rpm
Mount and examine
Unoxidized slide

↓

Sieve and/or sonify if needed.
Re-examine residue

↓

Oxidize if necessary
(e.g. Schultze, nitric acid, clorox)

↓

Distilled water wash

↓

Distilled water wash

↓

Distilled water wash

↓

Stain if necessary
(e.g. Safranin O)

↓

Mount residue

↓

Label and dry slides in oven
(duration and temperature
depends on mounting medium)

↓

Store final residue

↓

(Note that distilled water wash is given when using centrifuge)

↓

(Adapted from Wood *et al.,* (1996)

CHAPTER **10**

Pollen Physiology (Palynophysiology)

Pollen physiology has attracted the attention of plant breeders and horticulturists for plant improvement programmes ever since the discovery of the pollen tube by Giovanni Batista Amici (1924), an Italian astronomer and mathematician, while examining the papillate stigma of *Portulaca oleracea*. Physiological studies have mainly centred round pollen germination, storage, and on the artificial induction of pollen sterility in cultivated plants, to be of benefit in large-scale hybridization programmes. Very scanty information is available in pollen germination of Indian trees. An interesting development in the pollen studies is the study of the influence of pollen physiology on the mode of pollination and breeding systems of plants. In pollen physiology, the various aspects normally studied are: pollen chemistry, pollen storage, pollen viability, pollen germination *in vitro* and *in vivo*.

POLLEN GERMINATION

Each mature pollen grain upon germination produces two nuclei by mitotic division of the haploid nucleus of the dehisced grain. One of these nuclei (vegetative) controls the growth of the pollen tube as it progresses down the stylar tissue, the remaining nucleus (gametic) follows down the tube as it grown towards the micropyle of the ovule, at which point the gametic nucleus bypasses the vegetative one, and enters the micropyle to fuse with the haploid ovule nucleus, thus restoring the diploid state with regard to chromosome and gene complement, to form the zygote from which develops the embryo plant upon seed germination **(Fig. 10.1)**.

The arrival of a pollen grain upon the stigmatic surface of the gynoecium (female part of the flower) is stimulated by chemicals secreted by the stigmatic tissues (usually a sugar compound) to produce the pollen tube via one or more of the exine pores.

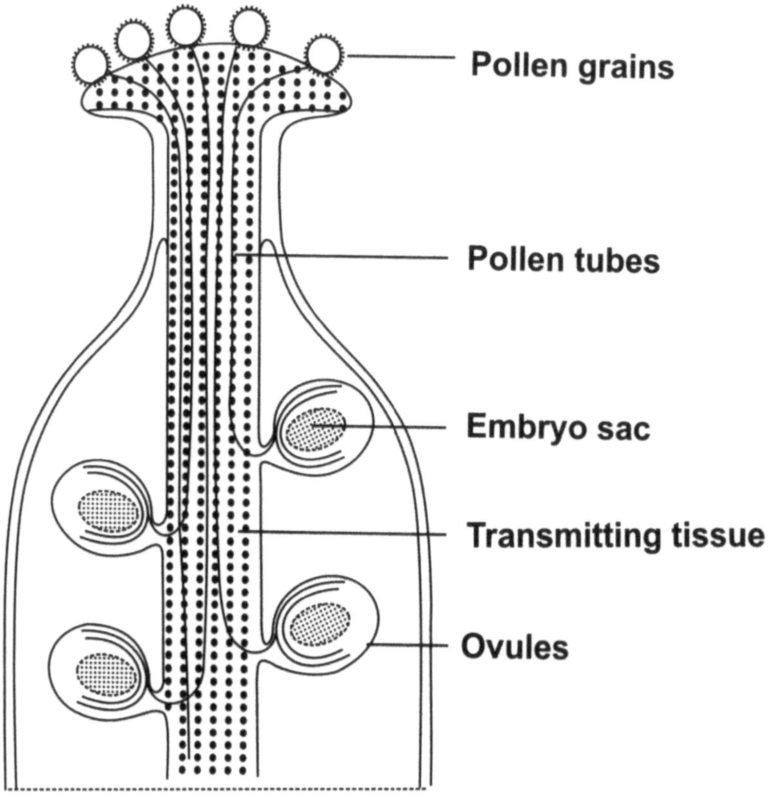

Fig. 10.1 *In vivo* pollen germination. Pollen tubes grow through the transmitting tissue and enter the ovules (all with embryo sac).

Successful pollination and fertilization are essential for sexual reproduction in higher plants. The germination of pollen requires a varying range of growth media like water, sugars, inorganic salts, hormones and vitamins. However, some species are known to germinate in distilled water without addition of any nutrients.

As the studies of pollen germination *in vivo* are difficult, our knowledge of the physiology and biochemistry of pollen germination and pollen tube growth is based on *in vitro* germination studies. In addition to moisture, carbohydrates, boron and calcium are generally essential prerequisites for artificial (*in vitro*) pollen germination. Pollen grains of most of the species require an optimum concentration of sugar solution and sucrose is the best carbohydrate source for pollen germination and tube growth.

Extract from flower parts, growth substances like kinetin, gibberellic acid, indole compounds, etc. and even some antibiotics (e.g. penicillin) are known to have a stimulating effect on pollen germination and tube growth. The temperature and pH have a profound effect on germination of the pollen grains. Ideally a temperature of 20°C to 30°C and a pH range of 5.5 to 7.0 have been found to be optimum for pollen germination.

The time interval between pollen hydration and tube initiation is termed as the lag-phase. The duration of the lag-phase varies from species to species. In some species it is limited to a few minutes while in others it may extend to many hours. The pollen tube emerges through the germ pore. It generally emerges as an extension of the intine. The intine, therefore, has to become less rigid before the pollen tube can emerge. One of the special features of the pollen tube is that growth is confined to the tip only.

Pollen grains of different species of plants have specific requirements for their germination. In some cases pollen germinate in ordinary tap water, starch paste with one or two parts of water and on parchment paper soaked with sugar solution or even in moist air. Following these observations, attempts have been made to cultivate pollen *in vitro*. Cane sugars have been frequently employed as a medium with a good rate of success. Various kinds of sugars (alone or with agar), nutrient elements (both major and minor) and some growth substances and vitamins have been used in germinating pollen grains of different species of plants.

Pollen germination studies involve assessment of the role of various substances mentioned below:

1. Role of sugars (nutritive)
2. Role of boron (a stimulant of pollen germination and pollen tube growth).
3. Effect of certain growth substances (like Indole Acetic Acid).
4. Role of lipids

Various steps are involved in the process of pollen germination. Pollen hydration, germination and penetration of the stigma by pollen tubes are influenced by the exudate on wet stigmas and by the pollen wall in species with dry stigmas. The exudate is known to allow pollen tubes to grow directly into the stigma, whereas the pollen wall establishes the contact with the stigma. This is followed by the pollen tubes growing into the papillae, which are covered by a cuticle.

Wolters-Arts *et al.* (1998) while working on the nature and role of exudate in tobacco plants, showed that lipids are the essential factor needed for pollen tubes to penetrate the stigma. They further concluded that lipids in the exudate direct pollen tube growth by controlling the flow of water to pollen in species with dry and wet stigmas.

METHOD OF STUDYING *IN VIVO* GERMINATION OF POLLEN GRAINS

In vivo germination of pollen grains in hybrids and their respective female parents is studied on the first, second and third day after opening of the flower. In the above experiment, the stigmas along with some portion of the styles are taken out of the flowers after the first, second and third day of anthesis. They were then subjected to the following treatments:

1. At first the stigmas along with part of style are kept in aceto-alcohol (acetic acid:alcohol 1:1) at 60°C for one and half hour.
2. They are washed with distilled water and macerated in 1% KOH solution at 60°C for an hour.
3. The stigmas were washed with distilled water and kept in it at 60°C for one hour and then again washed twice or thrice with water.
4. After washing, the stigmas are kept at 30°C in lactic acid for 10 minutes.
5. They are then stained in cotton blue. Slides are prepared by mounting the material in glycerine by exerting sufficient pressure and examined under the microscope for assessing *in vivo* germination of pollen

According to Heslop-Harrison *et al.* (1973) proteins are stored at two sites in pollen grain walls: in exine derived from tapetum, and intine synthesized by pollen grain itself during the course of development. These observations were made in Malvales. Such sporophytic and gametophytic fractions of pollen wall proteins are supposed to play an important role in pollen stigma interaction and they are, therefore, mostly concerned with compatibility reactions (Heslop-Harrison 1971).

As pollen grains come in contact with the stigma, it is soon subjected to hydration. Simultaneously the pollen wall proteins are released on the stigma, first the exine proteins and then the entire proteins (Knox and Heslop-Harrison 1975a).

Pollen grains are usually shed under dehydrated conditions (water-content $<20\%$), and the metabolic rate is very low (Wilson *et al.*, 1979) and it is expected that endogenous substrates would be gradually used up. Stored pollen grains require a higher sugar concentration for germination *in vitro* than fresh pollen (Johri and Vasil 1961).

POLLEN VIABILITY

Pollen grains contain the male gamete and may be regarded as microspores. Their viability is relatively short as the cytoplasmic contents have limited food reserves to maintain the living state, and as with all gametes, both plant and animal, viability is designed to be maintained to cover transfer and pollination.

Environmental factors, particularly temperature and humidity, greatly affect pollen viability. Retention of the pollen viability after shedding is termed as pollen longevity, and varies significantly from species to species. There is a close correlation between the cytology of pollen (two or three-celled) and loss of viability. Two-celled pollen grains generally retain viability for a longer period, than three-celled pollen grains (Brewbaker 1957). Two-celled pollen grain is more amenable to storage compared to three-celled pollen.

The most commonly used method for assessing pollen viability is *in vitro* germination of the pollen grains. It is rapid, reasonably simple, and fully quantitative. The percentage of *in vitro* germination can be correlated with its ability to set fruits and seeds following *in vivo* pollination (Visser 1955; Akihama *et al.*, 1978). Whereas in *in vivo* conditions, the most authentic and accurate test for viability is expressed as the ability of pollen to effect fertilization resulting in seed and fruit set. This fruit and seed set test has many limitations (Heslop-Harrison *et al.*, 1984). It is laborious and more time consuming.

METHODS OF DETERMINATION OF POLLEN VIABILITY

Triphenyl tetrazolium chloride (TTC) reacts with viable pollen and gives red colouration. Tetrazolium chloride and its derivatives have been used as vital stains for testing viability of tissues especially in seeds. Triphenyl tetrazolium chloride (TTC) is commonly known as tetrazolium salt. It is a white to pale yellow crystalline powder that darkens on exposure to light and is readily soluble in water.

In the presence of viable tissue the colourless solutions of 2, 3, 5-triphenyl tetrazolium chloride forms the insoluble red triphenyl formazon. This reduction reaction is caused by the dehydrogenase enzymes present in living tissues.

The TTC stain is kept in 10% stock solutions and diluted to the ratio – 1 part of stain: 5 parts of 60% sucrose solution at the time of treatment. However, a disadvantage of sucrose is the possibility that it may become contaminated with bacteria or other microorganisms and be decomposed by them. To guard against this contamination the two solutions should be kept separately until the slides are prepared.

Procedure

A drop of the final dilution is placed on a slide, and pollen from just dehisced anthers are dusted on the surface and a cover slip is placed on it. Two hours are required for the development of red colour. The proportion of grains developing a red colour was determined by counts made with a microscope using the low power objective, after 2 hours. The percentage of coloured grains is estimated. Coloured grains were considered viable and grains without colour were considered non viable. Thus, the pollen at the known time of anthesis can be tested for viability after several days.

METHODS FOR TESTING GERMINATION OF STORED POLLEN *IN VITRO* FOR VIABILITY OF STORED POLLEN

The anthers are collected in Petri dishes and allowed to dehisce in the laboratory. The whole pollen sample was divided equally into required number of small lots, transferred to small glass vials, which were loosely closed with cotton plugs and kept at the following temperatures and humidity combinations:

At room temperature 25°C – 29°C with 0, 5, 10, 25 or 30% RH

At freeze temperature (9°C – 12°C) with 0, 5, 10, 25 or 30% RH

The humidity was maintained in desiccators by a mixture of sulphuric acid and water. The viability tests of stored pollen are carried out at intervals of 24 hours. Viability is tested by germinating the pollen grains *in vitro*, in castor oil, which is found to be the best medium for germination of cotton pollen.

For this, pollen grains just after anthesis are sown on a drop of refined castor oil and spread on a microslide in the form of a circle, 1-5 cm in diameter. The pollen grains are carefully stirred so that the pollen grains are coated with oil completely. The slide with the pollen is placed on a stand in a Petri dish in which there is a thin layer of water. The dish is partly covered to allow free circulation of air and kept in total darkness at 38 c for 24 hours. The germination percentage and tube lengths are recorded after 24 hours.

Pollen Storage

As an aspect of conservation, the technique of preserving shortlived gametes-pollen and spores – by means of freeze-drying is used. This technique involves storing, samples of pollen in small glass vials, which are then placed in containers of liquid nitrogen. A state of 'suspended animation' ensues throughout the period of storage. Upon gradual return to normal temperatures, the pollen may be persuaded to fertilize female ovules, which like pollen can receive similar freeze-drying. Pollen banks have a role to play in the possible production of species on the verge of extinction.

Successful pollen storage is a very convenient tool in the hands of tree breeders for improving trees by hybridization, occurring in different regions and also of those blooming in different seasons. The longevity of pollen grains enables the introduction of new characters in plants by using stored living pollen. Thus, there is an immense importance of pollen storage in tree improvement programmes.

Jain and Shivanna *et al.* (1990) and other scientists have successfully stored pollen grains of several taxa in organic solvents such as benzene, petroleum ether, ethanol, acetone and chloroform. Certain pollen diluents have been found to be very effective in increasing pollen longevity. Powdered egg albumin, stale and powdered milk are useful in storing pollen as dry and powdery diluents.

The pollen grains are often mixed with diluents such as wheat flour, corn flour, powdered charcoal, talc and *Lycopodium* spores before storage (Johri and Vasil 1961). The main purpose of diluents is to increase the bulk of pollen so as to avoid wastage of the pollen sample during artificial pollination. Pollen grains of many taxa have been successfully stored in ultra-low temperature using dry ice (-80°C) or liquid nitrogen (-196°C), in suitable cylinders. The refrigerant has to be replenished at regular intervals. This technique is known as *Freeze-Drying* or *Lyophilization*. Generally, a mixture of solid CO_2 and acetone is used for freezing.

Lower temperature and humidity (0°-8°C and 0-50 RH) prolong the life of pollen grains of different plants (Johri and Vasil 1961; King, 1965; Holman and Brubaker 1926). Cross-pollination largely depends on insects. Poor weather conditions limit the activity of insects, and significantly reduce the yield because of insufficient pollination. To overcome this problem, attempts were made to carry manual pollinations with the help of stored pollen (Brown and Perking 1969; Legge 1976; Williams and Legge 1979). In the U.S.A. and U.K. there are several commercial firms, which sell stored pollen for supplementary pollinations. The ultimate aim of storage of the pollen grains by plant breeders is to establish 'pollen banks' through which the viable pollen of the desired species can be obtained at any time. Pollen banks are often established for the supply of viable pollen for mass scale hybridization. Pollen storage of viable pollen in pollen banks is done by various methods, such as storage by controlling temperature and humidity, storage by freezing and dehydration, influence of gases and pressure, storage with pollen diluents.

POLLEN CRYOBANK AT THE INDIAN INSTITUTE OF HORTICULTURAL RESEARCH (IIHR), BANGALORE, INDIA

Long term cold storage (cryopreservation) of nuclear genetic diversity in the form of pollen in several plants of horticultural, medicinal and forest importance can be successfully achieved (Ganeshan *et al.*, (2005). The ultimate result of this methodology is the establishment of the Pollen Cryobank at the Indian Institute of Horticultural Research in Bangalore, India which maintains more than 600 types of pollen belonging to 45 species and 15 families under cryogenic conditions.

Sustainable use of cryopreserved nuclear genetic diversity has been successfully demonstrated through field pollination with cryopreserved pollen. Thus, the cryopreserved pollen from the pollen bank can be put to use by plant breeders, hybrid seed producers, genebanks for various crops of economic importance. The IIHR maintains in the database, information pertaining to longevity, viability and media used for pollen germination, yearwise listing of collected pollen samples and yearwise lists of pollen samples cryopreserved in different years. Some of these cryopreserved pollen in the pollen bank maintained by the pollen storage laboratory of the Division of Plant Genetic Resources, IIHR includes pollen of Day lily, mango, citrus, Carnation, Gladiolus, Vinca rosea, grape, papaya, pomegranate, tomato, lemon, onion, rose, Solanums and capsicum.

Successful storage of pollen grains of some tree species by manipulation of temperature and humidity is shown in Table 10.1 (courtesy Sharma).

Table 10.1 *Successful storage of the pollen grains of some tree species by manipulation of temperature and humidity.*

Tree species	*Storage Temperature (°C)*	*Relative humidity (%)*	*Duration of storage*	*References*
Artocarpus sp.	0	25	8 months	Sinha 1972
Betula sp.	5	0 (vacuum)	2.5 years	Jensen 1964
Juglans nigra	4	-	3 months	Hall and Farmer 1971
Mangifera indica	0	0	14 months	Singh 1960
Pinus banksiana	2	25-75	1 year	Johnson 1943
P.nigra	5	0 (vacuum)	2.5 years	Jensen 1964
P.strobes	18	25	14 months	Duffield and Snow 1941
P.resinosa	0-4	25-50	14 months	Duffield and Snow 1941
P.sylvestris	2	25-75	1 year	Johnson 1943
Prunus americana	2-8	50	2.5 years	Nebel and Ruttle 1937
P.communis	2	0	1.5 years	King and Hesse 1938
Pyrus communis	2	0	3 years	Visser 1955
Populus sp.	18	Desiccated over silica gel for 12-16 h before storage	Many years	Knox *et al.* 1972 b.

The self-incompatibility is considered to be one of the main causes for the rapid evolution of angiosperms. Pollen is known to have a higher energy investment per gram of organic tissue than other plant parts. Pollen of anemophilous spp. have lower caloric contents than pollen of entomophilous species. In addition to its main function of pollination and fertilization, pollen attracts and nurtures a variety of pollinators. Enhanced pollen energy contents could attract pollen consumers for their energy requirements.

Though physiological studies have been undertaken mainly on pollen germination, the biochemical and biophysical factors responsible for successful germination are not as yet clearly understood.

CHAPTER 11

Minor Applications of Pollen Studies

There is no doubt that pollen studies have many applications, which can be classified into major and minor applications for the sake of convenience and proper understanding. Major applications are incorporated in Chapter 12 on Melissopalynology, Chapters 13, 14, 15, 16, 17, 18 and 19 on various aspects of Aerobiology, Chapter 20 on Forensic Palynology and Chapter 21 on use of pollen studies in exploration of fossil fuels Palaeopalynology and reconstruction of past vegetation (Palaeoecology).

In the present chapter, minor applications of pollen studies will be presented briefly under various captions such as Palynotaxonomy, Copropalynology followed by the role of palynology in environmental pollution (bio pollution).

A: PALYNOLOGY IN RELATION TO TAXONOMY (PALYNOTAXONOMY), ECOLOGY AND EVOLUTION

Taxonomy is the scientific discipline of classification. The desire to name and classify objects, whether biological or physical, in some logical order is a basic and universal human characteristic. From time immemorial, since indeed the dawn of comprehensible language, objects which are seen and felt have been given a name. The ordering of objects into some logical order, on whatever basis, came later. Some of the earliest attempts to classify objects date back classical times – Aristotle and Lucretius were two of the earliest 'taxonomists' involved in this effort.

Taxonomy, as a scientific method, based on careful observation and logical thinking, did not develop until the 16th-17th centuries, largely due to discoveries by explorers in the wake of early empire builders. It was not until the 17th century that a significant and, to some extent, enduring system of naming individual organisms and an attempt to organize a classification scheme which reflected relationships was effected. Modification to the most important classification scheme propounded by the Swedish botanist Linnaeus, has since tried to reflect an evolutionary approach, in which naturally occurring features play a dominant role to produce a natural classification as opposed to Linnaeus' highly artificial scheme. Linnaeus based his scheme on floral structure primarily. A century earlier Anthony Van Luewenhoeck published his observations on microscopic organisms observed through his primitive microscopic lens. His published drawings included pollen. However, the nature and form of pollen was not included in Linnaeus' scheme.

To some botanists and zoologists, taxonomy is a science; to others it is an art. Whichever it is, taxonomists come in two forms 'lumpers' and 'splitters'. The former tend to recognize fewer categories or taxons (classification units), while the latter consider many more taxons are necessary to reflect adequately the organization of that which is natural to the organisms concerned. As a consequence, taxonomy is a source of much disagreement and controversy, at times bitter. Taxonomy is an ongoing activity, partly due to the constant discovery of organisms hitherto unknown to science, to add to the huge volume of specimens collected and only waiting to be named, described and classified. Natural hybridization, especially in plants renders classification a slow and often difficult process. The Irish botanist Pugsley devoted a lifetime to the study of the genus *Hieracium* (hawkweeds) in the Asteraceae. He was by no means a solitary figure in lifetime devotion to the study of one small group of organisms. Hybridization in hawkweeds was especially difficult to unravel due to their notorious anomalous sexual reproductive mechanisms – not always involving pollen.

As pollen are the product of living plants, which are biological systems subject to individual genetic variation, they too, are subject to variation. Over time plants evolve, undergo minute changes and occasionally major changes (mutations), which must be viable for survival. Such changes may be reflected in pollen structure, physiology and genetic composition.

Taxonomic studies of pollen is an important aspect of palynology. Pollen is classified on morphological grounds. The advent of Scanning Election Microscope (SEM) has for several decades contributed to better understanding of pollen structure and its relationship to other species. Pollen classification is of necessity tied to the classification of the plant producing it. Pollen is specific to the plant, which produces it, so the plant or species can be identified by its pollen.

In recent years several atlases of pollen grains (including also some spores, e.g. ferns) have appeared, most with keys to the morphology drawn up on a dichotomous basis. Photomicrographs of the pollen types keyed comprise the bulk of such atlases, in which pollen is usually magnified × 400. Most atlases nowadays have a section of electron micrographs taken by SEM at magnifications c. × 2,000 – 10,000.

As the taxonomy of pollen and similarly spores of selected groups of fungi, are, of necessity, related to the gross taxonomy of the organism producing them, such palynological taxonomy will, likewise, undergo change concomitant with the organism's taxonomic change or revision. However, as the role of palynological taxonomy is more or less confined to the general anatomy and morphology of the pollen grain or spore, as viewed in the surface, facilitates the identification of these structures becomes a prime object of the palynologist involved.

Once identified, the pollen grain forms the basis of much more information, such as its relationship to other similar pollen, which reflects the taxonomy of the parent; indicator of the habitat in which the parent occurred.

Pollen grains have themselves evolved with the parent plant species over time. Many palynologists research into pollen and spores of fossilized taxa. The evolutionary trends and pathways are the focus of interest in such studies. These can often be seen to have taxonomic links with more recent and present day species of pollen and spores.

Pollination

Pollen grains contain the male gametes required to achieve fertilization. The process of transfer of pollen grains to the receptive stigmas of female flowers is known as pollination.

Mature pollen grains awaiting suitable conditions for anther dehiscence and dispersal, normally have one or more pores through which pollen tubes grow and develop their passage down the length of the stigma towards the ovule below. Most pollen grains fall into one of two varieties of aperture arrangement; those with a single aperture often producing a bilateral symmetry in the grain or having three radially symmetrical equatorially arranged with a radial symmetry. Some grains have no aperture (e.g. *Populus*) whilst others have more than three (e.g. *Plantago* and *Chenopodium*), single apertured grains are exemplified by grasses (Poaceae).

Pollination is carried out in a variety of ways: the most common are wind (*Anemophily*) and insect (*Entomophily*). Wind-pollinated grains are usually dry and produced in vast numbers (e.g. grasses). Generally, windborne pollen are small in size, light, smooth walled and lacking pollenkitt. Hence, the pollen wall is not sticky. Classical examples of this type are known in the pollen grains of Betulaceae, Urticaceae and Plantaginaceae. In contrast pollen dispersed by insects require a completely different type of exine.

Insect-pollinated grains tend to be sticky, readily adhering to the bodies of insects. The greater probability of fertilization being achieved by this direct mode of transfer, number of pollen grains produced by entomophilous flowers tend to be much smaller (e.g. other forms of pollination occur – *Ornithophily* (bird pollinators) e.g. by tropical bird species – sunbirds and honeybirds; small crawling fauna e.g. slugs, spiders, millipedes, centipedes etc. (*Zoophily*) and aquatic plants e.g. pondweeds, duckweeds where surface currents transport pollen from male to female flowers (*Hydrophily*). Pollination by bats (Chiroptera) is widespread in tropical floras (e.g. baobabs (*Andensonia*) *Kigelia* in East Africa, *Freycinetia insignis, Symbolanthus, Purpurella, Cobaea,* etc.)

Many flower species have evolved intricate morphological structures and pigmentation of corollas to achieve pollination, some of the most sophisticated features are found in the Orchid family (Orchidaceae).

However, anemophily and entomophily are not mutually exclusive – many plants whose flowers are exposed prominently (e.g. *Acer, Tilia, Salix*) may be pollinated by both agents (Amphiphily). To ensure pollination and subsequent fertilization, in the event of failure of visitation by insects, some plants are structured so as to undergo self-pollination (e.g. *Ophrys, Epipactis, Neotinea,* all genera of Orchidaceae). There is a little added advantage to be gained by self-pollination apart from the production of a seed/seeds, which may ensure survival of the particular individual.

Functional Factors Influencing Pollen Form

Pollen are known to have a variety of forms and a variety of pollen wall ornamentation, aperture condition, etc. According to Punt (1984) each form of pollen and each type of ornamentation has a definite function. This is more precise with respective to pollination mechanism. It appears that the pollen wall particularly the exine gets modified to suit the transport mechanism of pollen. During transport by wind, pollen grains are influenced by several factors including prevailing climatic conditions. Small apertures in the exine offer the best protection against desiccating climatic conditions.

A rough pollen surface favours attachment to insects. The presence of pollenkitt on the wall is useful for adherence to the insect body. (For example, pollen of the families Asteraceae and Malvaceae). Water transported pollen are usually thin walled and have reduced apertures as in Hydrocharitaceae and Puppiaceae. With regard to transport of pollen by birds and bats, there is some evidence that sticky pollen is found in bird-pollinated flowers. Further, it has been reported that some humming birds have a special forked tongue to suck nectar and pollen from flowers and they may transport pollen on their tongue, whereas other humming birds transport pollen grains on their head feathers.

The contents of pollen grains such as cytoplasm and the nuclei containing the genetic material need to be protected, particularly during their journey through the air. In general, the pollen needs to be protected against dehydration and physical strenuous conditions.

During pollination, pollen grains may be exposed to extreme dehydration. The ideal form of protection might appear to be a completely closed exine without openings to the outside (apertures). However, such a condition would affect germination and formation of pollen tubes. According to some palynologists inaperaturate pollen grains indicate a final level of phylogenetic series. Examples are seen in the pollen grains of *Phyllanthus* and *Dichapetalum*. In pollen grains of both these families, the endo apertures are still present, perhaps to maintain relatively thin areas for the exit of the pollen tube during germination.

Another classical protection mechanism against dehydration in the pollen grains is described as harmomegathy. This was first described by Wodehouse (1935). Some bird-pollinated plant pollen, and insect-pollinated pollen are known to have thick walls which allows physical protection to the pollen grains.

It is a known fact that the ultimate goal of a pollen grain is to germinate and carry the generative nuclei safely through the pollen tube to the ovule. In this process, the first step involves rehydration with liquids from the stigma. These liquids enter the pollen grains through the large open apertures in the pollen wall. In many primitive angiosperms, such as Magnoliaceae the pollen grains are characterized by large open apertures. In the course of angiosperm evolution better-protected enzymes have evolved. In this context, it is worth mentioning that the protection of the pollen cytoplasm of the pollen contents and ease of germination are two factors with contradictory requirements. Reduction in the size of apertures is usually compensated by an increase in the number of apertures resulting into a pantoporate condition, which is considered to be highly advanced as seen in the pollen grains of the families Caryophyllaceae, Convolvulaceae, Malvaceae and Moraceae. Genera such as *Urtica, Plantago* and *Chenopodium* are also Pantoporate. However, the most important factor in addition to environmental factors in deciding the form of pollen grains is the genetic factor, which is really of overriding importance.

POLLEN APERTURES: FORM AND FUNCTION

Thanikaimoni in his classical paper published in 1986, had reviewed various aspects of pollen apertures in relation to form and pollen function.

Morphologically, the aperture is an opening or thin area of the exine where the intine is usually thick. Physiologically, it is a germination zone or a harmomegathus. The latter is described as a device to accommodate volume changes in the pollen due to dehydration or rehydration.

An aperture is classified as simple or composite. A simple aperture can be described as an opening or thinning of the exine overlying a thick intine. Whereas a composite aperture is an opening or thinning of the exine overlying a thick intine differentiated into an outer unit known as the ecto-aperture and the inner unit referred as the endo-aperture, which are not congruent. In certain cases there is a local separation of the exine layers to form a cavity called vestibulum as in porate pollen. The apertural intine of the pollen of angiosperms and gymnosperms is characterized by the abundance of cytoplasmic tracts where as the apertural intine of the spores of pteridophytes and bryophytes lack them.

Operculum

It is defined as a thick exinous lid covering the aperture. It is distinguished from the thick non-apertural exine by the underlying thick intine. It is generally surrounded by a thinning of the exine which demarcates it from the non-apertural region. Operculae are of various shapes: circular, elliptic, annular or bridge like. The last one is also referred as ponto-operculum. The pollen grains of *Chamaerops humilis* are described as monosulcate ponto-operculate.

The number of apertures and their position such as distal, meridional, equatorial, global are genetically determined. Apertural forms may be a function of evolution. The phylogenetic diversifications of apertural congruation have been established in several angiosperm families on the basis of correlation of character states.

In the family Asclepiadaceae, there is no callose around the individual microspore or around the whole pollinium. Sporopollenin is formed from the secretory tapetum. The exine around the microspore inside the pollinium is considerably reduced. The special germination area in certain genera of Asclepidiadaceae represents a common aperture for all the monads of a pollinium.

The occurrence of different apertural types in various angiosperm genera are listed below:

1. Monosulcate – e.g. Arecaceae, Liliaceae
2. Monosulcate – operaulate – e.g. *Nymphaea*
3. Monosulcate – ponto-operculate e.g. Annonaceae
4. Diporate – tetraporate – periporate e.g. Amaryllidaceae, Bromeliaceae
5. Colpate – colporate – e.g. Menispermaceae
6. Colporate e.g. Asteraceae, 6-Stephanocolpate e.g. Passifloraceae, Pedaliaceae
7. Tricolporate e.g. Sarraceniaceae
8. Stephanocolporate – e.g. *Buxus* Tripororate – e.g. Betulaceae

9. Tricolporate – e.g. Myrtaceae, Lecythidaceae
10. Stephanocolporate e.g. *Utricularia*
11. Dicolpate e.g. *Pedicularis*
12. Tricolporate – e.g. Rubiaceae
13. Triprorate – e.g. Onagraceae

Thus, it can be concluded that the principle functions of the pollen aperture are protection, harmomegathy (accommodation of volumetric changes), ion exchange and germination.

Variations occurring in the morphology and number of apertures may be adaptations to the habitat and mode of pollination. The primary function of the exine, like that of the epidermal cell wall, is protection, which is offered through its impermeability and resistance to physicochemical and biological degradation. Bolick (1981) indicated that but for the sporopollenin coat, the pollen grains would plasmolyze very rapidly.

The pollenkitt also known as triphen or pollencoat is a complex mixture of lipid substances found on the surface of mature pollen grains. It is known to protect the pollen from the hazardous effects of the environment particularly from radiation hazards by virtue of its ultraviolet-absorbing pigments. This lipid layer also protects the pollen from desiccation.

Pollen grains and spores are susceptible to fungal attack. Generally the fungal mycelia penetrate the pollen through the exine apertures. Lloyddu (1974) explained that biomechanics, the study of biological form from the functional view point, is an examination of the principles of engineering design as implemented in living organisms. Rowley and Skvarla (1976) considered the aperture as an area, which responds to metabolic requirements for transfer of ions during growth and maturation of the pollen grains.

According to Heslop-Harrison (1975, 1979a b) the principal role of the apertures in pollen grains is to serve as an outlet at the time of germination. The different layers of apertural intine are of functional significance. During the germination of pollen grains, the outer layer of intine gets disrupted with the release of the underlying enzymes. The inner layer of intine becomes the precursor of the pollen tube. The enzyme load carried through the pollen tube degrades the cuticle of the stigmatic papilla.

It appears that the multiple apertures have the selective value of rapid germination and early fertilization as they provide several exits for the pollen tube to emerge out.

Adaptations to Habitat

In case of pollen of Xerophytic plants, higher temperature of the arid zones seems to select thicker exine and /or fewer and smaller apertures to prevent excessive water loss (Lee 1978) as in some members of Umbelliferae and

Boraginaceae. Multiaperturate operculate pollen grains with thick exines have a selective advantage in combining the mechanism against desiccation and quick germination in the xerophytic species with a short life span of pollen as in Cactaceae.

In contrast, the hydrophytic plant pollen are characterized by less rigid, discontinuous or very thin exine permitting easy volume changes. Aquatic species of *Utricularia* are multiaperturate whereas the terrestrial species of *Utricularia* have only 3-5 apertures.

In general, the size and surface of the pollen grain as well as those of apertures are often selected or modified according to the mode of the pollinating agents such as air, water, insects or birds. The cohesion of monads into tetrads and polyads seems to be a functional asset to entomophily since they often behave as a single harmomegathic unit.

The reduction in thickness of exine may be linked to the origin of the taxa under aquatic (e.g. Zosteraceae) or highly moist (e.g. Heliconiaceae) habitats. According to Thanikaimoni (1986) there is no pollen in aquatic angiosperms without exine. Reports of exineless pollen are actually cases of pollen with thin exine which often gets broken into bits during acetolysis, and escape observation. Aquatic plant species that are pollinated by water have multiaperturate – global pollen with thin elastic exine that could stretch readily and accommodate the increase in size due to imbibition of water. In order to overcome this problem certain aquatic species like *Potamogeton* and *Pontederia* produce flowers above the water surface. According to Schwantiz (1967) the exine reticulum of marine species of *Ruppia* is responsible for keeping the pollen afloat.

In conclusion it can be pointed out that the principal functions of the exine apertures are protection, harmomegathy, ion exchange and germination. The diverse forms of pollen are the result of any one or a combination of three factors, such as either a particular mode of development is an ontogenetic factor or of evolution which is a phylogenetic factor or of adaptations to the habitat and pollination i.e., an ecologic factor.

B: THE ROLE OF PALYNOLOGY IN AGRICULTURE, HORTICULTURE AND PLANT BREEDING

For many centuries it was known but not understood, that the yellow dust (pollen), when transferred by insects or carried artificially by man, from the dehisced anthers of one flower to the stigmas of another flower, resulted in the formation of fruit. As time progressed from the late 18th century, through the 19th and 20th centuries, an understanding of the nature of this process of transfer, called pollination, took place. The majority of pollen-producing plants require the pollen to be transferred, by whatever agent, from one flower to a different flower of the same or very closely related species.

Few flowering plants use their own pollen (self pollination), and usually only resort to this if cross-pollination has not occurred. Perhaps the best-known examples of cross pollination between closely related species dates from the early, Neolithic in the Middle East, the so-called 'fertile crescent', where varieties of wild wheat were gradually improved in yield by cross pollination. This hybridization process continued and expanded to include a wide range and variety of crop plants in agriculture. The discovery of Gregor Mendel's (1860) experimental studies on cross pollination in peas laid the foundation of the science of genetics in the early 20th century, and coupled with the rapidly developing understanding of cell structure and function – cytology – the scientific discipline of cytogenetics, have both made an important impact on the process of plant breeding.

Whilst early techniques used in plant hybridization and breeding were applied to crop/food plants, the development of horticulture and forestry were stimulated and have made a considerable impact on the economic life of man.

Recent developments in techniques of gene manipulation and the recognition of the role of selected genes which could be considered to be to the advantage of a plant lacking them, have led to the implantation of such genes, e.g. those which could resist disease or pest attack, in crop plants. Such techniques are known as Gene Modification (GM). The ability to alter the genetic makeup of plants by exposure to radiation has long been known and the subject of experimentation. However, most radiation-induced changes – man-made mutation – are deleterious. Genetic modification is a more precise and predictable practice. The aim of GM modification is said to benefit and enhance crop production. It is, however, a very controversial issue, not the least as the long-term impact on the environment and ecosystem is unknown.

The role of plant breeding in horticulture has a long history. The twin approaches of hybridization and plant propagation by genetic means have over the years produced a profusion of plant varieties selected for form, colour and scent. Hybridization by means of cross-pollination produces the range of floral types, especially seen today. The long sought after 'black tulip', by means of cross pollination and selection, has so far eluded attempts of over 400 years of effort – to produce the so – called Queen of the Night, which is sold as the 'black' tulip, is dark purple as is the very recently bred Black Hyacinth. Roses have over the past century and a half have had a most successful outcome from cross pollination (together with vegetative grafting of stock (original plant) and scion (grafted floral part) to produce the tremendous range of varieties obtained. Grafting and cross-pollination have long been employed in developing fruits.

Western Europe, Australia, New Zealand, South Africa and the U.S.A. have all developed many varieties of fruits. Potatoes have received a lot of specialized attention, especially in southern Scotland, where climate, soil and elevation have all played a major part in the breeding of new varieties, for example, the Pentland range of potato developed at Roslin near Edinburgh.

Palynology has played an important role in the soft fruit industry: raspberries strawberries, gooseberries, blueberries, red and black currants, grapes, brambles (blackberries) to name only a few from the temperate zones, in addition to the enormous range of fruits (and vegetables) produced by plant breeders in the tropical and subtropical parts of the world such as grape fruit, bananas, mangoes, etc.

Many, but not all, of the above breeding programmes have involved artificial pollination under strictly controlled conditions.

A visit to any local market around the world will reveal the bewildering display of fruits, flowers and vegetables, including cereals and pulses. Rice, a staple diet of millions of people, especially among Asians, has been the subject of intensive plant breeding. Pollination has played a major role here as rice is herbaceous and precludes vegetative reproduction. Other cereal grasses such as maize (corn), millet, wheat, oats, barley, rye together with the great variety of peas and beans derived from leguminous plants, have all been subject to breeding programmes based on pollination techniques.

Pollination techniques used in plant breeding programmes are an aspect of applied palynology, which have proved immensely successful in supplying, most importantly, food of enhanced quality and nourishment for mankind.

C: COPROPALYNOLOGY

The term copropalynology is derived from the Greek word *Kopros* meaning dung, indicating the study of pollen and spores present in dung. It is one of the applied branches of palynology in which droppings (excrements) of various animals such as sheep, goats, kangaroos, giraffes, elephants, horses, bats, etc., are analyzed palynologically which provide interesting insights into their feeding habits. In literature, there is a classical example in this connection with regard to pollen analysis of bat droppings. Samples from limestone caves near Harir in Northern Iraq have shown that the excrements of bats inhabiting these caves contain a significant number of pollen grains. This is concerned with the bats, which eat moths, which visit flowers at night to suck nectar. Bats roost from the roofs of caves, hence their excrement falls on the floor and piles up in the course of hundreds of thousands of years into a thick layer of guano. In this substance, accumulating continuously like a type of peat, are pollen grains, spores and also wing scales of moths. Thus, the copropalynological analysis of guano can reveal the history of entomophilous plants of the district by studying the pollen grains that lodge on or in insects, which in their turn have passed through the bat's digestive tracts. French palynologists restrict the term 'copropalynology' to the study of pollen and spores in fossilized dung, and the term with 'actuopalynology' for contemporary studies.

Aerobiological Aspects of Pollen Contamination of Bats (*Chiroptera*)

Caulton (2006) worked extensively on interesting aerobiological aspects of pollen inadvertently caught by bats during their food foraging flights for insects many of which are involved in the process of pollination. The pollen ingested with the insect prey by the bats is eventually passed out of the bats' gut as a faecal component. The bats' bodies become contaminated with pollen released by plants including grasses. It was possible to recover anemophilous as well as entomophilous pollen of *Pinus spp*, *Tilia spp* and grasses from the bats' fur when grooming. The faecal matter collected from different species of bats in the U.K. also yielded different types of pollen both of anemophilous and entomophilous plants.

Applications of copropalynology

Copropalynology deals with not only the fossil faeces of past animals but also of humans from the Ice Age. Oeggl (2001) has given an excellent account of pollen analysis of the Iceman's colon content. This investigation throws light on the Iceman's (Neolithic) lifestyle, environment, diet and the place he inhabited or where his settlement was located. The origin of this investigation goes back to 1991, when a chance discovery was made in the permanent ice of the Tyrolean Alps. A naked human body was thawed out of the ice at an altitude of 3,210 m in the proximity of the Otztal summit.

Palynological analysis of the colon contents of the Iceman yielded a large number of pollen grains and diatoms which were observed under LM at different magnifications with the help of phase contrast and interface contrast. Thirty different types of pollen and two types of spores were recovered from the food residues. Pollen of *Pinus*, *Picea* and *Ostrya* type were also identified. The presence of *Ostrya* type of pollen indicates the Iceman's origin, as the genus *Ostrya* has a restricted distribution in the Alps. *Betula* pollen with their gametophytes intact were also observed. As many as 16 different diatom species were recovered from the food residue. The most dominant species comprised: *Achanthes minutissima*, *Diatoma chrenbergii*, *Fragilaria arcus*, *Gamphonema pumilum* and *Navicula radiosa*. The presence of these diatoms are indicative of fresh water diatom flora typical of mountain rivulets in the area of investigation.

Pollen of herbs identified from the food residue consisted of Chenopodiaceae members, *Plantago major tysx*, *Polygonum persicaria* type, *Rumex acetosella* and Urticaceae members. Pollen became incorporated into the intestinal residue in many ways: such as drinking water, breathing and swallowing bronchial mucus. Pollen archieved by this means in coprolites is rich in information about diet, seasonality of site occupation and paleoenvironmental conditions (Byrant 1974; Sobolik 1998).

Some general conclusions can be drawn on the basis of detailed pollen analysis of colon content of the Iceman.

1) His last meal consisted mainly of Einkorn (*Triticum monococcum*) with traces of meat which he consumed along with drinking water full of diatoms.
2) In the hours preceding the Iceman's death, perhaps he walked through a warmth – demanding deciduous vegetation dominated by Hophornbeam (*Ostrya carpiniforia*).
3) The predominance of pollen from an early flowering season indicates that the Iceman died in late spring or early summer.

D: ANTIBACTERIAL PROPERTIES OF POLLEN

It was hypothesized that pollen possess some protective mechanism against microbial attack. Antibacterial activity is determined by disc diffusion method. The disc diffusion assay (Rasoanaivo and Urverg 1993) is used to determine the growth inhibition of bacteria by the pollen extracts.

The plates are evaluated after incubation at 37°C for 18 hours. Antibacterial activity is expressed as the ratio of the inhibition zone (mm) produced by the pollen extract and the inhibition zone caused by the reference, (Vlietinck *et al.*, 1995).

The experimental work was carried out by Pannicker (2003) to assess the antibacterial properties of *Prosopis juliflora.* It was found that their pollen extracts inhibited the growth of bacteria such as *Staphylococcus aureus* (standard strain) and *Staphylococcus aureus* [MRSA], *Staphylococcus aureus* [VRSA] and *Escherichia coli.*

Further it was determined that the aqueous extracts and all other extracts of *Prosopis* pollen can produce reactive zone against *Staphylococcus aureus* [MRSA], in which commercial antibiotic has no effect, *Prosopis* pollen extracts thus gave better results than those produced by antibiotics like Netromycin and Vancomycin.

CHAPTER 12

Melissopalynology

Melissopalynology (or melittopalynology) is one of the applied branches of palynology which deals with the study of pollen in honey. The term 'melittopalynology' is derived from the Greek words *Melitta* and *Melissa* (the attic form) meaning " A bee" as well as honey (in Homer). The Latin word *Mel* (mellis) also means honey. The International Commission for Bee Research prefers the term 'melissopalynology' which was also used by Rex Sawyer–world authority, involving the study of pollen in honey, conventionally the majority of palynologists have been using the term 'melittopalynology'.

HISTORICAL ACCOUNT

In their excellent review of work on melissopalynology, Jones and Bryant (1996) had indicated that the sweet taste of honey had attracted the early primate ancestors of humans and later humans too. Honey being a nature's gift derived from the hard work of honeybees and plant pollen attracted the attention of users and scientists right from prehistoric times. Cave paintings dating back to 15,000 years ago in Altamira region, northern Spain depict the honey combs and the ladders needed to reach bee hives on tall trees and rock crevices. The probable ancestors of the order Hymenoptera to which honeybees belong, evolved some 200 million years ago. Rock paintings in Central India, regarding human activities showed honey collections, which date back to the Mesolithic age. The use of honey as a sweetener in wine making was known to the Greeks and Romans. Perhaps the earliest evidence of honey gathering comes from the ancient paintings in Egyptian temples, which depicted slaves collecting honey from man-made hives inhabited by domesticated honeybees. Bee keeping for honey was carried out by using dome shaped beehives (skeps) made of twisted straw during the medieval period in Europe **(Fig. 12.1)**.

Early records indicate that *Apis mellifera* L. subsp. *mellifera* (family Apidae) commonly known as the German honeybee was first introduced in the New World (i.e., the United States of America) by European settlers

Fig. 12.1 Beehive made from twisted straw during medieval period in Europe.

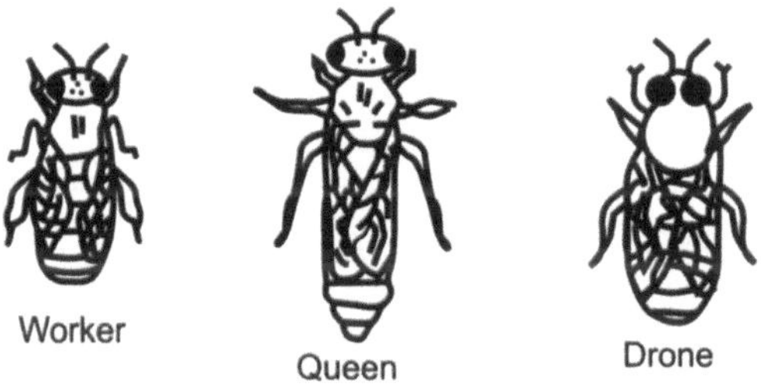

Fig. 12.2 Inhabitants of Bee colony: queen bee, drones and worker bees.

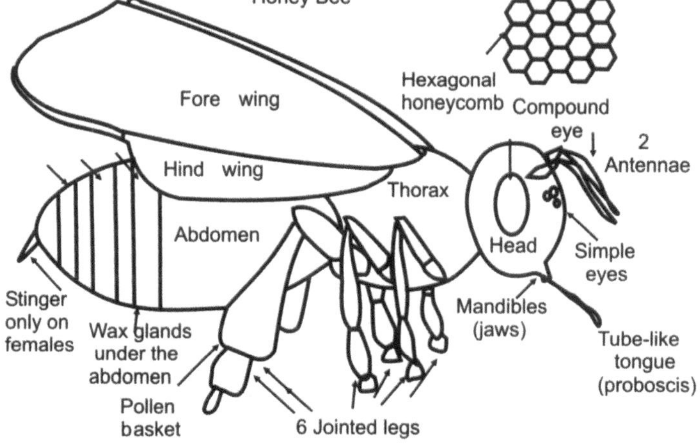

Fig. 12.3 Diagrammatic sketch of a typical honey bee showing various body parts.

in the early 17th Century. As early as 1908, Young published a short report on the pollen contents of domestic honey from the United States of America. This was followed by more exhaustive studies by Lieux (1978) who investigated 124 honey samples from Louisiana and Mississippi. Occurrence of *Prosopis* and *Acacia* pollen in honey samples was reported.

MELISSOPALYNOLOGY AND POLLEN FLORA

There is a very close relationship between honeybees, honey and pollen, which is referred to as bee botany. Pollination is a well-established fact. Bees and certain flowering plants have evolved a well-adjusted system of interdependence that ensures mutual benefits. Establishment of this mutual relationship between the flowering plants and their pollinators is one of the most significant events in organic evolution. Pollen analysis of honey is the most important aspect of melissopalynology. The assessment of the resource value of different plant species to bees is a major objective of melissopalynology.

Honey is a highly nutritious, sweet substance, which is produced by honeybees mainly from the nectar of flowers and from honeydew in temperate climates. The bees also use other raw materials, such as nectar from extra floral nectaries or sap from sugar cane.

In contrast, honeydew is derived from the secretions of the salivary glands and the gut of the plant-sucking insects. Thus certain plant-sucking insects of the order *Rhynchota* excrete a substance containing sugar called honeydew.

As per a recent report, the annual world production of honey is around 9,92,000 tonnes of which over 50% is produced by larger beekeeping countries like China (1,60,000), United States of America (75,000), Mexico (67,000), Canada (43,000), India (37,000), Argentina and Australia (35,000 each). In India, Tamil Nadu is the largest honey-producing state (2,373 tonnes) whereas Karnataka produces 750 tonnes of honey annually.

Pollen grains provide protein requirements for bees that are essential for building their body tissues especially during early embryonic growth. A colony of average size needs over 50 pounds (22.68 kg) of pollen per year.

THE CHEMICAL COMPOSITION OF POLLEN

Though chemical composition of pollen from different plant species varies over a fairly wide range, generally it contains: proteins 7-40%; fats 1.5-2.3%, carbohydrates 7-12%, minerals 2.8-10.6% and traces of enzymes, hormones, vitamins, pigments and other components.

Nectar, an aqueous sweet secretion by nectaries contains 5-80% sugars. Depending on the position, the nectaries may be floral or extra floral. Bees also feed on honeydew secreted by plant-feeding insects such as aphids. Nectar is eventually transformed into honey by an enzyme action and evaporation. Chemically, nectar is composed of a mixture of various sugars.

Pollen in Honey

A close observation on the life of honeybees indicates that four natural resources such as water, resin, nectar and pollen are required for survival of honeybees. Water is used to cool the hive and dilute honey fed to the larvae. Resin reinforces the hive and plug holes. Nectar serves as the major source of carbohydrates from which honeybees obtain their energy. Pollen is the major source of proteins, fat, minerals and vitamins.

It is estimated that a worker larva requires about 142 mg of honey for development and to make half a kg of honey, bees have to visit about 2 million flowers. On the average, one worker bee produces about half of a teaspoon of honey in its lifetime. Honeybees collect pollen from anthers in flowers by active movements of the legs and proboscis. During this process, pollen adhere to the bees' hairy legs and body which are scraped and deposited into the pollen basket. These pollen- packed baskets are transferred to comb cells.

The bees take precaution to prevent bacterial growth and pollen germination by applying a phytocidal acid to the pollen before it is deposited into the comb. These packed pollen are often referred as "bee bread" which is used for consumption by the bees. Pollen usually is trapped in nectar of the flower which is sucked by the bees. This is followed by enzyme action in the gut of the bees, and regurgitated in the form of honey. Pollen get into the honeycomb cells during grooming of her body.

Occasionally windborne pollen produced by anemophilous plants get into honey (e.g. *Pinus*-pine) however their quantity is far less than pollen of entomophilous plants in the honeycomb. It has been reported that *Salix* spp (willow) plants are often visited by bees, though they are primarily anemophilous.

On the basis of quality and quantity of pollen present in honey, bee-foraging plants are classified into major and minor floral sources. Honey of a dominant floral type has great consumer preference and commercial value. Thus, by knowing which plants are preferred by honey bees, bee keepers can locate their apiaries for maximum honey production. This information can be also used by the forestry departments to concentrate on growing more of such bee-favoured plants which ultimately leads to higher honey production.

A thorough knowledge of major floral resources is essential for a beekeeper to maintain healthy and productive bees and increase honey production.

Pollen Extraction from Honey

The extraction of pollen from honey is achieved by heating honey to 38°C in a water bath and stirring properly to ensure a uniform mixing of the pollen. Lieux (1981) recommends 20 gm of honey to be diluted with 100-200 ml of distilled water. The aqueous mixture is centrifuged for 10 minutes at 4,000 rpm to ensure complete sedimentation of the pollen. The pollen thus recovered from honey can be acetolyzed and stored in glycerine.

There have been very few innovations in the techniques used for the recovery of pollen grains from the honey samples. Jones and Bryant Jr (2001) have suggested alcohol dilution of honey samples. This suggestion has come forward after a long experimentation. It is well known that plants visited by honeybees can be determined by identifying and quantifying pollen in honey. Water is always used to dilute honey before pollen extraction. However, the amount of water used for dilution and the length of centrifugation varies from sample to sample. A liquid with a specific gravity less than water was examined for diluting honey by Jones and Bryant. They showed that pollen concentration values of alcohol-diluted honey samples were significantly higher than those of water-diluted honey samples. Alcohol-diluted samples also had significantly higher pollen concentration values than filtered samples. Higher pollen counts yielded significantly more pollen taxa than the lower pollen counts.

POLLEN ANALYSIS

Pollen is an essential tool in the analysis of honey. Pollen analysis helps in determining floral types and origin of honey as required by commercial import and export requirements by many countries (e.g. correct identification of pollen, categorization and correct labelling). Pollen analyses are an essential tool for honey standards and certification. Data obtained from pollen analysis determine the botanical and geographical origin of honey and the foraging ecology of honeybees. Pollen analysis also helps in determining if honey has been contaminated with poisonous pollen or adulterated. If honeybees die after foraging on poisonous plants sprayed with pesticides, pollen analysis of their pollen loads or honey indicates which taxa are responsible for this catastrophe (Bryant *et al.*, 1990).

Since pollen analyses of honey are concerned with pollen concentration values, pollen diversity and percentages, and floral resources used by honeybees, the alcohol dilution techniques proposed by Jones and Bryant should become the standard pollen extraction techniques for honey samples.

CONTAMINATED AND POISONOUS HONEYS

Honeys containing pollen from plants known to be poisonous or toxic are unsafe for human consumption. Individuals oversensitive to certain pollen may show allergic reactions to honeys contaminated with these airborne pollen. Pollen analysis helps to identify such honeys. Pollen of *Lasiosiphon* sp are not harmful to bees but honeys contaminated with these pollen are highly dangerous and poisonous for human consumption. In contrast pollen of *Euphorbia geniculata* is highly toxic, almost lethal to bees, but honeys contaminated with this pollen have no effects on human consumption.

Pollen grains of the poisonous plant, *Serjania letbalis* (Latin letbalis = deadly) belonging to the family Sapindaceae were found in a sample of stomach contents of a boy who had died after consuming such honey. Thus, the death may well be connected with these pollen grains. It is strange that the stingless honeybee *Trigona jaty*, which collect the pollen and make the honey, is however not harmed by the pollen of the poisonous plant. Many such deaths due to consumption of poisonous honey have been reported from the Cerrado Forest of São Paulo in Brazil.

There are different methods of analyzing pollen in honey. Basically all these methods involve the correct identification of pollen taxa and quantitying them. Agashe and Rangaswamy (2003) have identified a number of pollen recovered from honey and pollen loads in Karnataka, India. Some of them have been illustrated in **Figs. 12.4 and 12.5**.

Quantification

According to the International Commission for Bee Research (ICBR), pollen counts exceeding 12,00 per honey sample are needed to obtain percentages of taxa to an accuracy of within 1%. However, Moar (1985) recommends counts of 500-1,000 pollen grains per honey sample. These counts are sufficient to determine the precise geographic and botanical origin of each honey sample.

Codification of honey samples is done in many ways. The most accepted method was suggested by Louveaux *et al.* (1970, 1978). On the basis of pollen count of 200-300 per sample they may be categorized in different frequency classes.

a) Predominent pollen is the one which is represented by > 45% of total pollen count.
b) Secondary pollen occur between 16-45%.
c) Important minor pollen occur between 3-15%.
d) Minor pollen are below 3%.

Using the above frequency class system, honey sample can be codified as unifloral if it contains one predominant pollen type (e.g. clover honey, mustard, *Eucalyptus* honey). In case two pollen types are predominant both can be used in naming the honey (e.g. Baygall – tapelo honey). Honeys containing pollen grains predominantly belonging to single species or genus are called unifloral honeys. Such unifloral honeys are usually marketed under the name of their respective plant sources such as Clover honey, Jambul (*Eugenia*) honey, Mustard honey, etc.

If there are no predominant taxa, the honey should be classified as 'mixed floral' type. The frequency distribution of a pollen taxon in a series of honey samples is determined by dividing the number of samples in which a taxon occurs, by the total number of honey samples. Thus, a pollen taxon is classified as 'rare' if it is found in less than 10% of the samples, 'infrequent' if in 10-20%, 'frequent' if in 20-50% and 'very frequent' if in more than 50% of the samples.

Each unifloral honey has characteristic physico-chemical properties. Consumer preferences for unifloral honeys are based on these considerations. Pollen analysis in such cases serves to confirm the botanical sources of unifloral honeys.

In the case of multifloral honeys, pollen analysis indicates their botanical composition as represented by the spectrum of pollen variability. It can also reveal mixtures of different honeys as also their relative proportion approximately. Occurrence of both unifloral and multifloral honeys have been reported by Agashe and Rangaswamy (2003) from Karnataka, India **(Fig. 12.6)**.

GEOGRAPHICAL ORIGIN OF HONEYS

Local floras have characteristic plant associations that are reflected in the corresponding spectrum of pollen types represented in local honeys. This often enables to identify the geographical origin of honey samples. Honeys are often marketed under labels indicating their locality such as Kashmir honey, Nepal honey, Coorg honey, Mahabaleshwar honey and regions such as Mediterranean, African and Caribbean, Australian and New Zealand, Mexican, Candian, Asiatic Russia, etc. (Sawyer 1988) with incidental consumer preferences that can be protected through pollen analysis. As early as 1895, Pfister had suggested that pollen could be used to determine the geographic origin of honey. Five hundred to one thousand pollen grains per honey sample should be counted to determine the precise phytogeographic province.

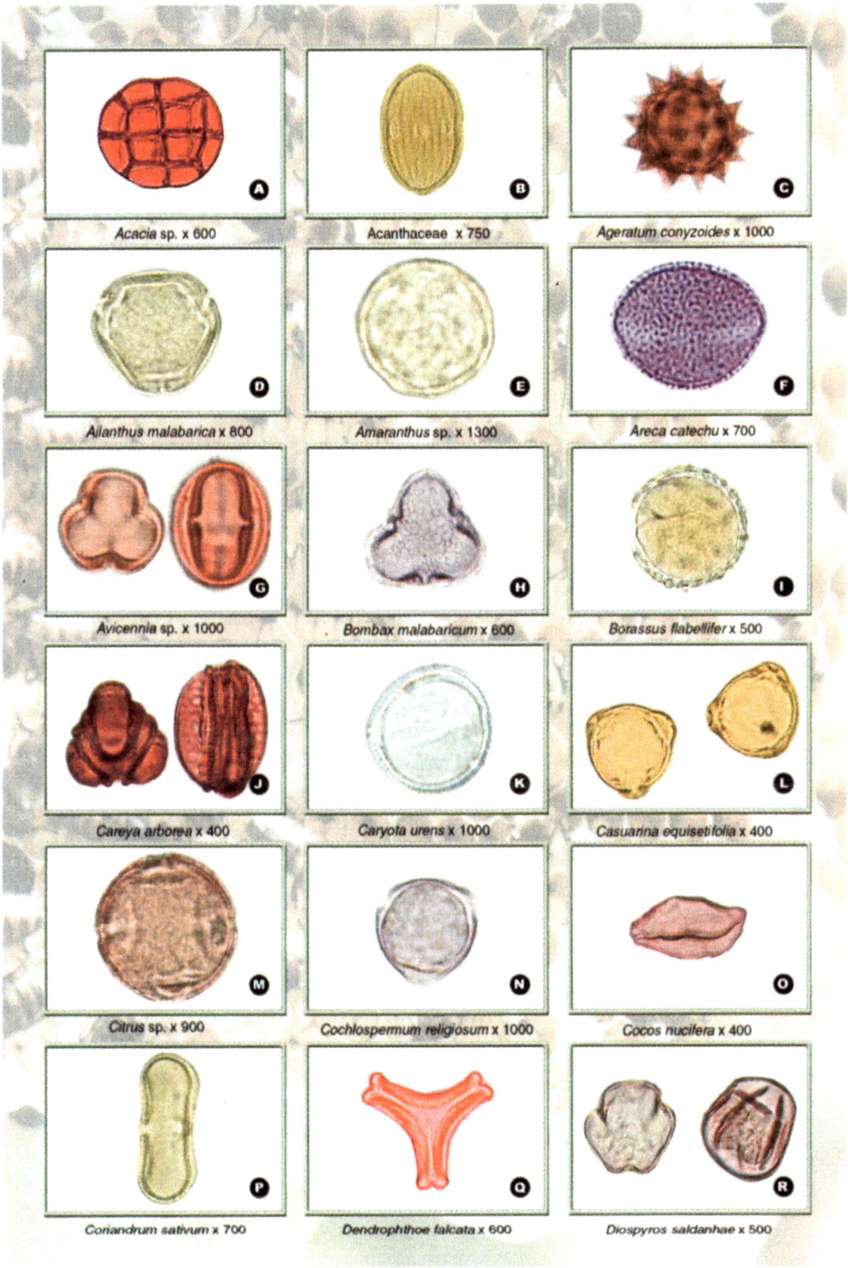

Fig. 12.4 Photographic plate showing major pollen types recovered from honey and pollen loads from Dakshina Kannada District, Karnataka, India.

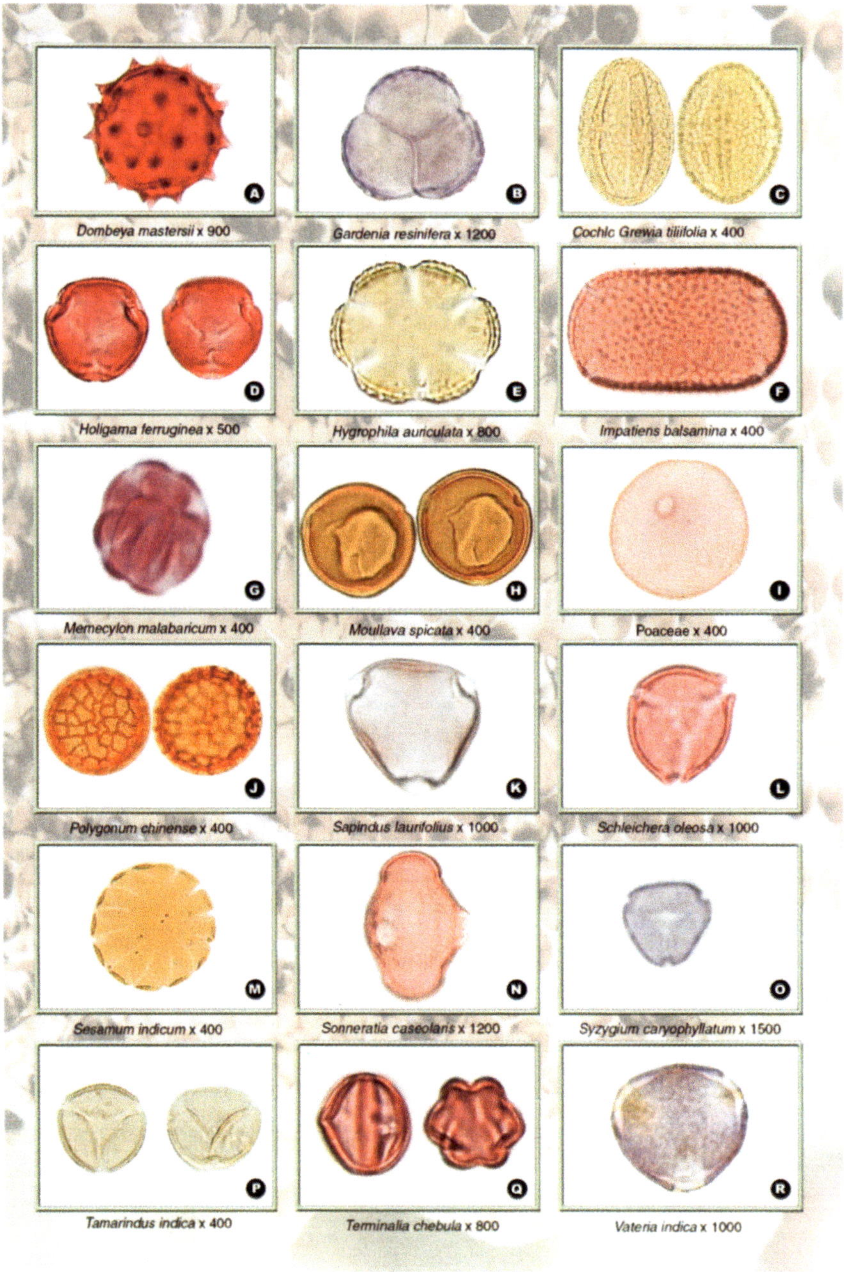

Fig. 12.5 Photographic plate showing major pollen types recovered from honey and pollen loads from Dakshina Kannada District, Karnataka, India.

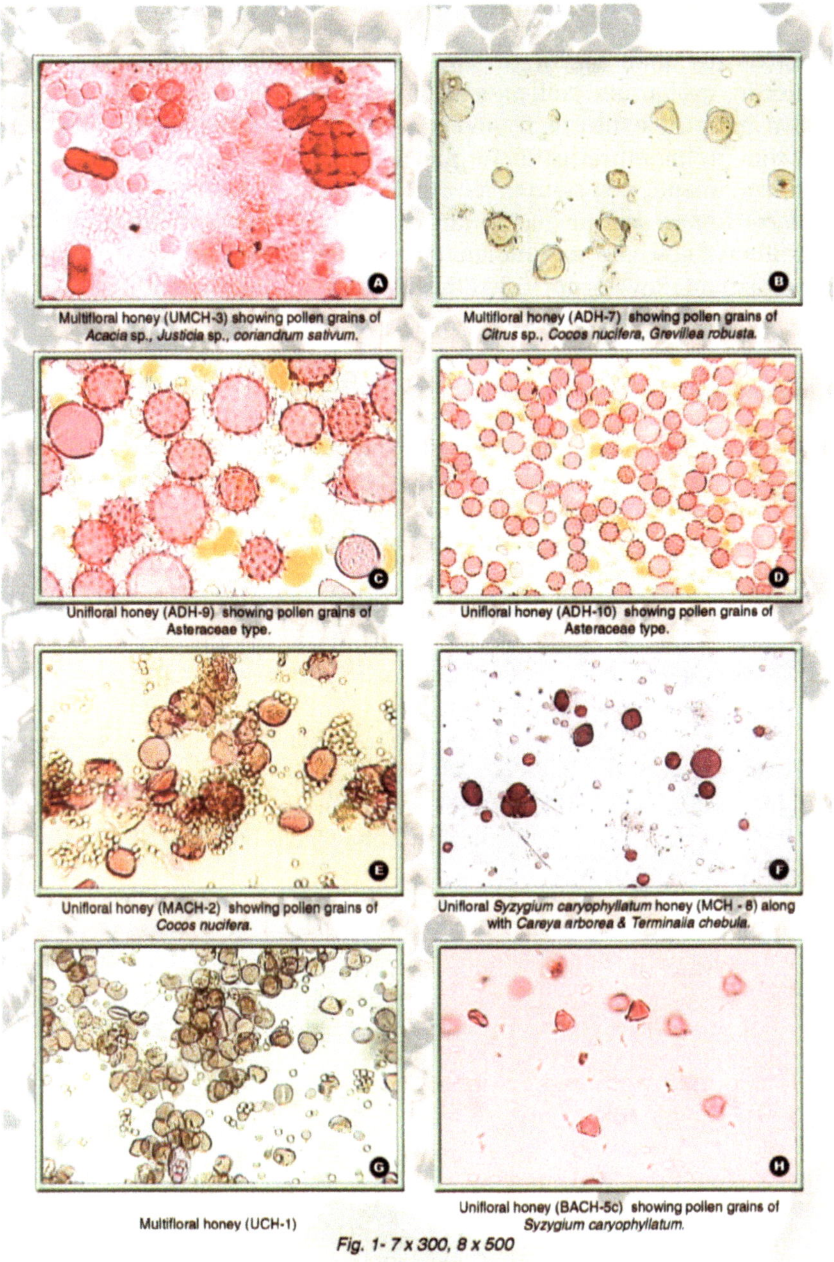

Fig. 12.6 Photographic plate showing occurrence of unifloral and multifloral honeys (**Figs. 12.4, 12.5, 12.6** from Agashe and Rangaswamy, 2003).

GEOGRAPHICAL FORAGING RANGE OF HONEYBEES

Bees can fly about 24 km/h. They feed on nectar, pollen or honey. For collecting nectar and pollen, bees visit flower after flower during which pollen get stuck to their body and inadvertently perform the most important function in the reproduction of plants namely pollination which leads to fertilization and seed setting. Normally the European honeybees i.e. *Apis mellifera* forage within half a km of the hive, however under extreme conditions these bees are known to forage up to 8 km from the hive. Many tropical plant flowers are visited for foraging by *Apis dorsata* **(Fig. 12.7)** and *Apis cerana* **(Fig. 12.8)**.

Honeybees

In order to understand the interaction between honeybees and plants it is important to know some facts of honeybees, their social life in the hive and a few statistical facts about their role in honey production and consumption.

Honeybee *Apis* belongs to the order Hymenoptera. There are four different species of honeybees.

Apis dorsata is diploid (Rock bee). It forms combs on rocks, branches of big trees, walls of buildings. Honey produced by this species is about 30 kg per comb.

Apis florea is also diploid (Little bee).

Apis indica tetraploid (*A. Sarana indica,* Indian bee). Honey produced by this species is about 6-8 kg per comb.

Apis mellifera is tetraploid (European bee).

Apis laboriosa The world's largest honeybee – one hive yields 40 lit of honey and 10 kg of wax.

Honeybees form highly organized permanent colonies of 50,000 or more disciplined individuals that live in a hive. Honeybees are social insects with a marked division of labour between the various types of bees in the colony. A bee colony includes a queen, drones and workers **(Fig.12.2)**. The queen is the only sexually developed female in the hive. She can be easily identified in a colony on account of her size. The queen bee, one per hive, is double the length and 2.8 times heavier than worker bee. She is the largest bee in the colony and is always surrounded by worker bees. A two-day-old larva is selected by workers to be reared as the queen. She will emerge from the queen chamber of the hive 11 days later to mate in the air with approximately 18 drones or male bees. During this mating she will receive several million sperm cells, which last her entire life span of nearly two years.

The queen starts to lay eggs about 10 days after mating. A young queen bee can lay as many as 3,000 eggs in a single day, which are deposited

Acacia sp.

Borreria sp.

Chromolaena odorata

Dalbergia sissoo

Gliricidia sepium

Helianthus annus

Hamelia patens

Ipomoea staphylina

Fig. 12.7 Visits by *Apis dorsata* (wild or giant honey bees) for foraging on different tropical plants flowers.

Apis cerana colony installed near mangrove vegetation.

A view of *Apis mellifera* comb showing queen bee.

Apis cerana foraging on *Guizotia abyssinica*.

Apis cerana foraging on *Psidium guajava*.

Apis mellifera foraging on *cocos nucifera*.

Stingless bees foraging on *Sonneratia caseolaris*.

A typical mangrove ecosystem at Nada.

Apis cerana foraging on *Ageratum conyzoides*.

Fig. 12.8 Visits by *Apis cerana indica* (Indian honey bee) for foraging on different tropical plants flowers. (A) *Apis cerana* colony established in a Bee box installed near mangrove vegetation adjacent to a back water river near Nada, Dakshina Kannada District, Karnataka, India. (F) Bees foraging on a mangrove flower of *Sonneratia caseolaris*. (G) A typical mangrove ecosystem near Nada, Dakishina Kannada District Karnataka, India.

individually in the hive cells. The queen bees never sting. Drones are stout male bees that likewise are without stings. They are the least active individuals in a colony and live with the sole purpose of mating with queen after which they die. Drones do not gather pollen or nectar from flowers. If the colony is short of food, drones are often removed from the hive.

Workers, the smallest bees in the colony are sterile females and the most active in the hive. A typical colony can have 50,000 to 60,000 worker bees. They are short lived with a life expectancy of approximately 28-35 days. Worker bees are known to perform various duties in the colony such as feeding the queen and larvae, guarding the hive entrance, protecting the queen and keeping the hive cool by fanning their wings. Their most important assignment is to scout and collect nectar and pollen by foraging flowers. In addition, worker bees produce wax through wax glands required for building combs which consist of hexagonal cells which are only 2/1000 inch (=c. 0.005 mm) thick, but support 25 times their own weight. Some of the worker bees act as scout bees. The main duties of the scout bees are to survey the area within the foraging range around the colony, to locate the presence of floral types available at a time, decide the best floral type which has high nutritive value and palatability, return to the hive and communicate all this information to other worker bees of the same colony. It is on the basis of this information from scouts that the activities of nectar and pollen gatherers begin.

Scout bees communicate their information about floral sources to foraging worker bees by certain methods. They also augment this further by depositing glandular aromatic secretious along the route. Thus, providing an odour trail which workers can follow due to their highly developed olfactory sensation. In all members of the genus *Apis*, scouts communicate their information on direction as well as distance of a specific floral source by means of certain types of dances on the comb surfaces.

In diploid species of *A. florea, A. dorsata* scouts perform communication dances on the horizontal upper surface of their combs and during their dancing gyrations point their heads towards the food source site. In advanced tetraploid *A. indica* and *A. mellifera* who build multiple combs enclosed in the dark, scouts perform communication dances on the vertical comb surface in dark. Each dancing pattern has a specific meaning for the bees in terms of both direction and distance of the floral source.

Dances serve the same purpose as language for their inter – communication. After the direction is thus conveyed, distance between the hive and floral source is indicated by means of the rhythm of 'round' or 'wag tail' dances which usually follow the pattern of number 8, which imply shorter or longer distances. Thus, bees have developed languages as well as dialects for inter communicating. The 'round dance' indicates the presence of food source near the hive. The tail wagging dance or 'waggle dance' indicates distance and directions of a food source further away. Thus, bees talk to their own kind by means of a dance language. Karl von Frisch received the Nobel Prize for his brilliant research on 'Bee Dancing'.

Honeybees' wings stroke 11,400 times per minute resulting into making distinctive buzzing sounds. The diagrammatic sketch in **Fig. 12.3** shows various parts of a typical bee. Like all insects, bees have three pairs of legs, a three-part body, a pair of antennae, compound eyes, jointed legs and a hard exoskeleton. The three body parts are head, thorax and abdomen with a terminal stinger (only in the sterile female).

Nectar is the major source of carbohydrates from which honeybees obtain their energy. Nectar is collected by foraging worker bees and is carried back to the hive in their stomachs. Upon returning to their hive, the nectar and pollen is transferred to hive workers for processing into honey. Enzymes from the bee's hypopharyngeal glands are added to the nectar. These enzymes break down the nectar into simple form of sugars, which are easy for digestion. The nectar is deposited in the cells and fanned to further reduce water in it.

It is estimated that to make 1 kg of honey, honeybees have to visit about two million flowers, fly a total of 80,000 km and carry about 84,000 loads of nectar back to the hive. It is reported by Thomson (1986), the British entomologist, that during the main flower blooming season, bees from a single hive visit as many as 2,50,000 flowers during the course of a single day.

Liriodendron tulipifera (tulip tree) is known to produce about a teaspoon (c. 5 mls) of nectar. During nectar gathering, a honeybee consumes 0.5 mg of ripe honey per km of flight. Feeding a bee larva from the egg to maturity requires about 142 mg of honey. The protein source needed for rearing a worker from the larval to the adult stage requires approximately 120-145 mg of pollen. An average size bee colony collects about 20-57 kg of pollen a year.

Sources of Pollen

Entomophilous species of *Salix* (willow), *Quercus* (oak), *Celtis* (hackberry) and several Poaceae (grasses) members as well as some of the wind-pollinated types of Asteraceae are considered important pollen sources for foraging honeybees. Certain families of flowering plants such as Rosaceae, Leguminosae, Tiliaceae (entomophilous), Asteraceae (Compositae), Scrophulariaceae, Labiatae, Aceraceae, Verbenaceae, Poaceae are known to be chief sources of pollen for honeybees. The following plants are known to provide good-pollen sources for honeybees in India: *Eucalyplus* sp *Psidium guajava, Syzigium sp., Mimosa pudica, Cocos nucifera, Pongamia pinnata,*

Sesamum indicum, Ricinus communis, Geizottia abyssinica, Helianthes annuus, Ailanthes excelsa and *Coriandrum sativum.*

APPLIED ASPECTS OF HONEY POLLEN

Pollen embedded in honey or pure pollen collected by honeybees as pollen loads and deposited in bee boxes find a number of applications. Beekeepers provide a mesh trap at the entrance of a bee box with a tray below in which pollen loads are deposited. The pollen loads consist of pure pollen collected by honeybees, which are packed in spoon shaped pollen baskets in the hind legs of bees. While entering the bee box, the hind legs of bees become stuck in the mesh trap and during this process pollen loads are detached and fall in the tray.

APPLIED ASPECTS OF BEE POLLEN

Bee gathered pollen is nature's most complete food, rich in vitamins, minerals, amino acids and a complete source of protein. It can give us that extra edge to stay healthy. An all-round nutritional supplement, bee pollen is ideal for daily use.

Bee pollen is often referred as the 'perfect food'. Studies from all over the world indicate that the pollen collected by bees from the stamen of flowers is worth its weight in gold. Bee pollen contains 22 amino acids (and higher amounts of the eight essential ones than most high-protein foods), 27 mineral salts, the full range of vitamins, hormones, carbohydrates, and more than 5,000 enzymes and coenzymes necessary for digestion and healing. A little known fact is that bee pollen is also rich in the bioflavonoids, important for capillary strength, and in vitamin B12. It is, in fact, one of the few vegetable sources of this vitamin. Preliminary observation indicates that bee pollen may prevent cancer. The Sloan-Kettering Institute for Cancer Research in New York City has been studying the effects of bee pollen, royal jelly and bee venom on cancer.

Hippocrates (c. 460 – c. 375 BC), the Father of Medicine, believed that bee pollen contributed to a long life. The Russian researcher Prof. Nicolai Vasilievich Tsitsin, a biologist and experimental botanist at the Longevity Institute, tried to discover why so many natives of Georgia, formerly of the Soviet Union, reportedly lived to be upwards of 125 years old. Most of these modern Methuselahs who live in dry, desert-like climates, are beekeepers, who every day eat raw, unprocessed honey with bee pollen. "All of the 200 or more people past 125 years of age in Georgia, without exception, state that their principal food is pollen and honey–mostly pollen." Prof. Tsitsin Naum Petrovich Joirich, chief scientist at the Longevity Academy in Vladivostak, said that "long lives are attained by bee pollen users. Bee pollen is one of the original treasure houses of nutrition and medicine. Each grain contains every important substance necessary to life."

Bee pollen reportedly can keep the skin youthful. Dr. Lars Erik Essen, a dermatologist in Halsingborg, Sweden, said that pollen exerts a powerful biological influence in preventing premature aging of cells and in stimulating growth of new skin tissue. Dr. Essen stated that bee pollen can help deliver more blood to the skin cells, guard against dehydration and smooth away shallow wrinkles.

Experiments reveal that bee pollen is an amazing biological stimulant with healing properties. In his book *Sexual Nutrition,* Morton Walker described the effects of bee pollen on both animals and humans with a variety of medical disorders. Treatment with bee pollen improved energy levels, relieved constipation and diarrhoea and acted as a tranquilizer for hyperactive patients. Other effects were increased blood haemoglobin and stress reduction at the cellular level.

Bee pollen has helped manage menstrual pain and irregularities, as shown by the double-blind study of Dr. Bogdan Tekavcic, chief of the Ljubljana Cen ' for Gynaecology in Yugoslavia. For two months, half of the women in the stı dy were given a mixture of bee pollen and royal jelly, and the other half a placebo. Almost all the women taking bee products demonstrated vast iı.ıprovement or total disappearance of menstrual pain. The placebo group showed little or no change.

The bee-gathered pollen is pressed into tablets or sold loose, but some companies add a binder of honey and concoct pollen bars–said to have been a favourite of Ronald Reagan (former American President). In Engelholm, Sweden, the Cernelle Company has been making pollen extract products since 1952. Cernelle manufactures a pollen toothpaste, pollen face cream, pollen animal feed and scores of other concoctions. These pure pollen, i.e., usually pollen of same species of plants are consumed in different forms for physical fitness and instant energy. Joggers often eat pollen tablets for general fitness. Pollen are extensively used in health food supplements as their medicinal use was known to man since prehistoric times. Pollen tablets as a nutritional supplement are available in the market **(Fig. 12.13)**. Pollen in liquid form is also available commercially. Pollen is known to possess antibiotic properties and also recommended for certain problems of intestinal functions, respiratory system and endocrine system.In certain European countries homeopathic doctors recommend pure pollen which act as strength-giving substances.

The Olympic sprinter Steve Riddick, who sported a 'Bee Power' T shirt after winning the gold in the 400-meter relays in 1976, used to consume pollen tablets. The World-class marathoner Gary Fanelli, used to take 10 pollen tablets before a race and often pop one while running.

APPLICATIONS OF RAW HONEY

'Honey is not a product but a mystic universe!' Its riches are beyond description. It contains and encloses the entire nature: the sun, the flowers, the freshness, youth, euphoria, the pleasure of living, in short a food for the Gods, the earthy paradise.

There is a treasury in every drop of honey? Honey contains over 25 different sugars, each one having a different function in the human metabolism. Among them are fructose (as dextro-fructose; fruit sugars), glucose, levulose, trehalose, meletoze, dextrose (dextro-glucose; grape sugars), maltose, kojibiose, isomaltose, nigerose, trehalose, genitobiose, laminaribiose, etc. Many of these sugars do not occur in nectar but are formed by the bees during ripening.

'Honey can be called the substance of divinity': the sweet nectar of the fragrant flowers which the bees collect, elaborate and then after depositing it in the cells of comb, evaporate by vibrating their wings for endless hours until it turns into the consistency of honey. Before it is evaporated, this nectar contains approximately 90% of water.

Honey is a unique food. It is a living, organic, instant energy building food, an antioxidant containing all the essential minerals necessary for life: seven vitamins of the B complex group, amino acids and enzymes. Honey goes into the blood stream in 15 minutes. Honey produces more energy than its weight and in doing so it uses part of the accumulated fat and cholesterol. It therefore is excellent in any weight reduction plan.

Honey is the oldest sweet known to mankind with a long, extremely interesting and fascinating history. Honey contains many minerals, vitamins, enzymes and antibacterial agent that acts like penicillin, and other antibiotics, killing microbes and germs that is correctly called a medicine; 'the elixir of life'. On account of its antiseptic properties it is often applied to open wounds. In fact, several reports are available indicating the use of honey to preserve dead human bodies like a mummy. It has been used as medicine for millenniums and is still used in prescriptions for cold. Honey also serves as s a mild laxative which helps digestion.

Honey is a perfect food. The bible mentions honey 68 times. "The promised land that flows milk and honey". It is advisable to use a glass of water with a little honey dissolved in it with a few drops of lemon juice as a thirst quencher instead of carbonated drinks, sodas, colas and others which destroy the teeth and the lining of the stomach resulting in cavities and ulcers.

In one of the ancient books of medicine from India, 'Yajurveda' (The book of life) as well as the book of 'Manu' it has been written that it is possible to prolong human life up to 500 years with a regimented diet of pollen-filled comb and honey.

Hippocrates, prescribed honey in the treatment of colds, coughs, rheumatism, insect, serpent and animal bites, wounds and a host of other ailments. He considered honey of a great nutritive value and a must for beauty. Pythagoras (c. 532 BC), the ancient Greek mathematician, philosopher and author, in his books affirms that honey has remarkable therapeutic properties.

The great sage of Central Asia, Abou Ali Ibn Sina, also known as Avicenna (c. 980 – 1037), was using honey and other products of hive successfully. In his medical book, a number of prescriptions based on honey are given. He mentions that honey fortifies the heart, recreates vigour, helps digestion, improves the appetite, reinforces the memory, helps and improves sexual deficiency, lifts the spirit and is reservoir of eternal youth.

In view of the applications of pollen analysis to bee problems, melissopalynology has been gradually assuming considerable importance. Apicultural laboratories in many countries have been developing their permanent reference collections of honey pollen slides. The International Commission for Bee-Botany with its Head Quarters in Switzerland has been attempting to coordinate and stimulate the work on melissopalynology in various countries.

HONEYBEES AND POLLINATION

Pollination is the process involving transfer of pollen grains from anthers to the stigmatic surface of gynoecium. Pollination is important to plants as the means by which the next generation is produced. The process of pollination requires a pollinator, which is the agent that moves pollen from the male anthers of a flower to the female stigma of a flower to accomplish fertilization or syngamy of the female gamete in the ovule of the flower by the male gamete from the pollen grain. The Pollenizer, which is different from the Pollinator is the plant that is a source of pollen for the pollination process.

POLLINATION – BEE FORAGING PLANTS IN EUROPE AND NORTH AMERICA

The majority of angiosperms (flowering plants) are entomophilous. Among the insects involved, the most important to man are the hymenoptera (ants, bees and wasps). The relationship between flowers and insects has been an evolutionary development. In both cases specific plants are dependent on specific insects to ensure survival for which pollination is the first step. Without successful pollination – no fertilization will occur, no seed will form and no fruit develop to disperse or achieve seed dispersal.

A visiting insect is rewarded with nectar and pollen and the flower for its part achieves pollination and subsequent fertilization. Insects are attracted to flowers by a variety of strategies to seek nectar and ensure pollination, including colourful corollas, bracts, honey guides and specialized morphology. In obtaining nectar, cross pollination is achieved as the insect goes from flower to flower of the same species.

Bees travel from flower to flower, collecting nectar (converted to honey later), and in the process pick up pollen grains. The bee collects the pollen by rubbing against the anthers. The pollen collects on the hind legs, in dense hairs referred to as a pollen basket. As the bee flies from flower to flower, the pollen grains are transferred onto the stigma of the female flower part. When bees are rearing large quantities of brood, they will deliberately gather pollen to meet the nutritional needs of the brood. A honeybee which is deliberately gathering pollen is up to 10 times more efficient as a pollinator than one that is primarily gathering nectar and is only incidentally gathering pollen.

Pollination, where pollinators are insects, is referred as entomophily. Pollination by animals is called as zoophily, whereas anemophily involves pollination by wind. Hydrophily involves pollination by water.

Some plants are self fertile or self-compatible, and can pollenize themselves. Other plants have chemical or physical barriers to self-pollination, and need to be cross-pollinated. In pollination management, a good pollenizer is a plant that provides compatible, viable and plentiful pollen, and blooms at the same time as the plant that is pollenized. Pollination can be by cross-pollination with a pollinator and an external pollenizer, self-pollination without any pollinator, or self-pollenization with a pollinator.

In cross-pollination (Syngamy): pollen is delivered to a flower of a different plant. In self-pollenization (Autogamy): pollen moves to the female part of the same flower, or to another flower on the same individual plant. This is sometimes referred to as self-pollination, but this is not synonymous with autogamy. The term self-pollination is restricted to those plants that accomplish pollination without an external pollinator (the stamens actually grow into contact with the pistil to transfer the pollen). Most peach varieties are autogamous, but not truly self-pollinated, as it is generally an insect pollinator that moves the pollen from anther to stigma.

In cleistogamy, pollination occurs before the flower opens and is always self-pollination. Some cleistogamous flowers never open, in contrast to chasmogamous flowers that open and then are pollinated. Cleistogamous flowers must of necessity be self compatible or self-fertile plants. Other plants are self-incompatible.

To attract pollinators, some flowers, such as the sunflower (*Helianthus*), when viewed under ultraviolet light as seen by honeybees, have a darker centre, where the pollen is located. There may also be patterns upon the petals. These are called nectar honey guides.

The other trend is the decline of pollinator populations, due to pesticide misuse and overuse, new diseases and parasites of bees, clearcut logging, decline of beekeeping, suburban development, with introduction of exotic flora to gardens, removal of hedges and other habitats from farms. It is estimated that one hive per acre will sufficiently pollinate watermelons. Patterns of climate e.g., excessive rainfall, prolonged drought-changes witnessed in recent years have influenced the decline of the industry as well as mite and viral diseases increase among colonies where bee keeping is a 'cottage industry' only.

Pollination Management

Honeybees play a major role in improving the yield due to cross-pollination in addition to their byproducts like honey and bee wax. Bees are the most efficient pollinators on account of following assets:

1. Bee colonies can be moved to areas, which require pollination.
2. Each colony contains large populations of foragers to work on crop plants.
3. Bees usually work on only one type of flower on each trip (floral fidelity).
4. As honeybees are efficient pollinating agents, by using pheromone sprays bees can be attracted to pollinate the crops to significantly increase the yield.

It is the horticultural practice that enhances the pollination of a crop, to improve yield or quality. The largest managed pollination event in the world is in California almonds, where nearly, half (about one million hives) of the U.S. honeybees are transported by trucks to the almond orchards each spring. It has been reported that New York's apple crop requires about 30,000 beehives; Maine's blueberry crop uses about 50,000 beehives each year. Bees are also brought to commercial plantings of cucumbers, squash, melons, strawberries, and many other crops. Thus, the management techniques of a beekeeper providing pollination service are different and somewhat incompatible, compared to a beekeeper who is trying to produce honey. The farmers only option in the current economy is to bring beehives to the field during blossom time.

In many tropical countries in the forest honeydew may not be available for honeybees on account of high density of ants. The ants feed on the honey dew and probably do not allow the bees to collect it. The knowledge of honeybee-foraging plants provides important information to bee keepers for apiary sites where floral resources are plentiful. Their information will boost the bee keeping industry. In turn this will lead to increase in honey production and other byproducts such as beeswax and propolis for local consumption and export. The melissopalynoligists can make recommendations for planting and conservation of bee-foraging plants.

The U.S. solution to the pollinator shortage, so far, has been for commercial beekeepers to become pollination contractors and to migrate. Beekeepers follow the bloom from south to north, to provide pollination for many different crops.

In Europe some 16 honey types are recognized, derived from *Acer, Brassica, Calluna, Castanea, Cirsium, Epilobium* (*Chamaenerion*), *Erica, Filipendula, Ligustium, Myosotis, Prunus/Pyrus, Salix, Sambucus, Trifolium* and *Vicia.*

Three honeydew honeys are derived from *Pinus, Quercus* and *Tilia.* In North America (U.S.A. and Canada), some 13 honey types are recognized derived from *Acer, Asclepius, Brassica, Citrus, Epilobium* (*Chamaenerion*), *Ilex, Liriodendron, Melilotus, Nyssa, Oxydendron, Rhus, Trifolium* and *Tilia.*

TRENDS OF MELISSOPALYNOLOGICAL WORK IN INDIA

Pollen analysis of pollen loads and honey samples have been carried out by various melissopalynologists such of Deodikar (1965), Chaturvedi (1973) from Lucknow and the adjacent area, by Garg and Nair (1993) from Kumaon in the Himalayan region, Ramajujam and Kalpana (1994) from Andhra Pradesh, Phadke (1962) from Mahabaleshwar, Rao and Lakshmi (1996) from the Western Ghat forests.

The Indian hive bee – *Apis cerana indica,* the rock bee *Apis dorsata,* the Himalayan bee *Apis laboriosa,* the dwarf bee *Apis florea* and several species of *Trigona* and related genera of the stingless bees are of major economic importance and are associated with the social and cultural life of many tribal populations in India.

Honeybees are lauded as the most important pollinators of crops. In spite of the rich diversity of angiosperm flora in India, very little work has so far been done on plant-pollinator relationship (Atluri *et al.,* 2003). It is an established fact that social or solitary, domesticated or wild bees totally depend on flowers for the survival on account of the food they provide in the form of pollen grains and nectar. Bees require an enormous quantity of pollen and nectar for nutrition for themselves and for their brood.

Generally, the bees are diurnal in their foraging activity; however they can also work in moonlit nights (Sihag 1984; Rao *et al.,* 2001). *Apis dorsata* or rockbee is the sole visitor and pollinator of the night blooming *Pterocarpus santalinus* belonging to the family *Fabaceae,* in its natural habitat. In this tree species, the flowers open at around midnight and are foraged during moonlit hours.

There have been several investigations in India, which have concentrated on pollination of agricultural crops by bees with a purpose of enhancing crop productivity. Bee pollination of the crops including the self-pollinated crops was proved to greatly enhance productivity of crops (Batra 1967; Kapil 1970; Deodikar and Suryanaraya 1977; Mishra 1989; Abrol 1993). Management of honeybees proved highly rewarding in several cultivated crops including sunflower (*Helianthus*). Studies at the Andhra University demonstrated an extremely important role of the carpenter bees in carrying out pollination in *Cassia spp*, Lamiaceae, Acanthaceae and Verbenaceae.

The stingless bee (*Trigona*) is known for its buzz collecting of pollen and is a proven and substantial contributor to pollination of at least nine crop species including *Mangifera indica* (Mango) and Cocos *nucifera* (Coconut).

The small stingless bees (*Meliponinae*) and honeybees (*Apinae*) are often described as cross-pollinators because of their opportunistic behaviour, low energy demand and short flight ranges. These bees are known to move '*en masse*' with the pollen adhered to them to other conspecific plants in the flower (Bawa 1980). This kind of *en masse* movement has been also demonstrated in the mass blooming *Terminalia tomentosa* where nectar respertion occurs simultaneously in all open flowers (Atluri *et al.*, 2000 b).

A web page: **Beekeeping in India** was started by Suryanarayana in 2001 to disseminate information on various aspects of beekeeping and bee research in India. The All India Bee Keepers Association has been engaged in beekeeping development activities right from its inception in 1937. The 'Indian Bee Journal' started 62 years ago is the only English journal in Asia devoted to tropical beekeeping. In addition, the association also publishes 'Bharatiya Madhvipalan' in Hindi and 'Bharatiya Jenu Vignyana Patrike' in Kannada.

MELISSOPALYNOLOGYCAL WORK IN KARNATAKA

Sheshagiri (1985), Agashe and Scinthia (1995), Agashe (1997), Rangaswamy (2001) carried out honey and pollen load analysis during a melissopalynological investigation at Thally, Dharampuri District, Bangalore and its environs, the coastal Karnataka districts of Mangalore and Udupi. They brought out important findings such as the occurrence of unifloral honeys from *Eucalyptus, Coriandrum sativum, Syzygium cumini, Psidium gujava, Pongamia pinnata* and *Phyllanthus.*

Melissopalynological investigation by Agashe and Rangaswamy from 2000-2003 in a major research project sponsored by Department of Science and Technology, New Delhi with emphasis on exploring the potential of bee-foraging activities on the mangroves of coastal Karnataka, India has been very significant. Bee boxes with *A. cerana indica* colonies were installed and operated in the mangrove belt near Kundapura, Nada, Udyavara,

Pithrodi and Kalmodi of Udupi district. Mangroves such as *Avicennia, Rhizophora, Kandelia* and *Sonneratia* produce large showy fragrant flowers with copious amount of nectar and pollen. Field observations made by them confirmed that honeybees visit these flowers frequently. Their pollen was also recovered from honey samples from these areas. The tremendous scope on exploiting mangroves for bee-foraging plants was indicated with evidence by them **(Enlarged view of 12.8E, F and G labelled as 12.8A, B and C respectively)**. The presence of *Avicennia sp, Rhizophora mucronata, Sonneratia caseolaris* and *Kandelia sp* pollen in honeys indicates the geographical origin of the honey.

Melissopalynological Work in Brazil, South America

Luz and Barth (2001) undertook a melissopalynological investigations in a mangrove area next to Rio de Janeiro, Brazil in South America. They noted that the evaluation of the apicultural importance of mangrove plant species that furnish nectar and pollen grains for the honeybees, may be a valuable option for the rational management of this ecosystem. Similar to other parts of the world, most of the mangrove areas around the Guanabara Bay of Rio de Janeiro have been covered with soil.

The purpose of this investigation by Luz and Barth was to evaluate mangrove and associated plant species for harvest by *Apis melllifera* with regard to pollen grains and honeys deposited inside the beehives. The nectariferous plant species noted by them were: *Croton sp. Eucalyptus sp, Eupatorium maximillianii, Gochantia polymorpha, Mimosa bimucronata, Mimosa pudica and Spondias sp.* Though typical mangrove species in the area were *Laguncularia racemosa* and *Avicennia tomentosa* **(Fig. 12.9)**. Their pollen were not commonly detected in the honey samples or pollen loads. They noticed that the honeybees travelled great distances in search of food. Pollen grains of *Eucalyptus* were found in the honey samples though the *Eucalyptus* plantations were 3 km away from beehives. According to Bastos (1995), bees can travel about 10 km in search of nectar and pollen grains from *Eucalyptus.*

The most abundant mangrove species in this area was *Laguncularia racemosa.* The flowers of this species have a sweet perfume and were very frequently visited by honeybees . The field observations revealed that bees stayed about five seconds on each flower for nectar collection. On account of its long flowering period its pollen were found in a number of honey samples and pollen loads, thus emphasizing its great importance for apiculture in Brazil. In contrast, the pollen grains of *Avicennia tomentosa* were not found in the honey samples.

It is a very significant fact that to date, except for the United States, several major honey producing regions require certification of their honey.

Apis mellifera foraging on *cocos nucifera*.

Stingless bees foraging on *Sonneratia caseolaris*.

A typical mangrove ecosystem at Nada.

Fig. 12.8 Enlarged view of Figs. 12.8-E, F & G labelled as 12.8-A, B & C, respectively.

Fig. 12.9 Representation of Nectariferous plant species in different months of the year 2001 in Mangrove area next to Rio de Janerio, Brazil, South America (From Luz and Barth 2001).

Plant Species & Pollen Types	Month										
	Jan	Feb	Apr	May	June	July	Aug	Sep	Oct	Nov	Dec
Albizzia lebbeck	X	X							X		X
Croton	X	X					X			X	X
Eupatorium			X								X
Gochnatia		X									X
Laguncularia racemosa			X						X		
Mimosa bimucronata			X	X	X			X			
Mimosa pudica	X	X			X						X
Palmae			X	X		X	X		X		X
Piptadenia communis									X	X	
Sapindaceae										X	
Spondias			X	X							

As per the law, certification includes verification of a honey's floral type, quality and precise place of origin. On account of the free trade between United States, Canada and Mexico, it has become possible to replace expensive domestic honey by inexpensive foreign honey.

A lot of melissopalynological work has been done in Mexico. Acosta-Castellano and Palacios-Chavez(2001) have investigated plants of apicultural interest in the Pluma Hidalgo zone, Oaxaca, Mexico which is known for cloud forests and semi-ever green montane tropical forests, in addition to coffee plantations. Palynological analysis carried out by them on the honey samples produced by *Apis mellifera ligustica* and pollen loads of these bees has indicated the occurrence of 31 types of pollen belonging to native tree species including families such as Actimidiaceae, Anacardiaceae, Astetaceae, Burseraceae, Fabaceae and Sapindaceae **(Fig. 12.11)**. In many aspects, pollen analysis results mentioned above resemble pollen analysis of honey samples from the Western Ghats of India.

On the basis of their experimental work Agashe and Rangaswamy have suggested the need for extensive mangrove cultivation and its sustainable management for apiculture development. They are of the opinion that by including pollen and nectar-yielding plants into social forestry programmes, it is possible to increase the total bee colony carrying-capacity of an area, which will enhance the total honey production. The same view has been supported recently by Debasis and Bera (2004) by analyzing honey samples collected from the mangrove area in Sunderbans, West Bengal. Out of nine honey samples collected from Sunderbans, eight were proved to be unifloral and one sample was of multifloral type, *Sonneratia apetala* was the dominant genus in the Sunderbans mangrove area and hence their pollen dominated in seven honey samples collected by them. In addition, other mangrove pollen recovered from honey samples belonged to *Bruguiera gymnorhiza* and *Acanthus ilicifolius*. It was concluded that

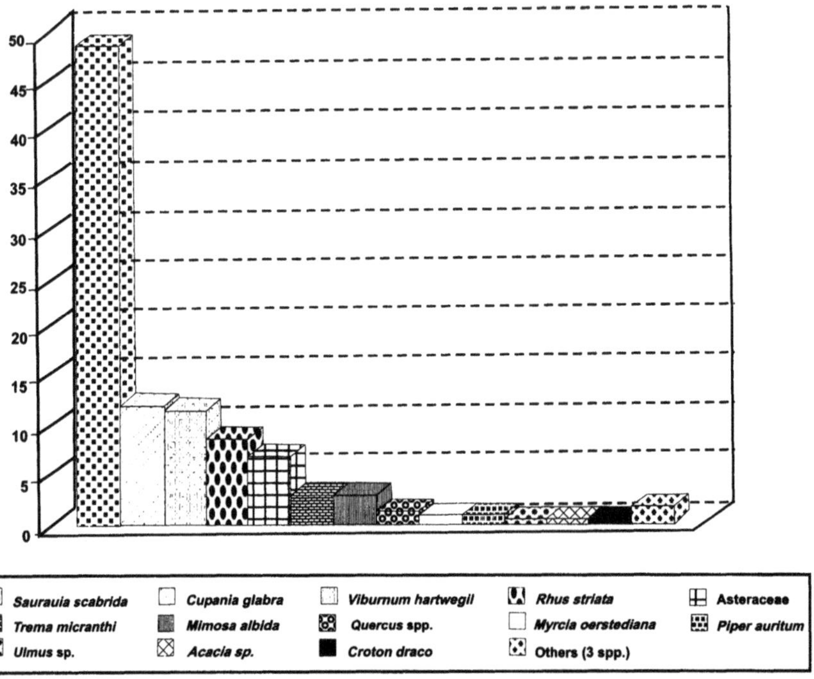

Fig. 12.11 Pollen analysis results of honey samples of pollen loads of *Apis mellifera ligustica* from Oaxaca, Mexico, South America (From Salvador Acosta-Castellano and Rodolfo Palacios-Chavez, 2001).

Sonneratia apetala was a major nectar and pollen source for honeybees *Apis dorsata, A. cerana indica* and *A. mellifera.* They observed that numerous honeybees foraged voraciously on *Sonneratia* flowers.

The pollen of *Sonneratia apetala* are isopolar, subprolate, 35-40 μm × 28-38 μm, triporate, pore surrounded by distinctly protruding thin collars, exine 4 μm thick, thicker at poles with a smooth pore membrane.

The major contribution of Agashe and Rangaswamy (2002) on the melissopalynological analysis of honeys from Dakshina Kannada and the Udupi districts of Karnataka, India was the compilation of pollination calendar of major bee-foraging plants **(Fig. 12.10)**. The beekeepers can select profusely flowering plants from this pollination calendar for migration of bee colonies.

Melissopalynological Work in Africa

Apicultural studies have been done extensively in Africa particularly in Ghana. Nectar and pollen sources for the African honeybee, *Apis-mellifera-adensonii* from different geographical regions of Ghana were investigated by Amoko and Picard (2001). The commonest nectar sources for the honeybees come from the plants *Elaeis guineense, Tridax procumbems, Panicum maximum, Ceiba pentandra, Mimosa pudica, Zea mays, Leucaena glauca, Securinega virosa, Morus mesozygia* and *Cocos nucifera* **(Fig. 12.12)**.

Certain plants do not produce nectar but their pollen is present in the honey samples. Among them, the predominant plants of Poaceae and Cyperaceae are: *Cynodon dactylon, Cyperus esculentus, Panicum maximum* and *Zea mays*.

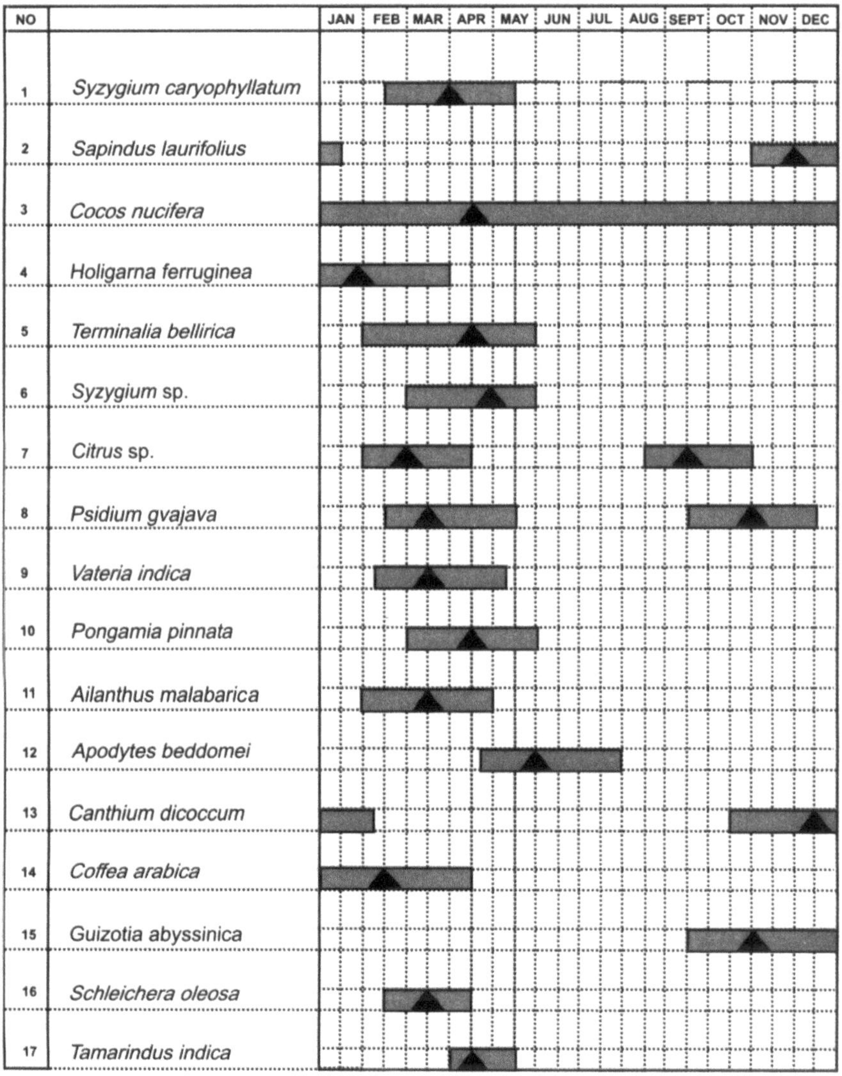

Fig. 12.10 Pollination calendar of major bee foraging plants from Dakshina Kannada and Udupi Districts, Karnataka, India. (From Rangaswamy, 2003).

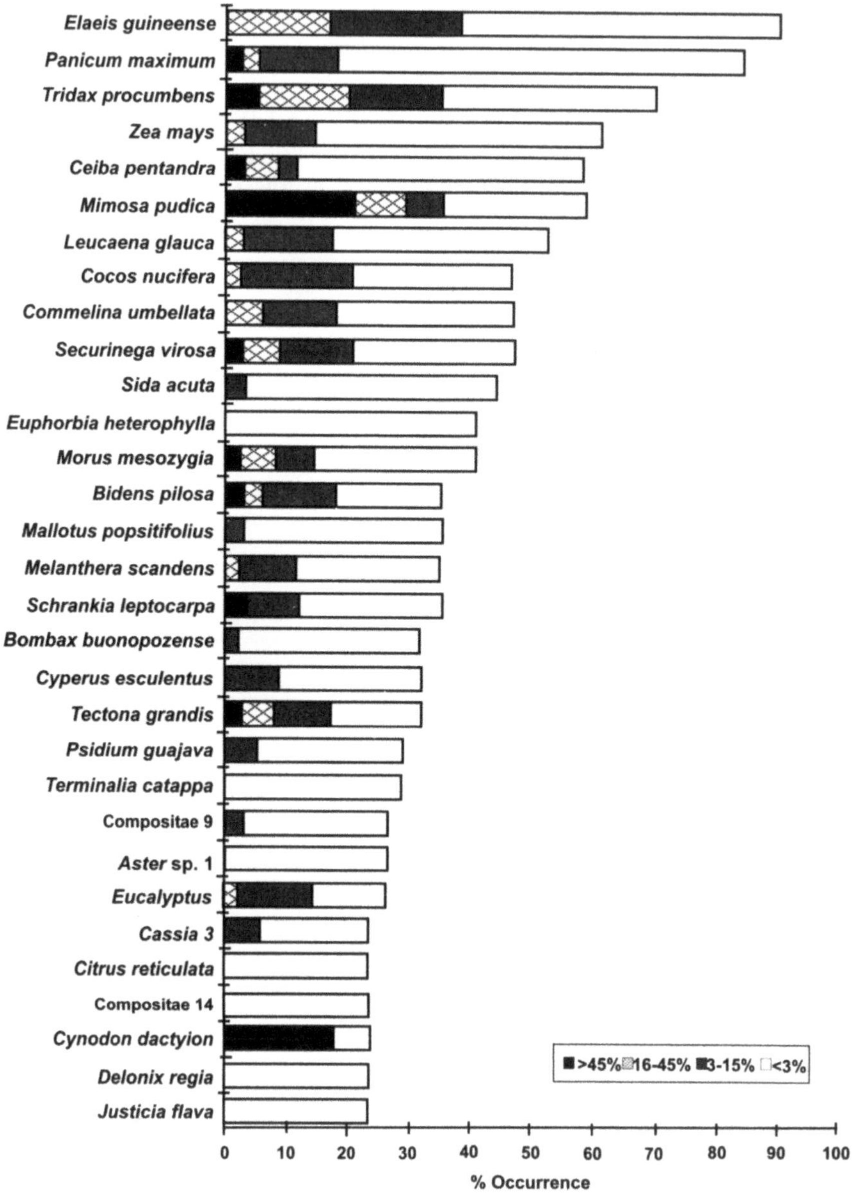

Fig. 12.12 Taxa occurring in more than 20% of the post September-December nectar flow honey samples and pollen sources of African honey bee *Apis mellifera adansonii* from Ghana, Africa. (From Joyce Amoko and Roberts Picard, 2001).

Fig. 12.13 Pollen tablets as nutritional supplement prepared from bee pollen which are commercially available. Here they are photographed from both sides of the packet (courtesy K.R. Shivanna).

CHAPTER **13**

Aerobiology: Aeropalynology - Part I

HISTORY OF AEROBIOLOGY

The term aerobiology was coined as early as 1930s by F. C. Meier who was the plant pathologist working in the Department of Agriculture, United States of America. However, this does not mean that aerobiological studies were not carried out prior to the 1930's. In fact, the preliminary aerobiological work and its applications to health and environmental pollution dates back to the period of the Vedas c. 3000 B.C.

Aerobiology involves the study of airborne bioparticles, that is, particles of biological origin (both from plants as well as animals). Subsequent to the 1930s, aerobiology was classified basically into indoor aerobiology and outdoor aerobiology. Some aerobiologists preferred to segregate the study of airborne pollen into a sub branch of aerobiology termed as 'aeropalynology'. The mycologists and plant pathologists preferred to study airborne fungal spores under a separate sub branch of aerobiology termed as 'aeromycology'.

Studies that deal primarily with pollen that are airborne may be called aeropalynology. Air contains many kinds of contaminants, organic and inorganic, having great diversity in size, shape, density and many other characteristics. According to Gregory (1961, 1973), aerobiology is usually understood to be the study of passively airborne microorganisms, their identity, behaviour, movements and survival. This field of science includes: identification, morphology, physiology, viability, longevity, sampling, concentrations, diurnal and seasonal patterns, phenology, emission, transport, dispersion, pollination, pollinosis and a host of other subjects. Louis Pasteur (1822-1895) proved in his classical experiments that air is the carrier of many common germs. Aerobiology can be defined as the study of microbial population of the atmosphere, now designated as air-spora.

Aerobiology involves study of airborne particles of plant and animal origin. These bioparticles get into the atmosphere after their release from the source. Pollen is nature's gift to mankind as it is responsible for pollination, fertilization, seed and fruit setting and multiplication of plants. However, some of the pollen after getting into air stream remain floating in the atmosphere before settling down on the ground. Some of these pollen on coming in contact with human beings induce allergic manifestations.

The primary objective of aerobiological studies is to monitor, determine and detect the occurrence of pollen and spores and their relative representation in the atmosphere. Once trapped in pollen traps or air sampling mechanism, they are microscopically scanned thoroughly in the laboratory. Though the atmosphere consists of several hundred types of pollen and fungal spores, in applied aerobiology only the ones having significance in allergy are concentrated upon.

On account of the tremendous applications of aeropalynology in public health and medicines, a new term has been added recently known as 'medical palynology'. This branch is concerned with the study of airborne pollen and fungal spores, which are responsible for causing allergic manifestations including the triggering effect leading to asthmatic attacks. (Mackay *et al.*, 1992) In addition, various aspects of immunotherapy are investigated involving hyposensitization of allergy patients by using pollen and fungal aeroallergen extracts.

According to a recent trend, the scope of aerobiology has been widened to incorporate different kinds of biological particles (air spora) for example, viruses, bacteria, microalgae, microfungi, lichen fragments, soredia, seeds, protozoan cysts, insects and insect parts, spiders. Abiotic particles or gases affecting living organisms are also included currently in the concept of aerobiology. Various processes that are involved in the aerobiological studies are depicted in the following Flow Chart No. 13.1.

Thus, the aerobiological pathway involves at least five major steps, which are: source, liberation, passive transport, deposition and impact on vegetation, water bodies and various substrates. It is obvious that these different steps are integrated with each other and they are affected by environmental factors, such as, meteorology, physics and atmospheric chemistry.

Aerobiology has become an interdisciplinary science of great significance and applications in different fields, such as, ecology, medicine, pathology, agriculture, forestry and meteorology. There are various ways of dispersal of pollen in the atmosphere, however, the most important factor is wind which transmits pollen grains and spores from the source to the target area. Hence, windborne pollen both of flowering plants or the angiosperms and gymnospermous plants are significant in aerobiological studies.

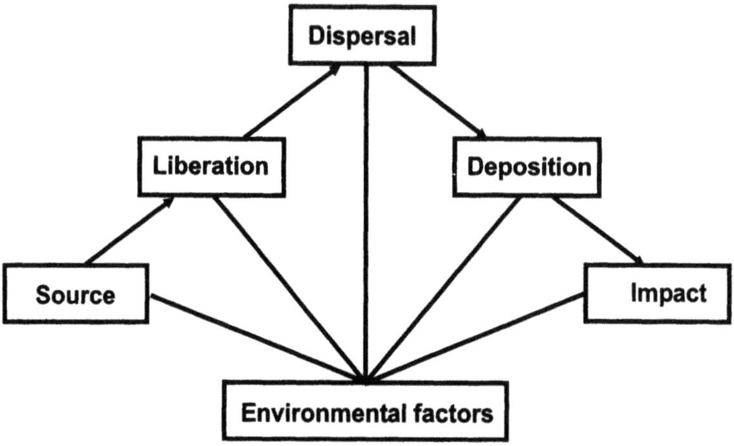

Flow Chart No. 13.1: Showing processes involved in aerobiological studies.

HISTORY OF AEROPALYNOLOGY: CONTRIBUTIONS OF SOME AEROBIOLOGISTS

Louis Pasteur (1822-1895)

Ariatti and Comtois (1993) rightly referred to Louis Pasteur as 'the first experimental aerobiologist'. As early as 1860, Louis Pasteur founded on precise scientific grounds, a new science: micrography. It was almost 70 years later (i.e., in 1930s) that F. C. Meier termed it as 'aerobiology'.

Louis Pasteur proved in his classical experiments that air is the carrier of many common germs. Louis Pasteur presented the true nature of the air spora: a limited cloud of solid dust dispersed by air movements and emptied by gravity. He carried out experiments in different seasons, localities and different heights. This intensive work was responsible for the discovery of important principles in aerobiology: such as 1) the absolute necessity of volumetric sampling; 2) the heterogenity of the air spora and 3) the aerobiological pathway: take-off, dispersal and deposition.

It is well known that it was Louis Pasteur who experimentally proved that microorganisms cause fermentation and diseases.

Fred Campbell Meier (1893-1938)

The term aerobiology was coined in the 1930s of the 20th Century, by the American plant pathologist, F. C. Meier. This term was probably parallel to hydrobiology. F. C. Meier worked for the U. S. Department of Agriculture for many years in various capacities. He was basically interested in plant diseases, which were distributed by airborne fungal spores. In order to

Fred Campbell Meier

investigate the atmosphere at various levels, in different areas and over considerable distances he worked with the famous aviator Charles A. Lindbergh known as first man to fly solo across the Atlantic in the aircraft named 'Spirit of St. Louis'.

On account of his close association with Lindbergh, he was able to carry out experiments in the air at higher altitudes by exposing a 'Sky-Hook', a special device for sampling airborne fungal spores, fragments of insects, wings. He carried out these sampling flights between Maine and Denmark a route, which included arctic areas. Initially he used adhesive coated microscope slides in above flights to trap suspended bioparticles. Later he used Petri dishes containing nutrient agar to trap and culture airborne fungal spores.

Meier compiled a vertical profile of fungal spores, ranging in height from 150 m to 11,000 m over Eastern United States. On the basis of his outstanding research work in a new direction, he was able to convince the National Research Council that it is worth extending this work even for other particles in air, such, as pollen known for causing hay fever. This also indicated his concern for human welfare.

He had visualized that further advances in such studies of the atmosphere could only be made by adopting an interdisciplinary approach, thus he was able to create an awareness of the importance of this kind of atmospheric study among botanists, meteorologists, zoologists, bacteriologists, plant pathologists and medical experts. In fact he gathered around him experts from the above disciplines. The ultimate result of this association of experts of different disciplines with common research interest and objective was the foundation of the Committee on Aerobiology, which met for the first time on November 12, 1937.

One of his close associates till the end was a medical adviser Dr. McKinley who was also a founder member of the Committee on Aerobiology. The committee tried to get the U. S. Government's support and other departments and collaborating agencies such as the U. S. Army, the Navy, the Coast Guards, National Research Council and also Pan American Airlines.

The end of such a dynamic research worker was very tragic. He set out on a flight from California to Manila with six other passengers including his close associate Dr. McKinley. The aeroplane, in which they were flying over the Pacific to carry out further aerobiological studies of the atmosphere, disappeared on July 29, 1938. It was later found out that the last radio message from the pilot came from an altitude of 2,750 m informing about the rainstorm. In spite of the intensive search, no trace was ever found of the aeroplane or its passengers and crew members.

Though F. C. Meier died in 1938, his ideas and inspiration and the term 'aerobiology' which he coined have survived and are flourishing today.

David Douglas Cunningham (1843-1914)

Chanda and Caulton (1990) have published a biographical profile according to which, David Douglas Cunningham's name will always be remembered by aerobiologists all over the world for his pioneering contribution to a thorough and systematic study of airborne microbes, for the first time in the tropics and probably one of the first of this kind of work in a global perspective. His monumental treatise *Microscopic Examinations of Air*, was published in 1873 by the Superintendent of Government Printing, Calcutta.

David Douglas Cunnigham

This illustrious scientist was born at Prestonpans near Edinburgh on September 29, 1843. He was only 20 years old when he entered the Medical Faculty of the University of Edinburgh and graduated with honours in medicine in 1867. In 1868 he joined the Indian Medical Service, passing out of the Army Medical School at Netley topping the list in autumn that year.

Before leaving for Germany, Cunningham worked for some time with Rev. M. J. Berkeley, Fellow of the Royal Society (F. R. S.) to learn his mycological techniques at Sibbertoft. He landed in Calcutta, then the capital of India, in 1869 and was attached on special duty to the Department of Sanitary Commissioner. He was engaged, in a series of pathological studies of great interest, especially in the context of Indian conditions. During this productive period Cunningham succeeded in bringing out *Microscopic Examination of Air* in 1873.

He was appointed Professor of Physiology in the Medical College of Bengal. His remarkable contributions to pathology, coupled with excellent teaching qualities brought him the laurel of being elected as a Fellow of the Royal Society in 1889.

Cunningham did intensive pathological studies in Calcutta under trying conditions. By 1897 his failing health forced him to return to England and was unable to go back to India. After attaining superannuation, Cunningham settled in Torquay, where he devoted the rest of his life to gardening, natural history and his books.

A series of field investigations into the airspora was in progress to find whether or not fluctuations in number and types of microbes present in the atmosphere were connected with outbreaks of such diseases as cholera, typhoid and malaria. In particular, Cunningham investigated the airspora utilizing a special model of the 'aeroconiscope' first devised by Maddox

(1870, 1871). The model consisted of a conical funnel, with the mouth directed into the wind by a vane, ending in a nozzle behind which a sticky microscope cover-glass was placed on which dust particles were impacted. Cunningham carried out his investigations in two Calcutta jails where cholera and other fevers were rife.

He sampled for a 24-hour period, and after microscopically examining the catches of airborne microorganisms, mainly fungal spores and pollen grains, published their illustrations in a series of coloured plates. However, he found no correlation between these microorganisms and the incidence of fever in the jails. He concluded that moist weather decreased inorganic dusts, but it appeared to increase the total number of fungal spores.

He first served as a Secretary of the Zoological Garden of Calcutta, later as President of the Committee of Management of the same institution. He was also a Fellow of the Linnean Society and Fellow of the University of Calcutta. In 1880, for a short period he acted as Superintendent of the Calcuttta Botanic Garden. After his retirement he was made an honorary physician to King George V, honorary surgeon to the Viceroy of India, and also a Companion of the Order of the Indian Empire. He passed away on December 31, 1914 at Torquay, Devon, U.K.

Charles H. Blackley (1820–1900)

Charles H. Blackley

The classical work of Charles H. Blackley entitled 'Experimental researches on the causes and nature of Catarrhus Aestivus' published in 1873, is always quoted by aerobiologists as one of the most generally accepted being the first text giving evidence of airborne pollen as the cause of hay fever. Blackley's most cited experiment dealt with the collection of airborne particles using kites.

Hay asthma had its birth place in England, and was first described by Bostock (1819), to whom we owe the designation of 'summer catarrah'. With regard to grass pollen allergy Blackley had stated that "the disease does not usually appear till the grass comes in flower: and as long as any flower remaining on the grass, the disease continues". If the influence arises from grass, it is not necessary it should be cut and dried, which means the presence of hay is not essential and the warmer the weather, and more advanced the vegetation, the earlier does it show itself.

Experiments on the Presumed Cause of Allergy

This first experiment in 1859 was accidental , a small cloud of pollen was detached from a bunch of *Poa nemoralis* placed in a vase and came in close proximity to his face. "I commenced sneezing violently, and I had a smart though short attack.... I was satisfied that my symptoms were due to the pollen"

Experiments with Chemicals and Odours

Blackley tried many volatile substances that "which produced head symptoms however, in no instances were there any symptoms in the least degree resembled those of Hay fever". Odours given by different flowers had sometimes a marked effect, but as for the preceding substances there were none of the symptoms of hay fever.

Blackley voluntarily tried the inhalation of odour from the microscopic fungi of *Chaetomium elatum* and involuntarily the one of *Penicillum glacum*. In this last case, Blackley had observed much earlier that straw dust could bring about an attack of sneezing.

In order to determine which fungi could be generated on damp straw, he placed wheat straw, slightly moistened in a close vessel at 38°C. In 24 hours, a white mycelium dotted with minute greenish black spots appeared (*Penicillium glocum*). After a few days, another crop of jet black coloured spots developed (*Chaetomium elatum*).

Charles Blackley concluded that "I have reason to believe that *Penicillium* generates symptoms not unlike those of hay fever in some respects, but differing materially in others, being much more like those of ordinary influenza".

Experiments with Dust and Pollen

Blackley was scientifically very observant. Once he reported on an attack caused by the dust cloud produced by a moving carriage on the road. Examination of the dust under the microscope revealed the presence of grass pollen grains. He concluded that "various channels by which a cause may reach a patient, in out of way places and at out of way times". On account of his ever alert and inquisitive mind, Blackley posed himself some questions: Can pollen produce the symptoms of hay fever? Does this property belong to all pollen? He also wondered if this condition is found in dried as well as in fresh pollen?

Subsequently the testing was done either by applying pollen to the mucous membrane, by inhaling it, by a decoction of pollen to the tongue or by inoculating the upper and lower limbs with fresh moistened pollen.

Very few scientists are aware that it was Blackley who first experimented with 'skin testing' for which he abraded a space of about quarter of an inch (c. 6 mm). Later pollen was applied after being placed on a piece of wet lint, the size of the abrasion, and was held in position by a strip of adhesive plaster. Scratching with a lancet, raised a weal such as seen in urticaria... Blackley concluded that the action of the different pollen grains was not related to their size, shape, roughness or to the poisonous character of the family. Later he hypothesized on the possible involvement of alkaloid in hay fever.

He undertook a series of experiments to establish the relationship if any, between the quantity of pollen found in the air and the intensity of symptoms.

Blackley used kites to collect airborne pollen at great heights. Of course, he found that kites were by no means as easily managed as first expected and he had many failures and disappointments.

He used a kite 6 inches in length and 3 inches in width. It had a central shaft (standard) and a semicircular top (bender). For covering the kite he used thin tissue, waterproofed with a mixture of boiled linseed oil and copal varnish. The kite carried a special slide holder. In this first experiment done in June 1868, his kite could reach an altitude varying from 90-150 m. On analysis of the slides he found the quantitative proportions of pollen in the upper strata were largely in excess of that of the lower strata (104:10). He repeated his experiment with a kite a year later when his kite with a slide could reach heights varying from 180-240 m capturing a total of 580 in contrast to 16-64 pollen at ground level. On an average he found 19 times more pollen in the upper level as compared to the ground level.

Philip Herries Gregory (1907-1986)

P. H. Gregory, primarily a phytopathologist was a Professor of Botany in London. His main interest was in fungi, particularly the dispersal of fungal spores. His fundamental studies on fungal spores are used by allergists.

Philip Herries Gregory

He gave particular emphasis to basidiospores of the dry rot fungus. He observed that many residences in London and other cities in the United Kingdom, which were damaged in bombings during the World War, were inadequately repaired and hence ideal for the growth and occurrence of the dry rot fungus *Serpula lacrymans* producing vast numbers of basidiospores. Gregory suggested the use of cascade impactors or Hirst's automatic volumetric spore traps instead of gravity deposition samplers to survey indoor fungal air spora.

In one of the surveys of a 16th century house, he found the average spore concentration during 24 hours at 360,000/m^3. In addition, he also surveyed air spora above a potato field (at 2 m height) by using a Hirst's sampler. He also emphasized the role of basidiospores in seasonal asthma. Philip Gregory conducted outstanding research on fungal aeroallergens and allergic diseases caused by them at the Rothmstead Experimental Research Station, Herpenden, England. This experimental station was a site where one could often see bales of hay outside his labortary, with thermometers stuck into them at various depths. These moist bales were loaded with thermophilic Actinomycete moulds. He carried out the studies on the dispersal of these mould spores (mostly *Aspergillus fumigatus*, possibly causing Farmers' Lung diseases) in a small wind tunnel at various speeds.

He promoted an inter-scientific or interdisciplinary approach, particularly with reference to aerobiology. He strove hard for the development of International Association of Aerobiology (IAA) with which he was associated from its inception. He was given the distinction of honorary member of the IAA at the First General Assembly meeting in August 1974 at the Hague. He had the opportunity to work with medical doctors such as Dr. A. William Frankland of United Kingdom. Philip Gregory is known for his outstanding book entitled 'Microbiology of the Atmosphere' published in 1973, (Gregory, 1973) which includes the different applied aspects of aerobiology.

He visited India during 1980-81, attending the Golden Jubilee Celebration of the Centre for Advanced Studies in Botany of the University of Madras. He is referred as the 'father' of modern plant pathology and also aerobiology. He was the president of British Mycological Society in 1951. He was elected as a Fellow of Royal Society (F. R. S.) in 1962. The Indian Phytopathological Society elected him as a fellow in 1985.

The Indian Aerobiological Society organizes the P. H. Gregory Award for the best paper presentation in its biannual conference, which is a fitting memory, and encouragement for young budding aerobiological researchers in India.

Professor John Malcolm Hirst, D. Sc. F. R. S. (1921-1997)

Prof. John Malcolm Hirst (1921-1997)

He was born in 1921 near Birmingham, England. He is remembered by aerobiologists as the designer of the Hirst type spore trap, which is known all over the world as a standard instrument for volumetric spore sampling.

He served in the Royal Navy until 1946 and later went to Reading University to study agricultural botany. In the summer of 1948, he met Philip Gregory at Rothamsted where he worked from 1950 onwards. He obtained a Doctor of Philosophy (Ph.D.) from University of London in 1955. He was elected a Fellow of the Royal Society in 1970. He retired from Long Ashton in 1984. Professor Hirst was approachable and ever ready to give help. He was President of the British Mycological Society in 1973.

Professor John Hirst became Founder President of the British Aerobiology Federation in 1990. He worked at the Institute of Arable Crop Research at Rothamsted, United Kingdom in the field of crop pathology and aerobiology. He died of cancer on December 30, 1997.

Hirst developed the trap when he needed a reliable method of retrieving airborne *Phytophthora* (late blight of potato) sporangia above a potato field. Hirst modified the Cascade impactor's second stage so that it could run continuously. The Hirst trap consisted of a slide drawn past the inlet slit at a constant speed over 24 hours by a clock mechanism and he mounted the trap on a wind vane to keep the slit facing into the wind.

Professor Hirst published his description of the Hirst trap in 1952 and immediately requests came to Rothamsted from other researchers to have duplicates made. It was decided to pass the design to an engineering firm and so production was begun by Casella Limited, later continued by Burkard Limited, when the seven-day trap was introduced. Within a few months Hirst traps were used by hospitals, studying hay fever and other allergic responses.

Professor T. Sreeramulu (1925-1974)

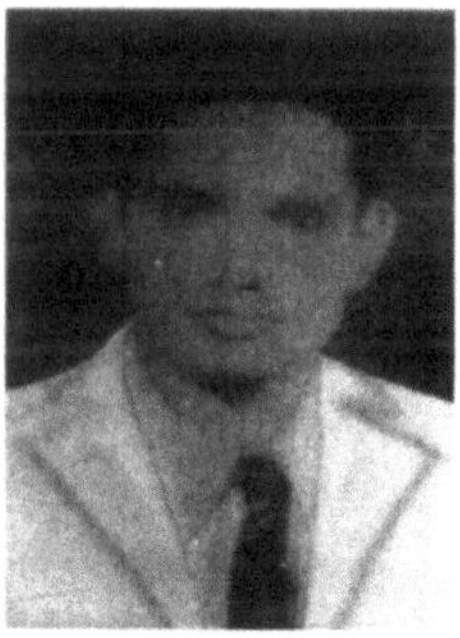

Professor T. Sreeramulu
(1925-1974)

Professor Tangirala Sreeramulu was born in Aranigadda, Krishna District in Andhra Pradesh, India on November 1, 1925. His early education was in Andhra Pradesh, but for his M.Sc., he went to Agra College, Agra. He specialized in mycology under the guidance of Professor K. C. Mehta, the first scientist to carry out long distance transport of rust spores (uredinales). He worked as a lecturer in the Department of Botany, Andhra University from 1948.

His research career started at Rothmstead Experimental Station, England in 1954 where he worked for his Doctor of Philosophy (Ph.D.) on fundamental problems on air spora under the guidance of Professor P. H. Gregory. On his return from the United Kingdom, he worked at the Postgraduate (PG) Centre of Andhra University at Guntur and later, as Professor and Head of the Department of Botany, Andhra University at Waltair (Visakhapatnam) until his untimely death on December 9, 1974.

In such a short life span he established aerobiological research on a firm footing and trained research students such as A. Ramalingam, C.

Subba Reddi, B. P. R. Vittal, K. V. Mallaiah, S. T. Tilak, who became outstanding aerobiologists in southern and western India.

His chief aeromycological contribution comprised of spore dispersal problems and airborne plant pathogens. He concentrated mainly on the occurrence of fungal spores pathogenic to rice and sugarcane crops.

He published more than 60 research papers among which included an outstanding paper on 'Spore Content of Air over the Mediterranean Sea' based on data collected by him on his voyage on the passenger ship 'S. S. STRATHMORE,' in which he returned from the United Kingdom to India. For collecting data on airborne spores he used Gregory's portable volumetric spore trap installed on the ship in which he travelled (Vittal 1974).

Besides being an excellent teacher and research worker. Professor Sreeramulu was a man of amicable disposition, affectionate temperament, witty but wise, and very helpful to his colleagues, friends and students. He always believed and strongly advocated the motto 'Publish or Perish.'

Dr. D. N. Shivpuri (1914-1990)

All the allergists and aerobiologists from India, owe a great sense of gratitude to Dr. D. N. Shivpuri for initiating and encouraging research work on aerobiology and its direct application to the diagnosis and treatment of allergy.

D.N. Shivpuri

Dr. Shivpuri was born in Rajouri, Jammu & Kashmir, India on August 10, 1914 and obtained M.B.B.S. in 1949 and M.D. in chest diseases in 1955 from Lucknow University. His professional life was devoted to research and practicing allergy and immunology at Delhi. He worked at the Vallabhai Patel Chest Institute, affiliated to the University of Delhi.

He had a profound knowledge of aerobiology, allergy and immunology. He guided several research students for their Doctor of Philosophy (Ph.D.) in Botany (aerobiology), prominent among them are Dr. M. K. Agarwal and Dr. A. B. Singh, who became outstanding research workers in aerobiology and immunology. The former worked at V. P. Chest Institute and later worked at the CSIR centre for biochemicals (Now Institute of Genomics) at Delhi. He also guided several doctors for their M. D. in chest diseases including allergy and immunology. He successfully brought together basic scientists and clinicians for fruitful interaction at scientific meetings by establishing the Indian College of Allergy and Applied Immunology (now the Indian College of Allergy, Asthma and Applied Immunology) in 1967, with its head quarters at the V. P. Chest Institute, Delhi. The college conducts annual conventions in different parts of India and training programmes for general physicians, which involve practical training in aeroallergens, diagnosis and treatment of allergy by using immunotherapy. The methodology for assessment of allergenicity to aeroallergens by skin tests, particularly the grading of skin sensitivity proposed by him in 1962 is widely followed by several clinicians all over India and abroad.

Dr. Shivpuri was also responsible for starting the publication *'Aspects of Allergy and Applied Immunology'* in 1967, which has been converted now as the Indian Journal of Allergy and Applied Immunology. He used to participate actively and carry on lively discussions on various problems in allergy and immunology at the annual conventions of the college. He died in 1990 in Delhi after a prolonged illness resulting from a road accident in London in 1985.

The ICAAI has rightly started an Oration Series in his honour by selecting one outstanding speaker in the field of allergy, immunology and allied subjects during annual conventions of the college.

Ruth M. Leuschner (Born on 20th September, 1922)

Ruth. M. Leuschner

There have been number of instances where scientists have made significant contributions inspite of the hardship they underwent during the early period of their careers.

In this context Ruth Leuschner is one of the perfect examples who is considered to be a renowned researcher in the field of aerobiology. In her long innings in the pursuit of scientific knowledge, she witnessed the development of aerobiology and has taken an energetic part in the dissemination of aerobiological information. She was also one of the founding members of the International Assocaition for Aerobiology (IAA) and served the association as Treasurer and Vice President.

Even at an advanced age of 85 she is very active and participates with unfailing enthusiasm in jointly sponsored aerobiological research and in various aerobiological seminars and conferences.

Dr. Leuschner during her early phase grew up and studied in a school in the old university of Basel in Switzerland. Surprisingly she was trained as a teacher of hand work and typing which helped her to teach in a school. She always had the ambition to study botany which was partly fulfilled due to encouragement of Prof. Gundo Boehm, a lifelong friend and colleague who was teaching physiology and light microscopy at the world famous University of Tubingen and University of Basel respectively. She first obtained a University Degree ('Higher Teaching Diploma') which was required to teaching assignments to the senior high school classes in botany, zoology with chemistry as a subsidiary subject. In addition to school teaching, she had to take a variety of other jobs to earn her living. One of the assignments she had taken was at the University library where her job commanded the task of sorting a vast collection of publications on pollen grains. During this period she developed a keen interest in the study of pollen grains on account of their endless variety of shapes and forms.

In fact the research project she completed for her teaching diploma was on 'Statistical study of apertures in the pollen grains of *Alnus* (alder)'. A chance meeting with Dr. Erika Stix at the Botanical Congress in Germany resulted in undertaking a research project for a dissertation on identification of airborne pollen in Basel. She obtained a Burkard pollen trap from Dr. Eric Stix. Many scientists like her had experienced many difficulties during their scientific careers. This is clear from the fact that there was no encouragement from the Deparment of Botany, where pollen morphology was considered more interesting than studying airborne pollen.

However, Dr. Rudolf Schuppli, Professor of Dermatology at the Cantonal hospital, Basel not only encouraged her to take up a survey of airborne pollen but also provided working space in the hospital along with a Burkard trap. Dr. Leuschner received a half time research grant from the Swiss National Foundation for scientific research in 1973.

With the support of Professor G. Boehm who provided financial assistance to purchase two Burkard traps, Leuschner initiated a network of pollen monitoring stations in Switzerland which later included several stations such as, Davos, Geneva, Neuchatel, Zurich, Samendan, Buschs SG, Nugano and Lucerne. She carried out comparative studies of pollen flight between Basel located at 273 m above sea level and Davos-Wolfgang which lies at 1,600 m above sea level. She contributed significantly to international publications including the Atlas of the European Allergenic Pollen (1974). She carried out aerobiological studies with the help of an individual pollen trap invented by Professor Boehm and used for correlating airborne pollen and allergy symptoms.

Ruth Leuschner has also published several popular articles and made radio broadcasts regarding airborne pollen. She has been writing articles regularly since 1979 which were published weekly in the Newspaper entitled 'Basler Zeitung'. The column which she used to write was captioned 'Orientation for allergy sufferers'. This column appeared in the newspaper regularly throughout the pollen season which educated the public about the prevalence of different pollen types in the atmosphere of Switzerland.

Jointly with Professor Boehm she successfully organized and chaired the Third International Conference on Aerobiology held in Basel in 1986. The proceedings of which was published under the title 'Advances in Aerobiology' in 1987.

In 1978, and subsequently in 1989 and 1990, she monitored and identified the pollen of ragweed i.e., *Ambrosia* in Basel and Nyon in Switzerland. She was a committee member of the Basel Botanical Society and also served as the editor of the journal Bauhinia. She was also the honorary president of the Swiss working group for aerobiology.

Leuschner has travelled extensively for academic work. She visited India thrice, the first time in 1990 to participate in the national conference of the Indian Aerobiological Society held at Pondicherry; in 1994 to participate in the Fifth International Conference on Aerobiology held at Bangalore, and in 1997 along with her sister, to deliver a lecture at the Department of Botany, Bangalore University, Bangalore.

Jennifer Jenkins has written an excellent biographical sketch of Ruth M Leuschner which appeared in Aerobiologia in 1996.

Gamal El-Ghazaly (1947-2001)

Gamal El - ghazaly

Gamal was born in Alexandria, Egypt on June 17, 1947. His early education and academic degrees were from Alexandria where he also studied palynology. His real interest in palynological research was kindled in Stockholm, Sweden where he obtained a Ph. D from the University of Stockholm, under the guidance of Dr. Siewert Nilsson, then Director of the famous Palynology Laboratory at Stockholm. His thesis was on 'Palynology of Hypoichoeiridineae and Scolymineae (compositae)'. Later he had a good break, when he received an 18 month fellowship to work with William Jenson at the University of California, Berkeley, where he studied wheat pollen development, with particular reference to the formation of microchannels in the exine.

Gamal started his teaching and research career as an Assistant Professor from 1983-1988 at the University of Aexandria. In 1988 he became the Chairman of the Department of Botany at the University of Qatar. At this time he brought a group of post graduate students for a field work tour to India. He visited the Departmentt of Botany, Bangalore University and had discussions with Prof. Agashe. He appeared to be more interested in full time research than teaching. He assumed the position of First Curator in the Palynological Laboratory of the Natural History Museum at Stockholm, Sweden, where he was granted Swedish nationality and also married a Greek laboratory technician. They settled in Stockholm and had two sons. On account of his up to date knowledge in taxonomy and palynology and his excellent editing ability, he became Editor of the 'World Pollen and Spore Flora' and later worked from 1998 as Editor-in-Chief of 'Grana', a highly recognized international journal of palynology.

On account of his knowledge of pollen morphology and other aspects of pollen studies, he took a keen interest and pursued aerobiological research at Stockholm. He published a comprehensive volume on airborne studies of pollen and fungal spores and the pollen calendar of Alexandria.

Gamal was very friendly, had a helpful attitude and had cordial relations with scientists from different parts of the globe. This also led to a number of collaborative projects with scientists from various countries.

He was especially interested in the structure and function of the enigmatic sporopollenin particles and the Ubisch bodies. He researched extensively on the localization and the release of allergens from tapetum and the pollen grains into the atmosphere. He participated as part of the faculty in the national and international courses in plant anatomy, plant taxonomy, aerobiology and melissopalynology. He published more than 73 research papers some of them jointly with authors from Sweden, Egypt, Qatar, Japan, the U.S.A. and Italy.

Most of his publications were on aeropalynology and the ecological aspects of pollen allergy.

His later research activities involved the ontogeny of pollen grains and tapetum, floral micromorphology in relation to taxonomy and phylogeny, localization and the release of allergens from pollen of Poaceae and Fabaceae.

One of his last unfulfilled scientific contribution prior to his accident and later sad demise was a project to write book on palynology jointly with his close friend Prof. Shripad N. Agashe and his mentor Prof. Siwert Nilsson.

Siwert Nilsson (1933-2002)

Siwert Nilsson

Siwert Nilsson was one of the greatest palynologists and aerobiologists of modern times. In a real sense, he was the true successor of Gunnar Erdtman, the father of palynology. Siwert Nilsson served and nurtured the famous Palynology Laboratory of the Natural History Museum at Stockholm, Sweden.

He was born on July 30, in the northern part of Sweden. He obtained his M.Sc. degree (Botany, Geography, Zoology) in 1959 from the Institute of Systematics at Uppsala University in Sweden. He was appointed assistant in 1959 to the Swedish Council of Natural Sciences in Palynological Laboratory in Stockholm under the leadership of Professor Gunnar Erdtman. He had a special interest in pollen morphological studies of the families Apocynaceae and Gentianaceae. He selected the latter family for his Ph.D. degree topic.

Siwert Nilsson took over the directorship of Palynological Laboratory in 1975 which was transferred to the Swedish Museum of Natural History in

Stockholm. He continued to be the director until his retirement on July 1, 1998. On account of his scientific contributions particularly in palynology, he was appointed as Professor in palynology. In addition to pollen morphological studies, Siwert initiated a research programme on aerobiology in 1973. Initially he carried out research in aerobiology in collaboration with scientists from meteorological and medical institutions in Stockholm.

The results of daily atmospheric pollen monitoring were provided to the press and radio in Stockholm for the benefit of the public. Siwert travelled widely for the academic work particularly to various European countries and also other regions. Palynological knowledge was spread by him to other scientists by organizing several courses: in Sweden at the Palynological Laboratory, University of Stockholm, the Karolinska Hospital, University of Bergen in Norway: in Turku, in Finland, in Kingston, in Jamaica: in Amazonas, in Brazil, in Pretoria and in Bloemfontein , in South Africa and in Havana, in Cuba. He was an excellent teacher and speaker for which he was appreciated throughout the world.

He guided a number of Ph.D. students from Sweden as well as other foreign universities. He published more than 100 research papers and he has several books to his credit. On account of his research interests and editing ability, he became the Editor-in-Chief of 'Grana' in 1985. He was also entrusted with the work of an editor for the 'World Pollen and Spore Flora'. Even after his retirement, Siwert Nilsson continued his scientific activity at the Palynological Laboratory in Stockholm as Professor Emeritus. In addition to Swedish, he had mastered other languages such as English, German and Spanish. He, along with his wife and daughter were very hospitable to visitors. Siwert Nilsson died suddenly on August 19, 2002. He will be always remembered as a fair and amiable leader, a helpful colleague and an affectionate friend.

Prof. Agashe had several occasions to interact with Siwert Nilsson both in Sweden and India. Their first meeting took place in Aurangabad, India, during the First National Conference on Aerobiology in 1982 organized by Prof. S.T. Tilak, another outstanding Indian aerobiologist. Subsequently Siwert Nilsson visited India several times and interacted with Prof. Agashe and his research students at the Aerobiology and Allergy Laborartory of the Department of Botany, Bangalore University, Bangalore which organized the 5th International Conference on Aerobiology at Bangalore in August 1994. Earlier Siwert Nilsson had organized the previous 4th International Conference on Aerobiology at Stockholm in August 1990.

He was a prolific writer and eloquent speaker. He had collaborative research projects in basic and applied palynology with scientists from different parts of the globe. He has published more than 100 research papers and published many books devoted to basic and applied palynology including aerobiology. He was President of the International Association of Aerobiology from 1974 to 1982. From 1985 to 2002 he was the Chief Editor of 'Grana', the world famous and internationally acclaimed palynological journal published from Stockholm, Sweden.

Dr. Eric Caulton, the co-author of this volume also had several occasions to meet and interact with Siwert Nilsson. He was very friendly with an ever helping attitude. Prof. Agashe always remembers the useful interactions with Siwert Nilsson in his laboratory at Stockholm in 1986. He had a memorable time when Siwert Nilsson drove him to Uppsala and showed the famous Carl Linnaeus' summer house, winter house, Museum and Botanical Gardens at Uppsala.

It is worth mentioning here that the idea for the present book on palynology jointly with Siwert Nilsson, Shripad N. Agashe and Gamal Gazaly was mooted by Shripad N. Agashe. Three of them had initiated the basic steps but due to untimely death of Siwert as well as his student Gamal Gazaly it could not be pursued. This project was later revived jointly by Prof. Agashe and Dr. Eric Caulton, and the present volume is the ultimate result of it.

AEROBIOLOGY OF POLLEN

Pollen constitutes a small part of the aeroplankton or air-spora present in the atmosphere. The most frequent particles of biological origin are microorganisms, especially the spores of fungi. For example: pollen represents only 2% of air-spora detected annually in Cardiff, United Kingdom. The others are fungal spores belonging to various groups: Fungi Imperfecti 43%, Basidomycetes 37%, Ascomycetes 17% and Phycomycetes less than 1%.

Particles when dispersed in air are termed as 'aerosols'. The presence of bacteria and viruses in aerosols is less easy to detect. Algae, leaf hairs, seeds, plant fragments and volatile materials including scents and terpenes also occur in aerosols. The terpenes are oily substances released from the leaves of trees in sunlight, and may form a blue haze in the atmosphere or aggregate and polymerize in sunlight, forming brown-black air soot. The atmosphere may also contain other particulates including bushfire ash, industrial ash spheres and cenospheres from incomplete fuel combustion.

Aerobiology is concerned with the behaviour of a suspension of particles, both viable and non-viable, whose transfer from one site to another is governed by atmospheric properties. These aerosols may travel short distances, or may be blown into the upper strata of the atmosphere and travel long distances before they are deposited. The atmosphere has been well described as a 'restless ocean of air.' It is divided into a number of zones: the troposphere, stratosphere, ozonosphere and mesosphere named in order of distance from the earth's surface. The troposphere houses nearly all the air necessary for life.

SETTLING OF POLLEN AND SPORES FROM ATMOSPHERE

Various characters are taken into account for settling of airborne pollen and spores on the substrates. The specific weight of pollen and spores and the velocity of fall are significant. It is known that the size, shape, volume, density, morphology and other properties together decide the air buoyancy, mode and rate of deposition of air-spora. Bacteria and viruses occasionally become dispersed along with pollen and spores. Air pollutants such as soot particles may act as rafts for pollen and spores and often pollen grains themselves function as carriers of airborne pollutants.

The air is a very important source for the dispersal and distribution of bioparticles. It is a viscous, mobile medium with a number of properties associated with the physical and chemical laws alongwith the processes of meteorology. The air is usually an unstable medium due to solar heating, temperature differences and friction, which form circulating air masses.

Several viable microorganisms such as bacteria and fungi have been reported from the stratosphere (Gregory 1973). Fulton (1966) had trapped *Alternaria* and *Cladosporium* at the altitudes up to 30,000 m. Agashe and Chatterjee (1987) had carried out an aeropalynological survey at different altitudes by using the aircraft sampling method. They reported the occurrence of several pollen grains including *Parthenium hysterophorous,* and several fungal spores at the attitude of 1,000 m in the atmosphere of Bangalore, India.

FIELD BOTANICAL STUDIES: POLLEN HERBARIUM

The prerequisite of all atmospheric pollen studies is the compilation of a pollen herbarium. Primarily a classification and description of pollen morphotypes responsible for allergic disorders will lead to a pollination calendar for identification of airborne pollen grains.

In any particular locality, the first step in the investigation of pollen types responsible for pollinosis consists of a thorough field botanical study of the various plants of that area.

A list of local plants classified into anemophilous, entomophilous and amphiphilous on the basis of the mode of pollination and with notes on their distribution, pollen production and phenology, should be prepared. Further, plants are classified on the basis of their habit: trees, shrubs or herbs. A 'weed' is a horticultural term and is a concept rather than a reality!

The area under investigation is normally visited fortnightly or more frequently to observe and record the different seasonally growing plants for the whole year. During field trips, bulk anthers are collected from anemophilous and entomophilous plants for preparing pollen reference slides. The anthers are stored in small vial tubes containing a few drops of acetic acid. The plants collected are identified and preserved in the form of a herbarium for future reference.

PHENOLOGY

It is the science of the relations of climate and periodic biological phenomena, such as the shedding of pollen. One of the useful aids in identification of pollen is the knowledge of pollination times in comparison to the dates when the samples were obtained. Frequent observations should be made on the development of flowers. Annual records over a period of a few years will enable one to predict for local areas, the onset of pollen types of the different species. It is advisable to include in the phenology list, the names of some plants that have conspicuous flowers, even though their pollen is not airborne. Such records will greatly help in the comparisons of the seasons from year to year. These flowers will act as indicators for a source of airborne pollen. It may be found that some plants flower approximately at the same time each year, while others vary greatly.

Aeropalynological survey and field botanical studies carried out in Bangalore, India has consistently shown that when nonallergenic *Taebubia argentia* starts flowering, it coincides with the flowering of allergenic *Holoptelea integrefolia*. Hence, *Taebubia argentia* acts here as a marker for *Holoptelea integrefolia*. A set of reference slides of the pollen of common plants of the locality, may be considered essential for such investigations.

For making pollen slides, the method suggested by Wodehouse (1935) is generally used in which fresh pollen is treated with alcohol, stained with methyl green or safranin, and mounted in glycerine jelly. In addition to field studies, further reliable information about important plants responsible for pollinosis is obtained. The compilation of permanent reference slides is referred as a pollen herbarium, which is useful for the identification of airborne pollen.

A pollen calendar is constructed on the basis of field botanical studies as well as aeropalynological surveys. The compilation of a pollen calendar is the ultimate objective of aerobiologists, as this is most useful for clinicians in diagnosis and treatment of allergy.

Pollen is produced by the seed plants that include gymnosperms and angiosperms. Pollen grains are male reproductive structures of seed plants. The transport of the male gametes (sperm) to the female gametes (eggs) where fertilization may occur is called pollination.

SIGNIFICANCE OF POLLINATION IN AEROALLERGEN STUDIES

Pollination is thus the transfer of pollen from male structures to female structures of the same species. Pollination is accomplished by several methods, the commonest method in flowering plants is by insects. In these cases, flowers may be showy, colourful, fragrant and otherwise attractive to the pollinating agents. The pollen may be large, sculptured and often with an adhesive coating. The pollen which are important in aerobiology are from plants in which wind is the pollinating agent. In these flowering plants, the flowers are usually small, inconspicuous, numerous and without odour. Such pollen grains are mostly small, smooth and non-adhesive. Airborne pollen is usually produced in large quantities. The flowers or clusters of flowers may be on long stalks that move with the wind. The filaments of the anthers may be long and flexuous.

Some airborne pollen are slightly adhesive and may be carried by both wind and insects as in *Tilia* (basswood or linden), *Acer* (maple), *Salix* (willow) and *Castanea* (chestnut), *Parthenium* (Congress grass). The pollen may stick together and are found in clumps on the samples e.g. *Ambrosia, Peltophorum, Parthenium*. It is mainly the windborne pollen grains that have a bearing in pollinosis, because they more easily come in contact with the hypersensitive tissues of human beings than other types of pollen.

Insect-borne pollen grains in general, are not produced in such abundance as the windborne ones, and for that reason, they are of lesser importance as causes of pollinosis. This does not mean however, that they can be completely ignored. It has been observed in a few instances that pollen from entomophilous plants like *Carica papaya* and *Argemone mexicana*, produced symptoms of pollinosis in patients (Shivpuri and Dua 1963, 1964; Agashe 1989).

POLLEN PRODUCTION

Several workers on pollen have made an attempt to assess the quantum of pollen discharge to the atmosphere (Agnihotri and Singh 1975; Khandelwal and Mittre 1973; Nair and Rastogi 1963; Mondal and Mandal 1998, Kessler & Harley 2004).

The quantity of pollen in the air depends on several factors, the most important being pollen production in the individual species. The amount of pollen production and methods of dispersal are very important factors, which are directly or indirectly involved in causing environmental pollution and allergy.

Nair and Rastogi (1963) had suggested a method of assessing pollen production within an anther of a flower. Unopened but mature flower buds are usually collected in the morning hours (6:30 to 8:30 am) Pollen grains were extracted from the anther by crushing and dispersing in 50 drops of 50% glycerine. One drop of this mixture of pollen in glycerine was placed on a microscope slide and covered with square cover glass of 18 × 18 mm dimension. The slide thus prepared was examined under the microscope and the pollen grains were counted. This was repeated by counting pollen grains in 10 drops of pollen dispersion. An average number of pollen was determined and multiplied by 50 to obtain the pollen production per anther.

In case of polyads, on account of their large size the anther was crushed and contents dispersed uniformly in 5 drops of 50% glycerine. Five drops of the mixture was placed on a slide covered with cover glass. The pollen count was determined for 20 anthers from different flowers and an average was taken to determine pollen production. Pollen production is usually studied in the flower buds. If the pollen grains are large, the Haemocytometer method (Oberle and Goertzen 1952; Nagarajan *et al.*, 1972) for estimating pollen count is not suitable. In such a case, counting of pollen grains is done by other methods.

In as early as 1940, Erdtman had suggested the simplest method of determination of pollen productivity. According to him anthers of a mature bud are crushed in 1.5 ml of 50% glycerine comprising 30 drops. Pollen grains are counted in five drops for each sample and an average is taken in one drop. This number is multiplied by 30 so as to get the total number of pollen grains per flower.

In general the pollen production is controlled not only by their size, but also by genetic and physiological factors. Average pollen production per flower in *Acacia mangium* is reported to be the highest (16,640), followed by *Acacia auriculiformis* (15,360), *Mimosa invisa* var. *invisa* (12,800), *Albizzia falcataria* (12,288). Out of these four plants, three are tree species, which are capable of contributing an enormous quantity of pollen to the atmosphere. The pollen of these three tree species have been proved to cause pollen allergy in human beings.

Agashe and Soucenadin (1992) had worked out pollen productivity in certain allergenically significant plants in Bangalore, India. Pollen productivity is listed in the following table.

Table 13.1 *Duration of flowering and pollen production of allergenically significant pollen in the atmosphere of Bangalore.*

Name of the plant	*Duration of flowering*	*Pollen grains/anther*
*Casuarina equisetifolia**	Jan-Mar, Sep-Oct	787
Eucalyptus	Jan-Dec	2740
*Ricinus communis**	Jan-Dec	491
*Amaranthus spinosus**	Jan-Dec	5709
*Holoptelea integrifolia***	Jan-Mar	1830

Plant species * Abundant ** Rare

Correlation between the size of the individual pollen grains and pollen production, that is the number of pollen produced per anther

Samanea saman has 32 grains in the polyad where the polyad size (85.15 μm) is roughly double that of the *Acacia* species (average size of

40.83 μm) and the pollen count in *Samanea saman* is 256 which is double that of *Acacia* species with the pollen count of 128. The size of the individual pollen in the polyad of *Acacia* species varies from 8-11 μm, where as the individual pollen size in *Samanea saman* varies from 20-30 μm, these results are in conformity with the view that there is a positive correlation between pollen size and pollen production per anther.

The pollen output or the pollen production is expressed in different ways. Different authors express the pollen production either as the absolute number of pollen grains per anther, flower, inflorescence shoot or the entire plant. However, majority of the palynologists prefer to interpret the pollen production in terms of number of pollen grains produced per anther.

A single anther of *Betula* is known to contain about 10,000 pollen grains, while a single catkin of *Betula* produces more than 5 million pollen grains. A shoot with inflorosence of *Cannabis sativa* (Hemp) may produce more than 500 million pollen grains. Conifers are also known to be high pollen producers. A 10-year-old branch system bearing male cones in *Pinus sylvestris* may produce about 350 million pollen grains. When the mass of pollen grains are dispersed from microsporangia of male cones of *Pinus*, a cream or light brown coloured cloud containing pollen is observed.

Some fungi are known to exceed angiosperms in the production of spores. In *Lycoperdon giaganteum* (giant puff ball) the estimated total number of spores produced amounted to about trillions (Buller 1909). The pollen production is also sometimes expressed in relative terms rather than absolutely. Hasselman (1919) estimated an annual production of about 75,000 tonnes of pollen of *Picea* (Spruce) in southern and central Sweden. A quantity of 28 to 60 kg of *Picea* pollen per hectare was reported.

LIBERATION OF POLLEN AND SPORES

The liberation process includes detachment of pollen grains or the spores from the mother plant followed by take off of the air-spora into the atmosphere. The liberation process may be active or passive. In his classical book "Fungal Spores–their liberation and dispersal" Ingold (1971) discussed in detail, various mechanisms of liberation and dispersal of fungal spores.

In gymnosperms and angiosperms the microsporangium or the anther wall ruptures on drying in different ways to discharge the pollen grains. In *Urtica dioica* drying and the tension phenomenon of the stamens are responsible for the sudden liberation of pollen grains.

DEHISCENCE OF ANTHER

Pollen is commonly released from the anther through a longitudinal slit-like opening in the anther wall. Other methods also occur. Dehiscence usually results from hygroscopic shrinkage of the anther wall. Change in humidity may cause repeated opening and closing of the pollen chambers. The pollen may be freed all at once or escape gradually.

The air dispersal of pollen and spores depends on various factors such as time of the day, variations in temperature and wind speed. On a bright sunny day, ascending air masses may bring bioparticles up to the convective layer.

The airborne pollen grains and spores may remain afloat in the atmosphere for a longer or shorter period depending on horizontal and turbulent winds, which prevent them from moving downwards gravimetrically (pollen rain). It is interesting to note that bio particles such as pollen and spores act as condensation nuclei in water drop formation. During this process, the precipitation helps in washing out atmospheric pollen and spores and thus effectively cleaning the atmosphere from air spora.

POLLEN RELEASE AND DISPERSAL

Since there is greater exposure to pollen due to the height of the trees they are more easily disseminated by wind or insects and the chance of pollen loss is therefore greater. The higher production is compensated by this loss. A more or less similar observation was drawn by Mondal and Mandal (1998), who stated that there is a tendency for gradual increase in pollen production from herbs to shrubs and then in trees.

It is also observed that pollen grains of a small size have wider distribution and are capable of contributing a significant amount of pollen to the atmosphere. On account of the small size, the pollen tend to float in the air for a longer duration than the larger-sized pollen which tend to settle down on the ground faster than the former.

Pollen release, and to a greater degree dispersal, are influenced by the prevailing weather, which in turn is influenced by the time of day. With a view of studying the process over a 24-hour period, Reddi *et al.* (1985) had selected plant species releasing pollen throughout the day and those shedding at different times of the day were chosen after making preliminary observations as to the temporal pattern of flowering. Natural populations of five plants, *Ailanthus excelsa, Amaranthus spinosus, Cyperus rotundus, Fimbristylis miliacea, Holoptelea integrifolia,* and *Mimosa pudica,* growing in and around Visakhapatnam and Anakapalle, India, were used for this study by them.

For quantifying the anthers dehisced each hour, 3-10 inflorescences, which had just begun to flower, distributed among different conspecific plants, were tagged. Each hour the number of anthers dehisced was noted and removed to avoid recounting them an hour later. During dark hours, light focused from a five-celled torch was used to facilitate observations.

Pollen concentrations were measured with rotorods (Perkins 1957) rotating at 2,200 rpm. A rotorod with the adhesive cellotape on the leading edges of its arms was run immediately above the source continuously for either one hour or half an hour, depending on the expected concentrations over the source. The details of the investigations are given below.

Ailanthus excelsa **(Simaroubaceae)**
The pollen release occurred from 4 pm to 10 am. Higher rates of pollen release were recorded between 4 pm and 7 pm, the maximum being attained at 6 pm. Pollen concentrations were evident throughout the 24-hour period and showed two peaks. One of these peaks occurring at 9 am was without appreciable concomitant pollen release. The activity of honeybees (*Apis cerana indica*) was very brisk between 4 pm and 7 pm and between 7 am and 10 am. Certainly, the bee activity had a major role in the dislodgement of pollen.

Amaranthus spinosus **(Amaranthaceae)**
The pollen release occurred over a five-hour period commencing from 8 am and ceasing after 12 noon. During the three hours from 8 am to 10 am, higher levels of pollen release were recorded, with the peak at 9 am. Pollen dispersal into the ambient atmosphere also began at the same time as pollen release. Aerial pollen was evident for 11 hours until 6 am. Higher concentrations occurred between 9 am and 11 am with the maximum being attained at 10 am.

Cyperus rotundus **(Cyperaceae)**
The pollen release occurred for a three-hour period from 4 am to 6 am with the peak at 5 am. Aerial concentrations of pollen also appeared at 4 am; they increased gradually to a peak at 8 am and then declined gradually to a zero at 4 pm. Furthermore, pollen remained in the air long after the cessation of pollen release.

Holoptelea integrifolia **(Ulmaceae)**
The pollen is released throughout the 24-hour period with no pronounced peak, pollen concentration levels in the air almost followed the rate of pollen release, but from 7 am to 10 am they reached higher levels with a steep rise at 8 am, which could be related to the bee activity and development of turbulent conditions due to increasing temperature and decreasing Rh. Though the conditions prevailing there after (until 3 pm) were also turbulent, the concentrations were low, probably due to increased pollen diffusion.

Mimosa pudica **(Mimosaceae)**
The pollen is released over a period of three hours from 8 am to 10 am with the peak occurring at 9 am. Pollen also appeared in the air from 8 am and the concentrations gradually rose until 10 am, but suddenly rushed to a peak at 11 am. In the next hour the concentrations abruptly declined, but during the subsequent hours there was a gradual decrease, reaching zero at 5 pm.

Based on the knowledge of the aerodynamics of particle transport and capture, it is now established that the extent of dispersal is controlled by an interaction of the terminal velocity of the pollen and wind velocity, and that the actual concentrations of pollen in the atmosphere at any point downwind of the source are influenced by eddying of the atmosphere (Whitehead 1969; Gregory 1973).

Both air temperature and relative humidity rise and fall once in a 24-hour period. From about dawn the temperature increases while Rh decreases towards a maximum and minimum respectively at midday, and thereafter the former decreases towards a minimum and the latter increases towards a maximum by about dawn. Moreover, wind speeds tend to diminish by night and increase by day. After sunrise, the solar energy increases air temperatures and air movement, and decreases Rh. Then the atmosphere becomes unstable and/or turbulent.

As Kramer (1979) emphasized, it is essential to distinguish between patterns of pollen or spore release and the occurrence of a spore of pollen type in the atmosphere, and the appropriate methodology must be adopted in the study of these two separate but integrated events in the aerobiology pathway.

LONG DISTANCE DISPERSAL OF POLLEN

Interesting observations on long distance dispersal of pollen were made by Erdtman (1937), above the mid-ocean between Europe and North America, he made these observations during his voyage from Gothenburg to New York and back. The occurrence of large quantities of pollen of *Pinus, Picea* and grasses in the coastal areas of Greenland at 600 to 1,000 km from the nearest forest has been recorded.

According to Moar (1985) *Casuarina* pollen on a glacier in the South Island of New Zealand was assumed to have been transported from Australia to New Zealand. The transport of spores of *Puccinia* and *Erysiphe* from the U.K. to Denmark and northern Germany has been reported.

It was reported that *Cedrus deodara* (cedar) pollen is transported by air from the Himalayan region up to Lucknow. This was confirmed scientifically as this pollen was trapped by the air samplers used during aeropalynological studies at Lucknow.

Air Sampling Principles Involved in the Deposition Process of Airborne Pollen and Spores

There are various methods available for monitoring pollen in the atmosphere. Although a great number of sampling devices are in use, all operate on only a few basic principles mentioned below. Each principle has both merits and demerits and some are more suitable than others for collecting particles in the pollen size range.

The five basic sampling principles are:

1) Gravitational settling
2) Impaction
3) Suction
4) Grab sampling
5) Impinging

GRAVITATIONAL SETTLING

This principle involves settling or sedimentation of airborne particles from the air due to gravitational pull. When the airborne particles reach the terminal velocity they fall down to the earth. This is because the gravitational force will be more than the terminal velocity. Exposure of a horizontal surface on which particles can settle by gravity is the simplest method of collecting airborne pollen.

In theory, particles simply settle at their terminal velocity and are retained by an adhesive on the sampling surface. The terminal velocity of a small smooth spherical particle is proportional to the square of the particle radius, the particle density and acceleration due to the gravity are inversely proportional to the viscosity of the air or another medium through which the particle falls (Stroke's equation). With regard to the deposition of airborne particles, Stoke's law states that the terminal velocity of smooth spheres in the size range of 1 to 100 μm in still air can be calculated with a good degree of accuracy. However, Stoke's law is perhaps not universally applicable to pollen grains and spores as they are not always smooth and spherical in shape. In completely calm or very still air (air lacking turbulence), this concept of gravitational settling is valid since gravity is the predominant depositing mechanism. If the air is not calm but remains turbulent, particles do not settle vertically but descend at an angle determined by their terminal velocity and the wind speed.

The collection efficiency of a 'gravity' sampler (Durham Sampler) can be a complex function of particle size, wind speed, wind direction, and turbulence. It is therefore, impossible, to define the volume of air sampled or to compute the concentration of particles in that air. Moreover, counts are not comparable from one time or place to another unless meteorological conditions are identical. At best such samples give an indication of the types of particles present and a very rough measure of their abundance.

If a horizontal sampling surface is exposed on the ground, it does give a measure of deposition per unit area on that particular surface. However, this gives little information on the concentration in the air above. If operated for prolonged periods, the high collection efficiency of these devices leads to overloading and versions have been designed which operate sequentially or intermittently.

IMPACTION

When the airborne particles are subjected to some obstacles in their way, they impact to the obstacle surface with some force and are deposited. Impaction is defined as collision by inertial forces of a small, airborne particle with an obstacle or surface in the air stream, usually at right angles to the mean direction of flow. Since wind speeds are generally much greater than gravitational settling rates, most small airborne particles travel a nearly horizontal course. Their mass and velocity give them an inertial force, which resists changes in speed and direction. When a particle approaches a physical obstacle, the air molecules surrounding the particle divert and flow around the obstacle. If the particle has sufficient inertia, it will continue on its original course or on a path somewhere between this and the path of the air molecules and may strike an obstacle.

In the atmosphere, the efficiency of impaction (the percentage of particles approaching an obstacle that actually strike it) is a direct function of the size, mass and velocity of the particle and an inverse function of the size of the obstacle. Besides efficiency of impaction, the efficiency of retention is also important. A particle, upon impact, may either stick to the obstacle or rebound from it and re-enter the air stream. A sampling surface must be coated with a good adhesive to insure adequate retention. Sampling efficiency is a product of impaction efficiency and retention efficiency and can be determined experimentally in a wind tunnel. Wind, which carries the spores when it comes across the cylinder, the rays deviate away and later converge behind the cylinder. The inertia of the particle helps in impaction of the particle to the cylinder.

Particles may impact on obstacles of any shape, but vertical cylinders are most commonly used as impaction samplers since they are horizontally symmetrical and their impaction efficiency can be calculated. The relationship between efficiency of impaction and cylinder size is illustrated in **Fig. 13.1** and is given by the equation below.

$$E = d/D$$

Where E = efficiency of impaction

D = cylinder diameter from which particles impact

d = crosswind diameter from which particles impact

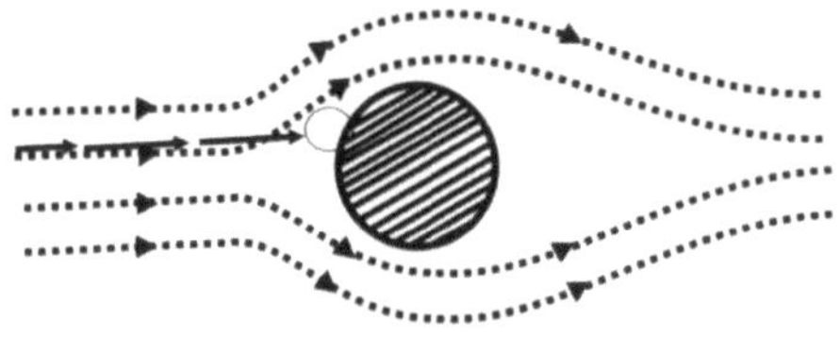

Fig. 13.1 Showing the relationship between efficiency of impaction and cylinder size.

Thus, ratio of d/D is larger for the smaller cylinder; indicating that a smaller cylinder is more efficient than a larger one, the other entire variable being equal.

SUCTION

Samplers in which air containing particulate matter to be sampled is drawn into an entrance by suction from a vacuum pump or another air-moving device which may be classified as suction. The samplers based on this principle are used for many air sampling purposes.

GRAB SAMPLING

Grab sampling consists of quickly capturing a volume of air by some means, hopefully without changing the concentration of particles contained therein. The sample is normally returned to the laboratory where the particles or other constituents of interest are removed from the air and counted or measured. A single grab sample is too limited in both time and space to be useful for pollen sampling, but repetitive samples may be useful in certain studies.

IMPINGING

Impingement is defined as collision of turbulent atmospheric motions of a small, airborne particle with a surface, usually not at right angles to the mean direction of flow. Liquid impingement results when the air stream is diverted into a liquid, which retains the particles as the air bubbles to the surface.

Many methods are used within such samplers for collecting the material of interest from the air stream. These methods include filtration, impaction, electrostatic and thermal precipitation and liquid impingement.

Description of Air Samplers (Pollen Traps) Including their Merits and Demerits

Many sampling devices operating on the principles described above have been used for sampling pollen and other airborne particles. The samplers used for sampling airborne pollen should have as many as possible of the following characteristics:

1) The samplers should have a reasonably high efficiency for the particles of interest under all normal operating conditions.
2) If the efficiency differs with wind speed or other factors, the manner of variation should be known.

3) The sampler should test a large enough volume of air to give a representative sample even when concentrations are low.
4) The efficiency should not change significantly due to overloading before the sampler has operated for a long enough period or has taken a large enough volume of air to give a satisfactory sample.
5) The volume of air sampled per unit time preferably should be constant; but if not, it should be capable of being calculated simply and accurately from associated data such as the wind speed.
6) Changing, storing and examination of the samples should be reasonably simple.
7) The sampler should be so designed and constructed that it will not be damaged by exposure to the normal range of atmospheric conditions.
8) The sampler should be commercially available or easily constructed, at a cost low enough to permit general use.
9) The sampler should be installed in a secure place to avoid tampering.

GRAVITATIONAL SETTLING SAMPLERS

This category includes all sampling devices in which the sampling surface is exposed in a horizontal position either on the ground or at some elevation. Capture of pollen and other airborne particles takes place by turbulent impingement as well as by gravitational settling. Retention of settled particles is not normally a problem with adhesives commonly used except that rain or heavy dew falling directly on the adhesive-coated surface may loosen some particles and float them away.

DURHAM SAMPLER

It was designed by Oren C. Durham in 1946, who was for many years Head Botanist at the Abbott Laboratories. Durham's sampler was adopted as the standard pollen sampler by the Pollen and Mold Committee of the American Academy of Allergy. It is still used sometimes by allergists, hospitals, and public health agencies on account of its easy availability.

The Durham sampler consists of a mount for positioning a glass microscope slide holder between two horizontal circular metallic disks. It is usually mounted on a metal rod, pipe or steel angle support at least several feet above the ground or on a roof top to have free flow of air all around **(Fig. 13.2)**.

The advantages of the Durham Sampler are:

1) The slides are easily loaded.
2) It is inexpensive.
3) It has no moving parts.
4) It requires no electric power supply.

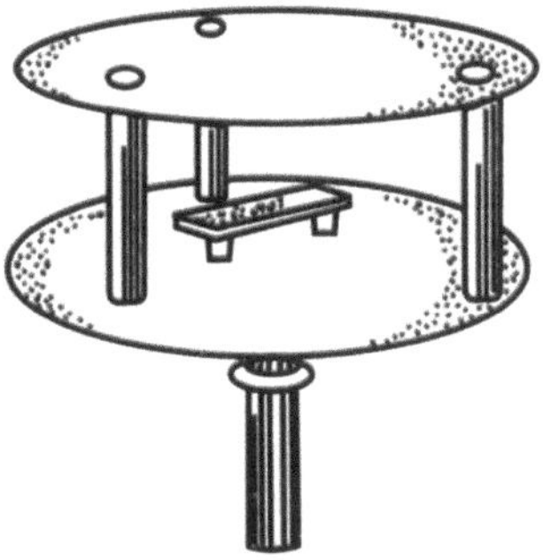

Fig. 13.2 Durham Sampler (Non-volumetric).

Disadvantages are:

1) The volume of air sampled is unknown, so the catch cannot be converted to a volumetric measure of concentration.
2) The efficiency cannot be determined.
3) The catch is relatively low.
4) The catch is a function of wind speed, turbulence and orientation of the sampler with respect to wind direction as well as concentration of pollen in the air.

Several attempts have been made to improve the sampler. Some workers have tilted slide samplers at a 45° angle in attempts to collect by both impaction and settling. However, none of these modifications appears to be a significant improvement over the original sampler.

Counts from the Durham Sampler are so greatly influenced by factors other than the concentration of pollen in the air that they indicate only in a general way the presence and abundance of pollen in the atmosphere. Tests have shown that it is impossible to determine the volume of air sampled; therefore, the data are qualitative at best.

Daily counts are highly misleading, as counts from different localities are not properly comparable unless the influencing meteorological parameters are the same. However, averages of the daily counts over a pollination season are useful for comparison with other localities, as the influencing factors usually tend to average out.

The daily counts, as taken with the Durham Sampler and reported by the news media, should be interpreted with an understanding of limitations of the sampler.

WIND IMPACTION SAMPLERS

These samplers suitable for catching airborne pollen may be divided into wind impaction and powered impaction samplers, both of which collect airborne particles on surfaces at right angles to the wind.

Flag Sampler

The flag sampler **(Fig. 13.3)** consists of a 5 cm length of 2.5 cm wide transparent cellulose tape wrapped around a 5 cm long straight pin, 1 mm in diameter (Harrington *et al.*, 1959). The tape is pressed together except near the tail where the ends are separated by a thin piece of folded paper to facilitate removal after exposure. The portion of tape around the pin is coated with an adhesive. The pin is inserted in a 1.8 cm long and 0.3 cm wide glass tubing sealed at the bottom. This serves as a bearing and allows the flag to move freely with the wind. After exposure, the tape is removed and mounted on a microscope slide for examination.

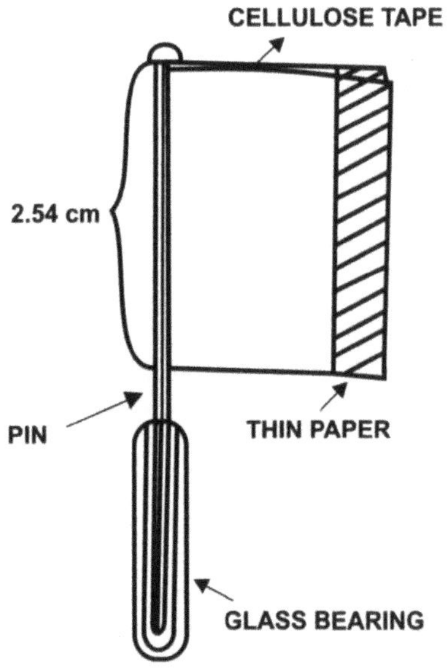

Fig. 13.3 Flag sampler (wind impaction sampler).

Advantages

1) Simple, inexpensive, no power needed, no moving parts.
2) Being small, it is portable and can be used for spot sampling.
3) Its efficiency can be computed if characteristics are known or determined experimentally for particles of interest.
4) The volume sampled can be determined if the wind speed is known.

Disadvantages

1) The efficiency and the volume sampled vary with wind speed; so if quantitative measurements are desired, the wind speed must be measured.

2) The efficiency varies with particle size and density.
3) The small size may lead to overloading during prolonged sampling periods if concentrations of airborne particulates are high.
4) Difficulty is sometimes experienced in separating the two halves of the tape and in placing the exposed portion flat on a microscope slide.

VERTICAL CYLINDER POLLEN TRAP

Glass vertical cylinders were shown to be reasonably efficient in sampling air-spora (Gregory 1951; Hirst 1959). The sampler was modified by Ramalingam (1968) to suit for routine aerobiological work in Indian conditions **(Fig. 13.4)**. Vertical cylinder pollen trap, a wind impaction air sampler, is found to be most suitable on account of its simple construction and efficiency. The spore-trapping surface is an adhesive-coated cellophane tape wrapped around a glass cylinder of 0.53 cm in diameter, suspended under a metallic shield. This device is usually installed on the roof of high-rise buildings to facilitate free flow of air around it. Irrespective of wind direction, the airborne pollen and spores get trapped on the vertical cylinder.

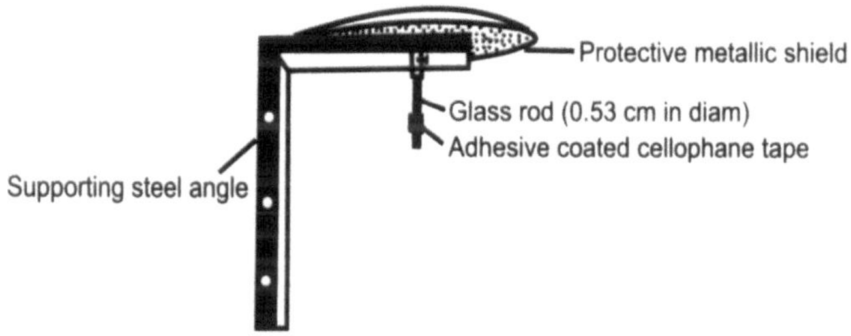

Fig. 13.4 Vertical Cylinder Pollen Trap (Non-volumetric wind impaction sampler).

Agashe and Anand alongwith their research students Jacob Abraham, Meenakshi Chaterjee and K. V. Nagalkshamma have extensively used verticle cylinder samplers for collecting data on airborne pollen and spores of the atmosphere in Bangalore, India, while on an ICMR funded research project in 1980.

The Cour Girouette Sampler (Cour, 1974)

The apparatus involves two methods impact and sedimentation/ gravitational. Two pieces of apparatus are involved during smpling: a height-adjustable pole carrying two vertically-aligned filtres **(Fig. 13.4-a)** and a fixed height sampler carrying a horizontally aligned filtre **(Fig. 13.4-b)**. When sampling the vertical filtres are raised to a height of 3 m above ground level and locked during the exposure period. The horizontal filtre

a.

b.

Fig. 13.4a & b a. The Cour Grovette Sampler: Showing a height-adjustable pole carrying two vertically-aligned filters b. The Cour Grovette Sampler: showing a fixed height sampler carrying a horizontally aligned filter.

is placed on a platform stand fixed at 1 m above ground level. Both pieces of apparatus are in close proximity.

The vertically aligned filters (2) face the oncoming wind by virtue of a large weather vane which is freely moving. Each of the two filtres carries a rain shield which also helps the filters to maximize the area of wind and airborne particles impacting on the filtres. The horizontal filter receives airborne particles descending under the force of gravity.

The filters are made of medical muslin soaked in silicone oil and fastened into acrylic frames exposing 400 cm^2 of filter surface for exposure. Optimum trapping efficiency requires the vertically-aligned filters to comprise five thickness/layers of muslin, whereas the horizontally aligned filters require eight. To distinguish the two types of filter when removed and undergoing treatment/storage, the frames of vertical filtres are blue whilst those of the horizontal filters are green.

Treatment of the exposed filters after removal of the acrylic frames involves treatment in Conc. H_2SO_4 to dissolve the muslin, followed by a prolonged period (usually overnight) in Hydrofluoric acid to dissolve the trapped silica particles, which, if not dissolved, can obscure the pollen and spores in subsequent examination. After repeated washings and centrifugations, the decanted residues can be volumetrically sampled for staining and slide preparation.

During the period of exposure temperature, humidity and wind speed are monitored. All are important parameters when interpreting the results of microscopic examination. Advantages of the Cour Girivette method lie in the efficiency of the techniques and apparatus involved and the superb quality of the slides produced. The principle disadvantages is the time required for treatment from the removal of the filters to the preparation of the slides for examination, which precludes the method for production of daily pollen counts. For longer periods of monitoring and forecasting (e.g. grape and olive harvests) the method is ideal and is widely used (Ribeiro *et al.*, 2007).

Rotating Impaction Samplers

Rotating impaction samplers are well suited for sampling airborne pollen and their use by allergists and aerobiologists is increasing. Although they require power and may have occasional mechanical or electrical problems, they measure pollen concentrations with acceptable accuracy.

ROTOROD SAMPLER (OLD MODEL)

Perkins (1957) developed a battery operated rotorod sampler, which is known to sample air at a constant rotational speed. Since the efficiency of the stationary impaction sampler is low and highly variable, the rotating impactor has been more advantageously used. The device relies upon the high efficiency with which small airborne particles are deposited on narrow rods oriented at right angles to high velocity winds **(Fig 13.5)**. It has been developed into a cheap, portable high efficient sampler with greater sensitivity. It is well fitted for use in the field and is relatively independent of the external wind speed.

In this sampler instead of moving the spores to the impacting surface in a current of air, the surface is rotated so that it strikes the spores. The volume of air swept can be calculated from the frontal area of the rod, the diameter through which it is turned and the number of revolutions for which it is run. According to Gregory (1951) this width should give more than 60-70% efficiency of deposition for 20 μ diameter spores at wind speeds about 4 mph (about 2 mm/sec). The unit is powered by a small battery operated DC motor in a protective case. It rotates at 2,500 rpm giving a linear speed of 10.6 mps and samples 120 litres/min of air. The arms are 6 cm long and 4 cm from the centre of rotation. Both arms are slightly bent inwards and are made from one piece of metal either brass or aluminium having 0.159 cm cross sectional area designed to slip over a special hub on the motor shaft. This sampler has a high efficiency for pollen-sized particles, but the arms are troublesome to handle without disturbing the sample and are difficult to place under a microscope. The sample can be viewed only in a reflected light.

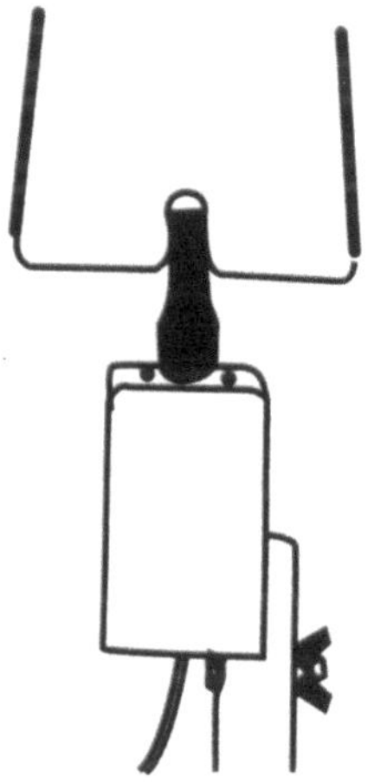

Fig. 13.5 Perkins Rotorod sampler.

Bainbridge and Lacey suggested sticking adhesive cellophane tapes on the metal rods (leading edge) and coating them with glycerine jelly. Later after exposure these exposed strips of tapes are mounted on the slide for microscopic examination.

The sampling rate is the volume of air swept over by the collecting surface per unit time. The volume of air can be calculated on the basis of the dimensions as explained below.

$$2 \text{ (arms)} \times 0.159 \text{ cm} \times 6 \text{ cm} \times 8 \times 2300 \times 10^3$$
$$= 48.0 \times 10^3 \times 2300 \text{ litres/min}$$
$$= \text{approximately } 100 \text{ litres/min}$$

The model has been tested for efficiency and it shows 85% efficiency. The sampling efficiency for particles greater than 15 μm is 100%. Wind speed has little effect on the efficiency unless it is close to the linear velocity of the collector. High winds, however, do increase, drag and load on the motor. The rotorod tested experimentally showed 60-90% efficiency.

The Rotorod Sampler is useful for a short period of sampling up to 2 hours. The sampler is volumetric and highly efficient and the efficiency remains unaffected at high wind speeds.

ROTOROD SAMPLER MODEL (40)

Rotorod Sampler (model 40) is volumetric, intermittent rotation impaction sampling device capable of quantitatively and qualitatively sampling airborne particles in the size range of 1 to 100 μm at sampling rates up to 120 litres of air per minute. The sampler consists of a constant speed motor of 2,400 rpm and two aerodynamically designed collector Lucite rods (1.3 mm in width), which are rotated by the sampler motor at 2,400 rpm. The retracting head holds two rods within the protective housing when the sampler is idle and when the sampler is activated; the rods are extended to a position perpendicular to the head. Rods are inserted in the pivot blocks and fastened with small thumbscrews (**Fig. 13.6**).

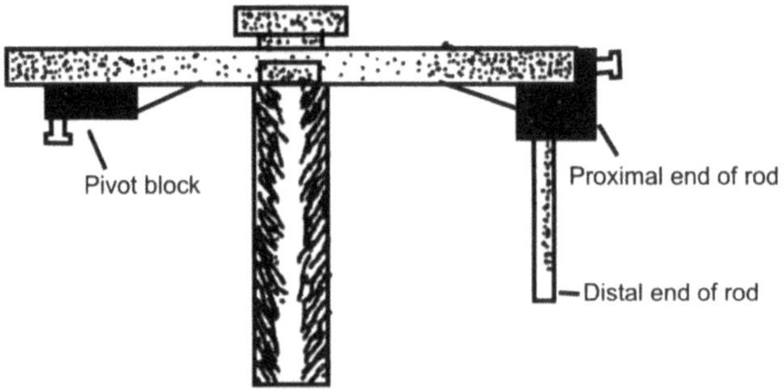

Fig. 13.6 Diagrammatic view of the Rotorod Sampler (Model 40).

The particles are impacted on one face of the rod, which has been smeared with adhesive (glycerine jelly). It is exposed for one minute after every nine minutes of rest time when it remains folded and static. The exposed rods are mounted on a grooved stage adapter, which consists of four parallel grooves of approximately the same width of the rod. By placing a coverslip carefully the rods are microscopically examined thoroughly under 40 × objective and 10 × eyepieces. After correct identification of the trapped airborne fungal spores, their percentage frequency is expressed as numbers per m^3 of air sampled.

Advantages

1) The volume of air sampled is known.
2) The efficiency is high and may be calculated or determined experimentally for specific particles.

Disadvantages

1) The efficiency varies with particle size and density.
2) It requires uninterrupted power supply.

SUCTION SAMPLERS

Hirst Spore Trap

The Hirst spore trap, (invented by Hirst in 1952), was the first suction type sampler readily available for sampling pollen and other spores. The vane tail keeps the 2.14 mm intake orifice facing the wind, and a rain shield protects the orifice from precipitation. It must be provided with an external vacuum pump (1/6 HP Motor).

The efficiency though variable with wind speed and with particle size, is reasonably high. Inside the housing containing the orifice, a greased microscope slide is drawn upward by a clockwork at a rate of 2 mm/hour. Particles in the air sampled are deposited by impaction on the slide, which is changed each day.

The suction trap provides data on rapid changes in the composition of air-spora. The spores in a measured volume of air are drawn through an orifice and are impacted on a sticky surface on a slowly moving microscope slide. The air is sucked through at the rate of 10 litres per minute, impinges on the microscope slide coated with solvent and vaseline, which form a sticky surface. The spore-free air passes out through the instrument into the pump. Thus, it leaves a trace at the end of 24 hours.

In the Hirst spore trap, there is a possibility of overloading of air spora on the collecting surface within a short period and so it requires constant checking.

Advantages of Hirst Spore Trap

1) It measures variation in concentration with time.
2) The efficiency is reasonably high.
3) The slides are easy to insert, handle and count.

Disadvantages

1) The efficiency varies with wind speed and with particle size and density.
2) Electric power and an external vacuum pump are required.

Anderson 2 – Stage Sampler

Andersen 2 – stage sampler is a multi-orifice cascade impactor. This unit is used whenever size distribution is not required and only respirable or nonrespirable segregation or total counts are needed. Viable particles above 0.8 microns can be collected on agar plates. The sampler is constructed of aluminium with two stages, which are held together with three dowel pins, and three teflon caps. Each impactor stage contains multiple precision-drilled orifices. When air is drawn through the sampler, multiple jets of air in each stage direct any airborne particles toward the surface of the agar collection surface for that stage. Each stage contains 200 tapered orifices. The diameter of the orifices on the first stage is 1.5 mm and 0.4 mm on the second stage.

Standard 100 × 15 mm Petri dishes with nutrient media were used as collecting surfaces on each stage. The exhaust section of each stage is approximately 19 mm large in diameter than the Petri dish, which allows unimpacted particles to go around the dish and into the next stage. A continuous vacuum pump is provided which will provide a constant sample flow of 1 ACFM. The sampler takes in 28.3 litres per minute of air through the opening at the top and impinges it successively on to the Petri plate containing nutrient media, placed below each sieve.

The number of fungal colony forming units (cfu) on each plate are counted and totalled. They are expressed as the number of colony forming unit per cubic meter (cfu/m^3) of air, as per the following conversion formula:

$$\frac{\text{Total number of colonies from two plates}}{\text{Total volume of air sampled}} \times 1000$$

The conversion factor was calculated in the following manner:

Suction rate = 28.3 litres per minute
If operated for 10 minutes; air taken in is = 0.283 m^3
If 0.283 m^3 has 1 colony
1 m^3 will have 1/0.0283
= 3.53 colonies

Conversion factor for estimating the number of cfu/m^3 of air for 10 minutes sampling time = 3.53.

This is the sampler, which separates small light spores from large heavy spores. The heavy spores are impacted on the first Petri dish and the lighter ones are swept away in the air stream and are impacted on the second Petri dish. Although this is helpful in the separation of different spore

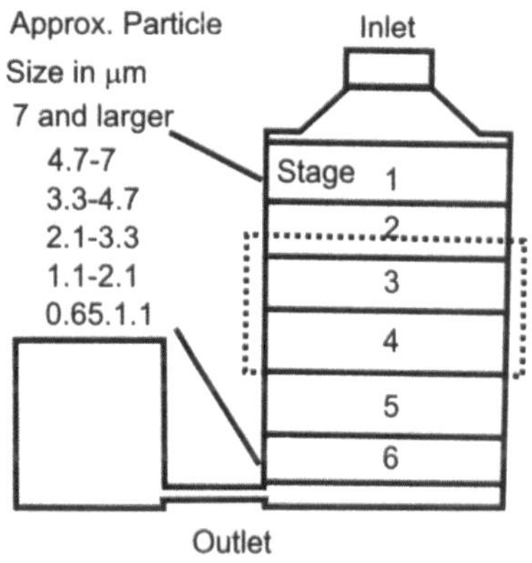

Fig. 13.7a Diagrammatic view of 6-stage Anderson sampler.

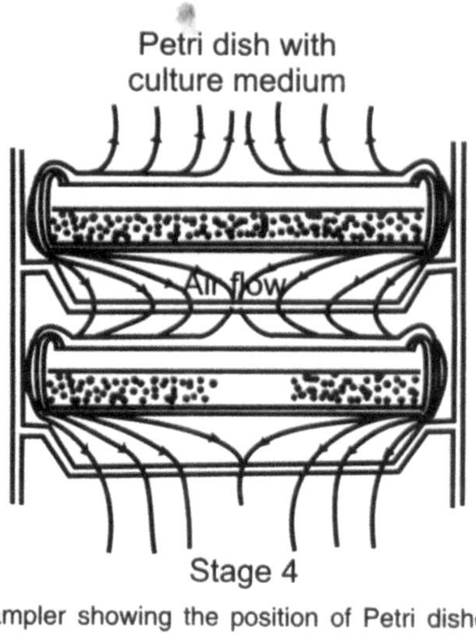

Fig.13.7b Anderson sampler showing the position of Petri dishes with culture medium.

types it does not indicate the true nature of the sample that is taken at a particular interval of time.

6-Stage Anderson Sampler

This is slightly modified sampler compared to the 2-stage Anderson sampler. In this sampler, the air after entering the circular orifice is drawn through a series of six circular plates each perforated with 400 holes. The spores are impacted through holes onto the sterile medium taken in Petri dishes. The plates in the series have progressively smaller holes. The largest particle gets deposited in the first while the smallest in the last Petri dish. In the Petri dishes different media are used for different size fractions. Air is sampled at the rate of 28.3 litres per minute. The sampler has proved useful for sampling particles less than 8 µm, which include the bacteria Actinomycetes and moulds. May (1945) modified this sampler by changing the pattern of holes to avoid wall losses occurring due to larger particles and recommended not to use the nose cone of the original model. Further, in the wind, the device for intake should be fitted with a large stagnation point shield **(Figs. 13.7 a, b)**.

BURKARD SEVEN-DAY RECORDING VOLUMETRIC SPORE TRAP

The Burkard seven-day recording volumetric spore trap is similar in principle to the Hirst spore trap. It has a built-in vacuum pump, and samples continuously for a week without attention on an adhesive-coated transparent tape on a clock-driven drum behind the entrance orifice **(Figs. 13.8-a, b)**. Particles are sucked into the orifice beneath the rain shield and impacted on adhesive-coated tape wrapped on a drum, which rotates behind the orifice over a weekly period. After exposure, the tape is cut into seven daily segments, which are mounted on microscope slides for examination. Each day's exposed tape segment can be further divided into 24 smaller parts corresponding hourly exposure with trapped pollen and spores. Other advantages and disadvantages are similar to those of the Hirst spore trap.

Protocol for Slide Preparation for Daily Pollen Count

1) Remove tape (exposed for previous 24 hours) from drum and align on perspex ruled block.
2) Cut 24 hour exposed section and transfer carefully to clean, dated slide. Placing a small drop of distilled water in the middle of the slide prior to lowering the tape will help the latter to adhere during inversion of slide.

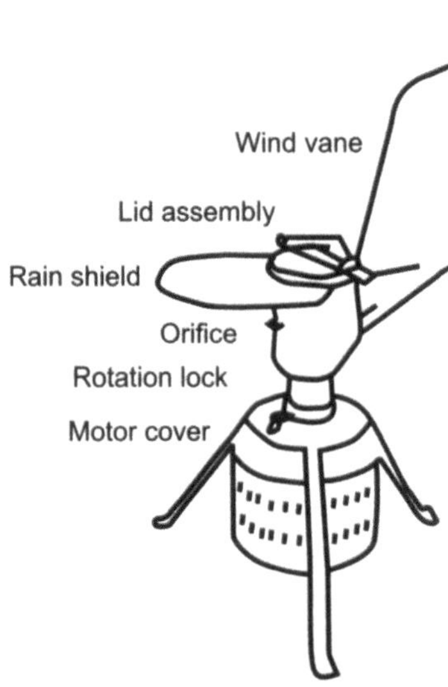

Fig. 13.8-a Burkard seven-day recording volumetric spore trap.

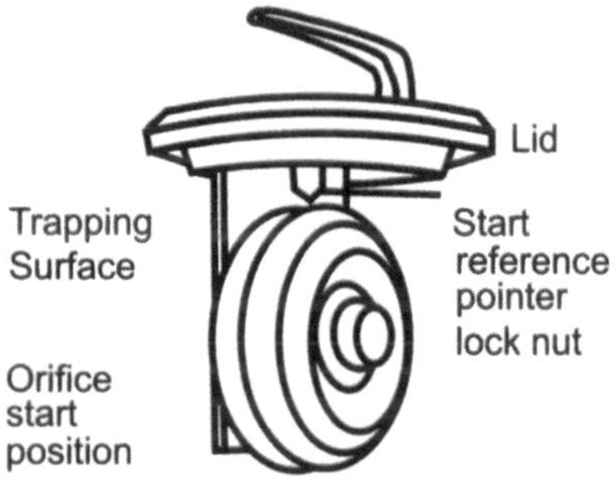

Fig. 13.8b Burkard seven-day recording volumetric spore trap showing details of lid with clock-driven drum with adhesive-coated tape.

3) Meantime phenolyzed basic fuchsin stain has to be kept for melting on a hot plate.
4) Transfer sufficient melted stain to cover area of cover slip when placed on slide. Pasteur pipette is used for this (cover slips are 20 × 60 mm No. 1 thickness).
5) Invert slide with tape onto cover slip bearing stain by means of a mounted needle. Apply gentle pressure to cover slip having inverted the slide with cover slip + tape on upper surface.
6) Place slide on microscope stage with date and start of exposure period on the left of the stage, at mid-point of the slide.
7) Move the slide to where the edge of the tape is aligned with the right hand of the microscope field.

8) Note the position on the horizontal vernier scale.
9) Move the tape 2 mm on to begin the first transect.
10) Move the slide vertically across the field examining × 400 magnification (× 10 eyepiece in conjunction with × 40 objective).
11) Identify and count each pollen grain as it appears. Grains are counted even if only partially visible in the left hand margin of the field, but not counted if partially covered in the right hand margin of the field (i.e., polar halves).
12) Record each taxon [usually at generic level, but for grasses at family (Poaceae) level].

TILAK AIR SAMPLER

It is a modified version of the Burkard sampler fabricated for Indian weather conditions by the outstanding Indian aerobiologist Professor S. T. Tilak. Instead of a vacuum pump, an exhaust fan is provided on the top position of the sampler. The apparatus runs on electric power supply (AC – 220 V) and provides a continuous sampling of air for eight days. The electric clock fitted in the instrument is synchronized with the drum. Air is sucked through the orifice of the projecting tube at the rate of 5 litres per minute and it impinges on the transparent cellotape, which is 1.5 cm in breadth and stuck on the slowly rotating drum. The drum completes one circle in eight days, thus giving the trace of catches for eight days.

The tape should be coated with glycerine mixed with vaseline or petroleum jelly. The mounting of cellotape is done in glycerine jelly. Scanning is done by dividing the tape into eight sections, which are mounted on eight separate slides. The tape can be further divided into hourly intervals for microscopic examination.

CALCULATIONS TO OBTAIN CONVERSION FACTOR

A) Sampled area : 8.4 cm × 1 cm = 8.4 sq. cm
84,00,00,000 sq. microns
(1 cm = 10,000 microns)

B) Scanned area : 20 microns (length) × 20 microns (width) × 24 hours = 9,600 sq. microns

C) Volume of air sampled per minute = 5 litres / min
In 24 hours = 5 × 24 × 60 = 7,200 L / 24 hour
$5 \times 0.001000028 = 0.005\ m^3$

To convert 1 litre into cubic metre

$$\frac{9600}{1,000,000} \times 7200 = 69.12 \text{ litres}$$

Therefore, conversion factor = $\frac{1000}{69.12}$ = 14.46 (conversion factor)

The number of spores, thus scanned, multiplied by conversion factor would give the number of spores in m^3 of air. For example 5 x 14 = spores = 70 spores m^3 of air.

Filters

Filters samplers particularly the high volume samplers, which can sample up to 1,415 litres per minute of air generally are used for sampling non-biological air pollutants.

AIRCRAFT SAMPLERS

Pollen and spores have also been sampled from aircrafts. Most of these samplers have nonisokinetic entrances, but some isokinetic samplers for use on light aircrafts have been developed. Upper air-spora of the atmosphere in Bangalore, India was studied by Agashe and Chatterjee (1986) by using the aircraft sampling technique. They sampled the air at 305 m, 610 m and 915 m above ground level for a period of seven months (October 1984 to April 1985) with nutrient agar Petri plates and glycerine-coated micro slides exposed to the air stream outside the cockpit of a light trainer aircraft flown at a cruising speed of 115 kmph. It is interesting to note that major types of fungal spores and pollen trapped at all the altitudes were: *Cladosporium, Penicillium, Alternaria, Aspergillus,* Smut spores, *Uredospores* and *Parthenium hysterophorus, Eucalyptus* species, *Casuarina equisetifolia, Ricinus Communis* and Poaceae. They observed that pollen grains were maximum at lower heights with minimum at 915 m.

Automatic Pollen Monitor: A New Air Sampler for Aerobiological Survey

Teranishi *et al.* (2006) have advocated the use of a new automatic pollen monitor (KP-1000, Kowa Co. Ltd., Nagoya, Japan) introduced at the University of Toyama, Faculty of Medicine in 2004. The basic principle of the measurement by this new pollen monitor is based on the auto fluorescence of the individual pollen grains excited by ultraviolet light. In order to study the efficiency and accuracy of this monitor, Teranishi *et al.*, compared the results obtained from this monitor with pollen counts determined by using the Hirst type (Burkard) pollen trap. Both results showed significant correlation. However, further improvement of the equipment is required for more precise and accurate counting before it is used universally.

Remote Sensing Techniques in Aerobiology

Anda *et al.*, (2006) had indicated the possible use of remote sensing to identify terrestrial growth of short ragweed: *Ambrosia artemisifolia* a notorious weed responsible for health hazards. They indicated that remote sensing technology is probably the most efficient tool for locating and detecting ragweed population. The data may be useful for the implemenatation policies aiming at controlling the spread of the obnoxious ragweed.

This endeavour of detecting short ragweed from space was possible due to the joint efforts of Ragweed Research Association and 'Centre National de la Recherche Scientifique' (CNRS). The spectral signature of short ragweed was identified for the first time in the world in 2003. This experiment has clearly demonstrated that areas of 100 sq m are detectable from space.

Of course the researchers were aware of the limitations of the remote sensing method in technical constraints which are necessary to understand before undertaking future programmes of space agencies.

A Person's Hair as a Pollen Trap

A very unique way of collecting pollen trapped in a person's hair was deviced by Penel and De Clercq (2006) in France. Every individual has a natural pollen trap such as hair or clothes. Hair is a very significant pollen trap since hair by its nature, 'follows' you everywhere, 24 hours a day. It was surprising to note that after hiking in a ragweed field in mid-September a person's hair yielded up to 140,000 pollen grains trapped. The pollen were recovered by a simple hair wash. Obviously a person's surroundings, particularly the vegetation have a clear impact on the number and types of pollen trapped in the hair.

Choice of Samplers

1) Particles < 5 µm in diameter and not requiring culture are best sampled by suction samplers. Anisokinetic conditions do not introduce serious errors for these small particles.
2) Particles between 5 and 15 µm in diameter are not sampled very efficiently by either suction or impaction samplers.
3) It is generally important to determine the collection efficiency for the particles being considered.

4) Above a diameter of 15 µm, particles are best sampled by rotating impactors using continuous, intermittent or sequential operations.
5) It is also important to consider the sampling location, sampling season, sampling period, and various stains for microscopic analysis and culture techniques.
6) Identification of airborne particles requires considerable training and experience, extensive consultation of the literature, reference collections and the assistance of specialists.

POLLEN CALENDAR

The ultimate aim of an aerobiologist is to compile a pollen/spore calendar, which will be useful to allergologists and the patients suffering from allergy. The aerobiological survey of an area involves aeropalynological surveys, identification of airborne pollen and spores, and determination of atmospheric pollen count. Although the atmosphere consists of an array of pollen and fungal spores, only a few of them are responsible for allergic manifestations. The knowledge of the occurrence and concentration of these allergenic pollen and spores, which can be inferred from a pollen calendar, is of great help to the clinicians. A detailed pollen calendar of a region is a prerequisite for the immunological treatment of pollen allergies (Caulton *et al.*, 1997).

Atmospheric surveys conducted in France and a comparative study from Montpellier and Font-Romen provided pollen calendars for *Alder*, Cupressaceae, *Pinus*, Poaceae, *Rumex* and Urticaceae. The pollen calendar for Switzerland is provided by Leuschner (1974). From Germany, Stix (1974) provided pollen calendars for Darmstadt and a pollen calendar for Huddinge (Nilsson and Palmberg Gotthard 1982) was formulated on a five years survey. Pollen calendars have also been published for Turin (Caramiello *et al.*, 1989), Alexandria (Ghazaly and Fawzy 1988) and Stockholm (Nilsson and Praglowski 1974; Engstrum and Nilsson 1979), Sweden, Scotland, United Kingdom (Caulton *et al.*, 1997). In India, pollen calendars have been complied for Secunderabad (Nayar and Ramanujam 1989), Bombay (Tripathi *et al.*, 1982b), Allahabad, U.P. (Nautiyal and Midha, 1984), Nagpur (Deshpande *et al.*, 1976), Chennai (Vittal *et al.*, 2001) and for Delhi (Singh 1987). Pollen calendar of Bangalore city was compiled for two years, 1982-83 and 1983-84 (Agashe and Abraham 1988, 1990), which showed the count and relative abundance of 12 major pollen types.

A sizeable number of residents in Bangalore, India suffer from allergic manifestations. Reports claim that 1.31% of the population suffers from asthma (Asthma Research Society 1979). From a retrospective statistical study done for three years on the number of cases of asthma admission in Victoria Hospital, Bangalore, India it was found that most of them showed symptoms during August to October, which correlated to the peak of pollen season of *Parthenium, Casuarina, Ricinus* and Amaranth-Chenopod (Agashe and Philip 1990).

The main objective of the continuous air sampling was to get the qualitative and day-to-day variations in the concentration of different pollen and fungal spore types. These data enabled compilation of the pollen calendar, which depicts the duration and concentration of various pollen types in the atmosphere.

Pollen calendars compiled by aerobiologists provide knowledge of the occurrence and concentration of the allergenic pollen, which is of great help to the clinicians for proper diagnosis. Hence, proper interaction among the aerobiologists and allergologists is essential to tackle the problems of suffering patients. The pollen types which are most significant in the atmosphere and in considerable concentration are selected for testing on the patients. The tests include skin prick tests and evaluation of serum IgE levels in the patients. In addition, nasal, conjunctival and bronchial provocation tests are also done. Once the offending allergens are detected by using the above procedures, the patients suffering from allergic manifestations are treated by subjecting them to immunotherapy. Thus, allergy patients get sufficient protection from the effects of airborne pollen and spores if they undergo immunotherapy.

As a prerequisite to the evaluation of allergencity by an allergologist, knowledge of a pollen calendar of the local region, is essential on two basic counts. First, only the relevant antigens need to be tested on the patients, the pollen of which is most predominant in the atmosphere. There are many antigen kits available in the market and most of them will not be specific to the locality. Hence, selection of the right antigens is ensured by consulting a pollen calendar. Second, many patients suffer from seasonal allergic manifestations to the seasonal occurrence of the pollen types. It is evident from the work known earlier that the magnitude as well as quality of the annual pollen load in the atmosphere vary significantly. Therefore, it is essential that an aerobiological survey of an area is conducted continuously over a number of years (Agashe 1993).

Pollen calendars should be compiled and updated every year. The annual pollen calendars give a picture of the change in trends of the peak and concentrations over the years. **Figs. 13.9 a and b** show the change in peak seasons for the pollen types over the period of 1981 to 1990 for Bangalore city, India, was recorded by Agashe (1994). The six most dominant pollen types, which are also allergenically significant have been studied. These are *Parthenium, Casuarina,* Poaceae, *Eucalyptus,* Amaranth-Chenopod and *Holoptelia. Parthenium,* an anemophilous pollen, exhibited maximum concentration during September to December and peak incidence was observed in July and August. *Casuarina* pollen was present throughout

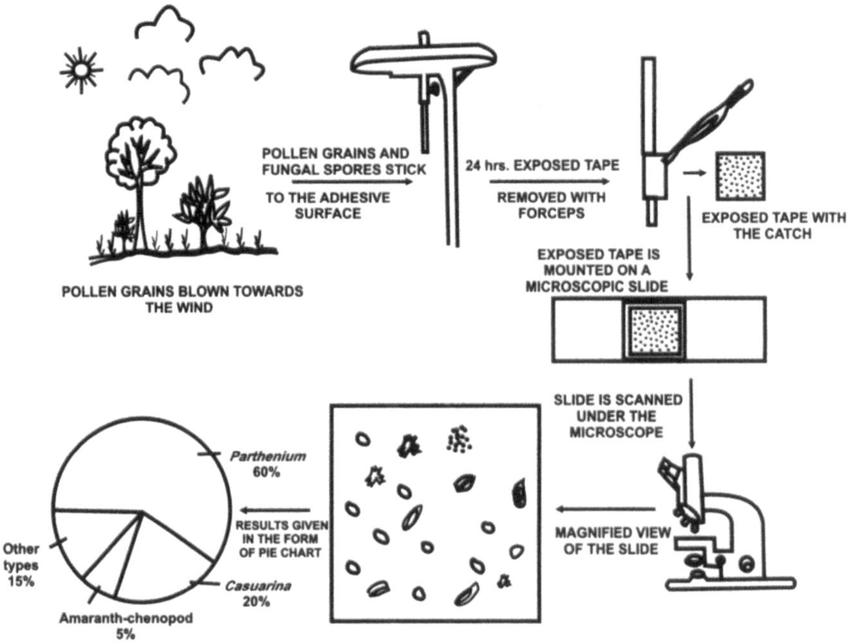

Flow Chart No. 13.2: Showing various steps involved in atmospheric pollen monitoring.

the year but showed two peak seasons. The main season was from February to March and the second peak was observed from September to November. The monthly maxima were observed mainly in February and September and occasionally in October, during the 10 years of survey. The Poaceae is a heterogenous group and the pollen was dominant between October and January. But the monthly peak differed over the years and was observed in July (1982), April (1983) and May (1984). Otherwise, the peak was observed in November or December. *Eucalyptus* pollen was also present throughout the year although the maximum concentration was noticed between October and December. The pollen of Amaranth-Chenopod was recorded throughout the year as different species of *Amaranthus*, flower at different times of the year. The prevalence of this pollen was maximum between July and October. *Holoptelia* pollen was dominant in the atmosphere between January and April.

The need for updating the pollen calendar is further substantiated by **Figs. 13.9 a, b** and **13.10 a, b** which show the pollen calendar of Bangalore, India for 1982-1983 and 1983-84 respectively (Agashe and Abraham 1990). It can be observed that the atmospheric pollen peak over the years have not varied considerably. But during 1989-1990, the occurrence of atmospheric pollen was prolonged for most of the pollen types. *Eucalyptus*, *Peltophorum*,

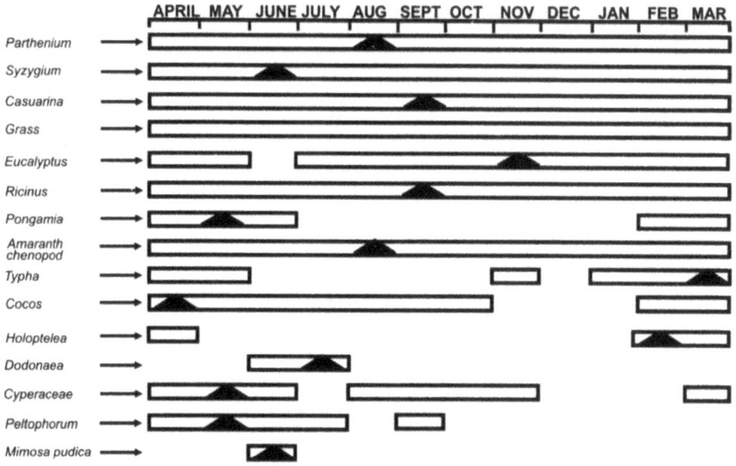

Fig.13.9a Pollen calendar of Bangalore City, India for the year 1982-1983.

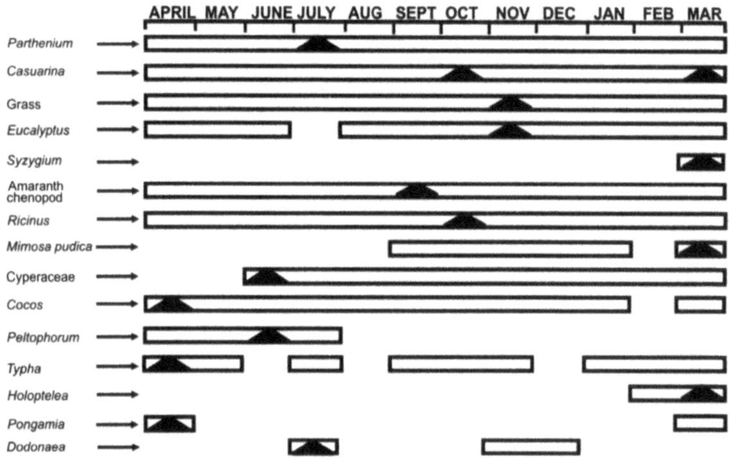

Fig. 13.9b Pollen calendar of Bangalore City, India for the year 1983-1984.

Cocos, Dodonaea, Mimosa, Cyperaceae and *Typha* had wider atmospheric pollen occurrence during 1989-1990 compared to 1982-1983. It was more pronounced in the case of *Mimosa* and *Dodonaea. Mimosa* was dominant only in June during 1982-1983 but was recorded from August to April in 1989-1990. The pollen of *Dodonaea* a fast-spreading weed in Bangalore was dominant only during June-July and February-March in 1982-1983 but was prevalent throughout the year except for a month of May during 1989-1990.

It is evident from the above-mentioned account that the magnitude and the quality of annual pollen load in the atmosphere can vary significantly.

As such, it is necessary that the aerobiological survey of an area be a continuous process over the years. It is imperative for an allergist to update the pollen calendar every year and to keep track of the variations in air-spora. Updating of the pollen calendar every year also gives a clear picture of the duration of occurrence of pollen in the atmosphere. The prolongation of occurrence of allergenically significant pollen in the atmosphere influences the prolongation of allergy symptoms in patients. This will help in a better correlation of allergy symptoms of patients with atmospheric pollen and proper diagnosis. Another practical use of a long-term survey of an area can be the construction of statistical models for prediction of the start and intensity of pollen season.

It is necessary to make a distinction between a pollen calendar and a pollination calendar.

Pollen calendars differ from Annual calendars, in that they are concerned with events relevant only to plant flowering and the release of pollen to the atmosphere from the source.

Pollen calendars and pollination calendars are very closely linked. In fact data generated from pollination calendars precede pollen calendars. Pollination calendars deal with flowering of plants leading to pollination in different months of a calendar year. Various aspects are studied in it. They are: initiation (or onset) of flowering, peak flowering, and duration and termination of flowering.

The pollen count on the other hand concerns the occurrence, abundance and decline of pollen and spores in the atmosphere. Pollination calendars are compiled by botanists, particularly taxonomists, based on physical observations. In contrast, Pollen Calendars are compiled by aerobiologists operating pollen traps or air samplers continuously, i.e., round the clock and round the year.

Pollen often remains floating in the atmosphere for a long period even after the pollination period is over. Thus, some of these pollen grains may cause allergy outside the known pollination period. Compilation of Pollen Calendars requires a lot of effort, patience, a thorough knowledge of plants in the field and analysis of pollen data in the laboratory. In fact, compilation of pollen calendars is the ultimate aim of an aerobiologist. Pollen calendars have great significance in pollen allergy as they serve as important guidelines to allergy practitioners with respect to the onset of allergenically significant pollen season, their peak and decline in the atmosphere. In this context Pollen Calendars serve as a bridge between aerobiologists and allergists, as both are dependent on each other and equally benefit from each other.

The pollen and fungal spore spectrum keeps changing in the atmosphere depending on several parameters, including weather factors. Recently Agashe (2005) has suggested that ideally, one has to aim for a four-way correlation, that is, correlation between flowering, pollen count, weather factors and allergic manifestations. Aerobiologists should concentrate on compiling four-way correlation as depicted below.

Flowering (Pollination calendar)
Pollen count (Pollen calendar)
Weather factors (meteorology)

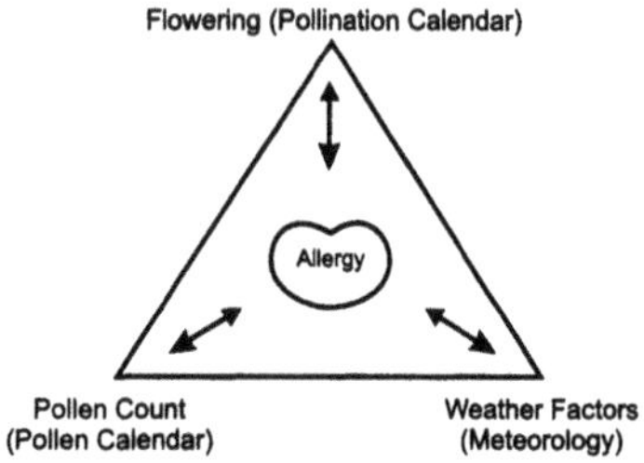

POLLEN CALENDAR OF BANGALORE CITY, INDIA

The magnitude as well as the quality of pollen load in the atmosphere vary from year to year. There can be significant variations in atmospheric pollen even between successive years. This aspect was highlighted by Agashe and Abraham (1990) while compiling the pollen calendar of Bangalore city for the two consecutive years, 1982-1983 and 1983-1984. It is seen that in these two pollen calendars, 15 major pollen types constituted over 96% of the total pollen load in the atmosphere of Bangalore. There were not many qualitative variations between the two calendars as the 15 major pollen types remained the same for both the years. However, quantitative variations between the two pollen calendars were noteworthy **(Figs. 13.9 a, b)**. There was a 19% increase in the number of pollen grains in the air during 1983-1984 (745/cm^2) as compared to 1982-1983 (6522/cm^2) *Parthenium* pollen which is known to be second highest pollen in the atmosphere of Bangalore showed a 26% increase during 1983-1984 where as *Syzygium* pollen count recorded an 81% decrease in 1983-1984. Pollen of *Mimosa pudica* (sensitive plant) was 15th among the dominant airborne pollen during 1982-1983, whereas during 1983-1984, it was 8th in the order of predominance of airborne pollen **(Figs. 13.10 a, b, c)**.

The high pollen months were June (1,141/ cm^2) and July (1,782 /cm^2) for the year 1982-1983 and 1983-1984 respectively. The maximum *Parthenium* incidence was in August during 1982-1983 whereas it was in July during 1983-1984. Similarly, peak pollen incidence of *Casuarina* shifted from February in 1982-1983 to March in 1983-1984.

It was also observed that pre-monsoon showers (May-June) of the year 1983 (324.9 mm) were slightly more and also more spread out than in the year 1982 (244.3 mm). This aided *Parthenium* weed to proliferate and flower profusely.

Pre-monsoon showers help for the prolific vegetative growth of *Parthenium* plants, which leads to profuse flowering in the monsoon and post monsoon season. Early rains (May-June) during the year invariably cause the airborne pollen of herbaceous taxa to peak earlier and late rains postpone the peak. It is very important that an aerobiological survey of an

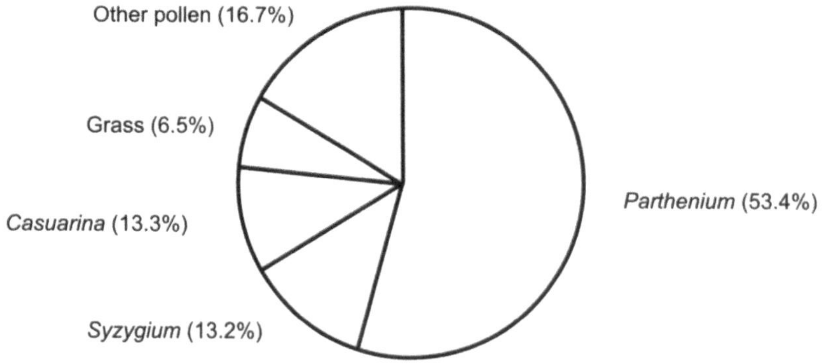

Fig. 13.10a Atmospheric pollen in Bangalore, India during 1982-1983.

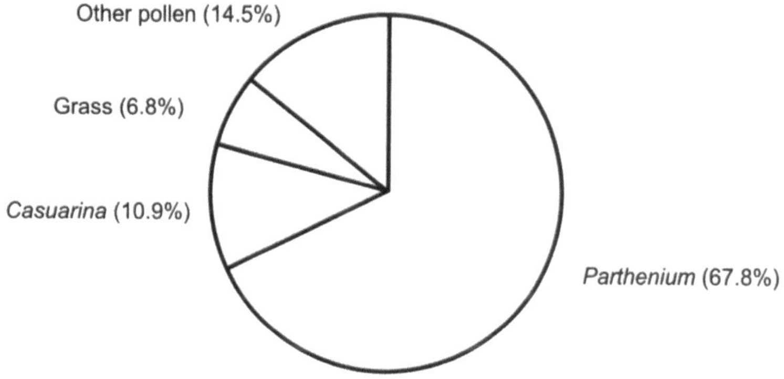

Fig. 13.10b Atmospheric pollen in Bangalore, India during 1983-1984.

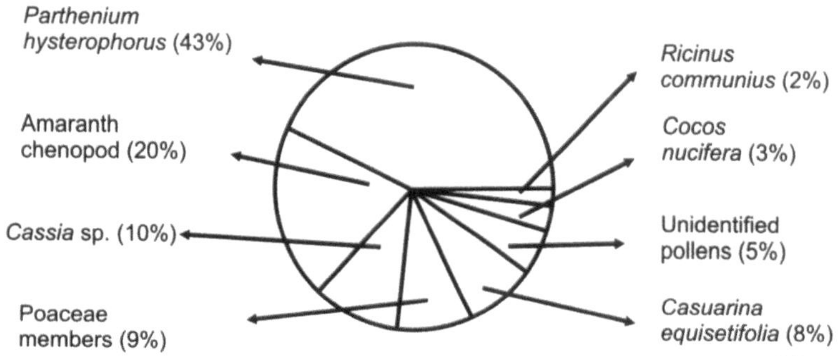

Fig. 13.10c Allergically significant pollen representation in the atmosphere of Bangalore City, India (Figs. 13.9a,13.9b,13.10a, 13.10b and 13.10c from Agashe and Jacob, Abraham, 1990).

area should be a continuous process over several years (Agashe 1996). It is also imperative for an allergist to update his/her knowledge of the pollen calendar every year and keep track of the variations in the air spora which have tremendous implications in allergic manifestations.

POLLEN CALENDAR OF ALLAHABAD, U.P., INDIA

Aeropalynological survey of Allahabad was carried out by Nautiyal and Midha (1984) during 1973-1979. The survey was carried out mostly by gravity slide method in different areas of Allahabad City. Airspora determined at and around the Allahabad Univesity, Botany Department from Oct 1973-Sep 1974 has been depicted in **Fig. 13.10d**.

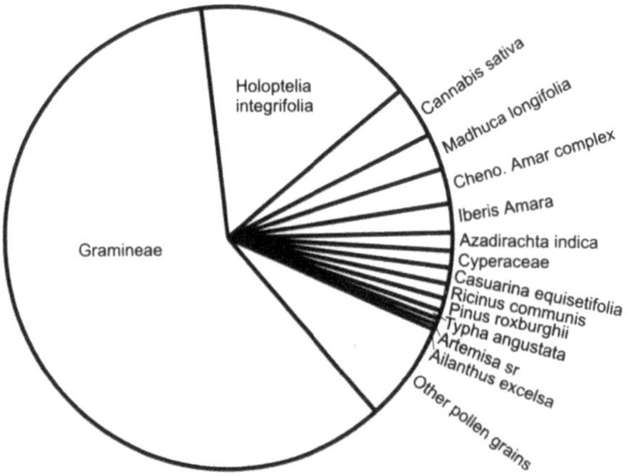

Fig. 13.10d Pollen calendar of Allahabad, U.P., India from Oct 1973-Sep 1974.

The pollen spectrum shows prepondance of grass pollen (Gramineae), followed by *Holoptelia integrifolia* and other pollen types including *Cannabis sativa, Madhuca indica,* Amaranthaceae, *Iberis amara, Azadirachta indica,* Cyperaceae, *Casuarina equisetifolia, Ricinus communis, Pinus roxburgii, Typha angustata, Artemisia sp., Ailanthus excelsa* and others. The aeropalynological survey was later carried out again from Feb 1978 to Jan 1979 in the same city. Surprisingly the aeropollen spectrum was dominated by *Pinus roxburghii* followed by *Holoptelea integrifolia,* Gramineae and other pollen as indicated in the Pie chart **(Fig. 13.10e)**.

Public awareness of aeroallergens: A successful attempt in Bangalore, India.

It has been proved substantially that environmental biopollution, i.e., pollution caused by bio particles or particles of biological origin indoor and outdoor has significant role to play in human health hazards. Some of these bioparticles are responsible for causing various types of allergies hence they are called as aeroallergens. The commonest 'aeroallergens' reported from Bangalore comprise of pollen produced by anemophilous plants such as *Parthenium hysterophorus, Holoptelea integrifolia, Casuarina equisetifolia,* Poaceae, *Ricinus communis, Eucalyptus* sp., Amaranthus-Chenopod. *Cocos nucifera, Peltophorum pterocarpum, Syzygium* sp., *Cassia*

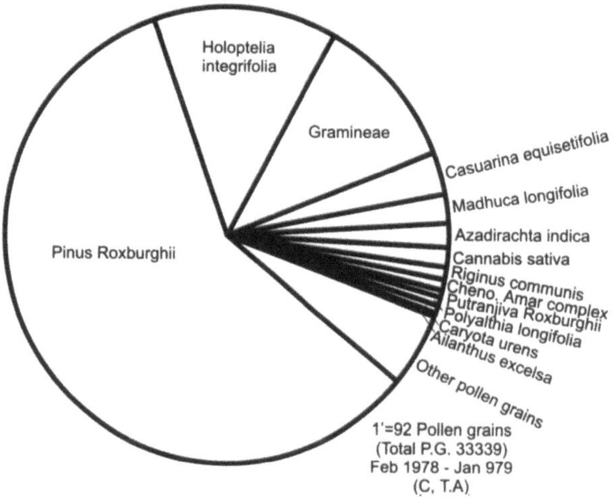

Fig. 13.10e Pollen calendar of Allahabad, U.P., India from Feb 1978 to Jan 1979.

sp., *Mimosa pudica*. In addition, the atmosphere is full of aeroallergens (mould spores) such as *Cladosporium, Periconia, Nigrospora, Alternaria, Helminthosporium,* Smut spores (Ustilaginales), *Aspergillus* and *Penicillium,* etc.

In a recent paper Agashe (2007) has emphasized the importance of predominant aeroallergens as biopollutants on human health particularly in Bangalore, India.

It has been proved experimentally and clinically that the following most common aeroallergens (biopollutants) are responsible for causing allergic disorders in the Bangalore atmosphere: Pollen of *Parthenium, Amaranthus, Minosa,* Grasses, *Ricinus, Prosopis, Albizzia, Holoptelea.* The fungal spores include *Cladosporium, Alternaria, Curvularia, Helminthosporium, Aspergillus.*

The above mentioned airborne pollen and fungal spores which act as biopollutants pose health hazards with respect to allergies. Their occurrence and relative abundance in the atmosphere is revealed in the pollen calendar compiled by the aerobiologists. However, it should be borne in mind that the relative abundance of pollen and spores is liable to change from year to year as it is dependent on biological factors as well as meteorological (weather) factors. Hence, it is recommended that pollen calendar should be updated every year. There appears to be a perfect correlation between higher pollen count, weather factors and prevalence of allergy.

Another classical example of importance of monitoring of aeroallergen biopollutants was thoroughly investigated in Bangalore, India. This is with regard to *Holoptelea integrefolia* known to produce highly allergic pollen and for a very short flowering duration. In fact there are a very few trees of this species in Bangalore.The plant flowers for only a fortnight either during late January or February each year. These trees are prolific pollen producers.

During its peak flowering Bangalore atmosphere is known to be occupied by as mush as 70% of total pollen population. Allergenicity of these airborne pollen has been clinically proved, particularly with regard to asthma.

The need for continuous monitoring of airborne pollen and fungal spores has been greatly emphasized by Agashe (2002). However, the most neglected aspect seems to be creating awareness of the results of air monitoring among the clinicians and allergy sufferers in India. In the Western countries daily pollen counts of airborne pollen and mould spores are publicized through the mass media such as radio, television and newspapers.

A fairly successful attempt has been made in this direction by Agashe and his co-research workers from the Aerobiology and Allergy Laboratory, Department of Botany, Bangalore University, Bangalore, for the first time in India and perhaps in Asia since December 1995. The awareness of pollen calendars, and atmospheric pollen and mould spore counts was created by them by publishing their results in the daily newspaper ('Times of India', Bangalore edition).

The daily pollen monitoring was done by operating the rotorod (Model - 40) sampler round the clock. The qualitative and quantitative analysis results were compiled on a daily basis and after consolidating these results for a period of one week were published in the 'Times of India' (Bangalore edition) once in a week under the captions 'Pollen Watch and Mould Watch' **(Fig. 13.11)**.

Agashe (1999) published a comparative account of modality of reporting atmospheric pollen count in newspapers in certain major cities in the United States of America cities and Bangalore, India **(Fig. 13.12)**.

ALLERGY STATUS IN BANGALORE, INDIA

Inspite of the fact that Bangalore has the unique distinction of gaining many adjectives including 'Air Conditioned City of India' , 'Garden City', etc. it is also gaining another dubious distinction as 'Allergy City'. It is said that this city has the highest atmospheric pollution next to Delhi, which is responsible for a large proportion of the population suffering from various types of allergies. On account of the numerous parks and gardens–grasses, weeds and trees, atmospheric pollen occur abundantly. This causes allergic symptoms in sensitive individuals.

There have been number of reports indicating that many people who are free from allergies develop the allergic symptoms the moment they enter Bangalore city limits. However, they feel much better and perhaps feel free from allergy symptoms the moment they leave the city.

The above concept has been depicted in the cartoon **Fig.13.13.** It should be noted that allergy sufferers who have to stay in Bangalore can get better

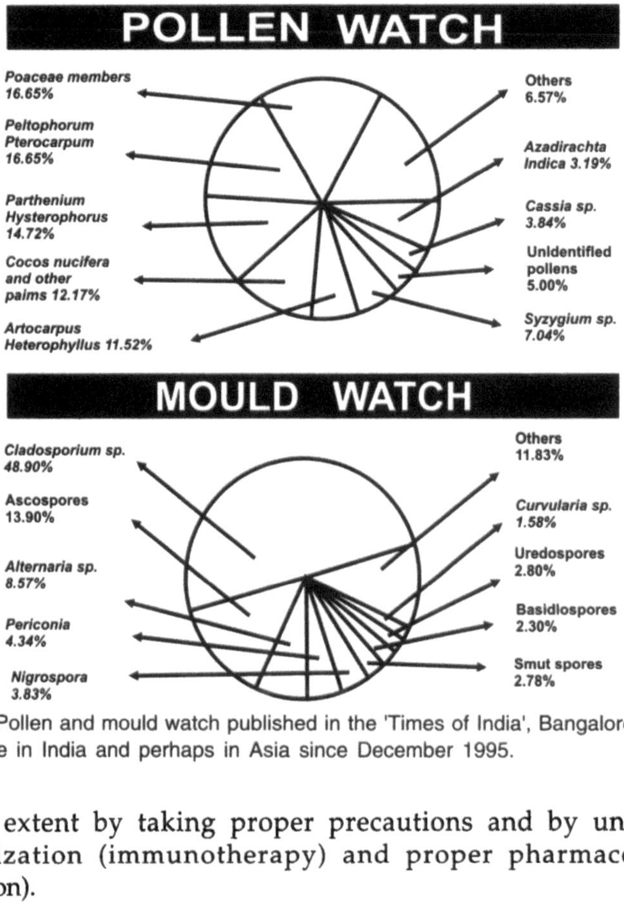

Fig. 13.11 Pollen and mould watch published in the 'Times of India', Bangalore Edition for the first time in India and perhaps in Asia since December 1995.

to some extent by taking proper precautions and by undergoing desensitization (immunotherapy) and proper pharmacotherapy (medication).

POLLEN NETWORKS IN NORTH AMERICA

Pollen monitoring in North America does not appear to have the close integration of sites and data obtained that is characteristic now for most West European countries. There may be a number of reasons for this situation. The sheer geographical size of the United States of America and Canada may preclude the organizational problems involved in the establishment of networks.

The uneven distribution of palynologists especially in Alaska, Canada and the more remote parts of the U.S.A. again, may mitigate against the production of national scale pollen data. Lastly, funding in North America, particularly in the U.S.A. is more tightly controlled by the large pharmaceutical sponsors, which often establish a fairly strict legal control over the publication and dissemination of data obtained from their

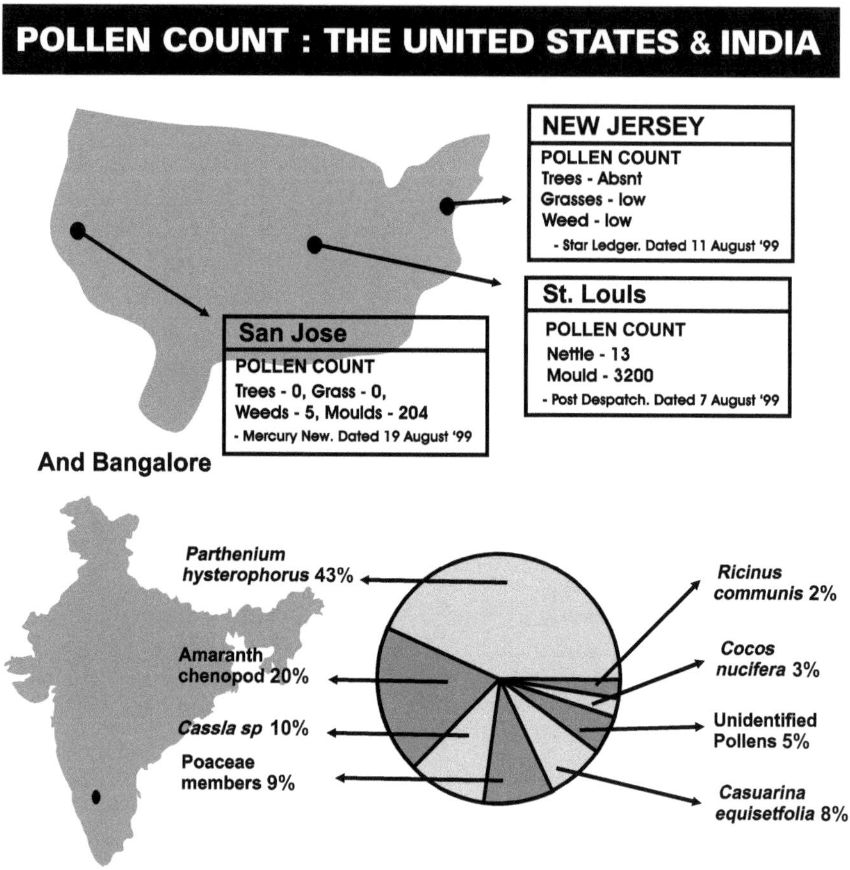

Fig. 13.12 Comparison of atmospheric pollen count reporting through mass medium such as Newspapers in some U.S.A. cities and Bangalore City, India.

sponsored sites. In short there appears to be a greater freedom of publication of data at international conferences and in the European scientific literature than appears to be the case in North America.

Nevertheless, considerable literature emanating from North America is extant with regard to all aspects of palynological research and at individual institutional levels – mainly universities.

By means of periodic national and international conferences and symposia held under the auspices of American, European and Asian organizations, dissemination of data from North America reaches the international scientific domain, oral communication, poster presentation and personal contact all play a vital and valuable role in the inter change of ideas and results between North America and the rest of the 'palynological world'.

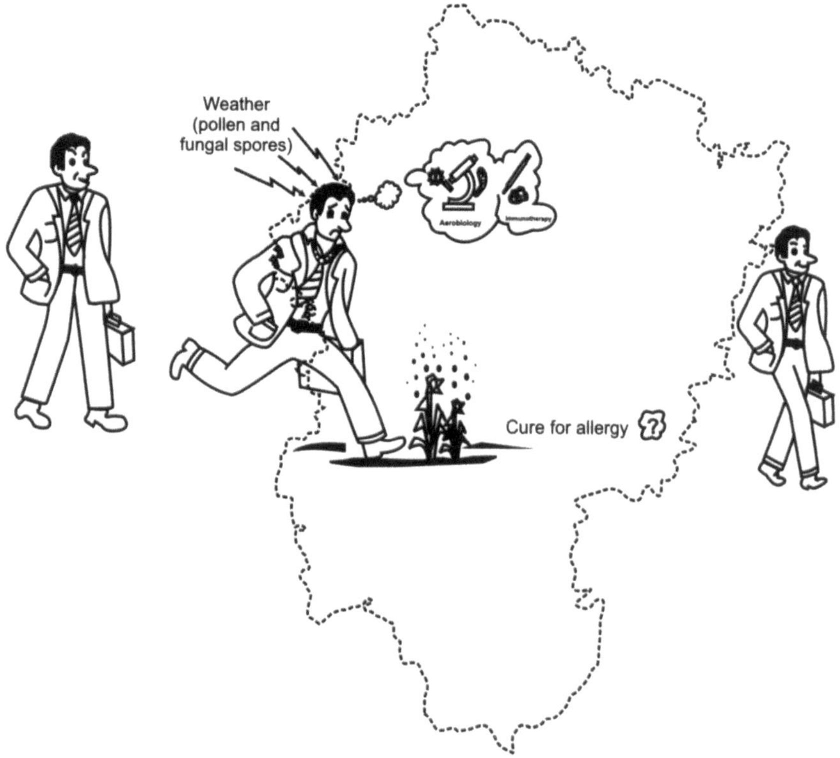

Fig. 13.13 Cartoon depicting status of allergy in Bangalore City, India.

The following cartoon **(Fig. 13.14)** modified from the one available in literature conveys two points in a humorous way. The first point it conveys is the fact that pollen count and weather have a close releationship. In many western countries daily pollen count is announced as part of the weather bulletin on T.V. In Sweden and other European countries, even the forecast for a probable atmospheric pollen count is broadcast along with weather forecast.

We have already explained in the foregoing chapter, the efforts that are involved in collecting and processing data to determine the pollen count. The second and most important humorous point indicates that **"It is not the way pollen count is determined"**.

It is hoped that this aspect of pollen count and its correlation with weather and its public awareness through mass media will be taken up seriously by research workers in Asian and African countries.

Fig. 13.14 Cartoon indicating reporting of atmospheric pollen count on T.V. and also depicting wrong way of determining atmospheric pollen count.

CHAPTER **14**

Aerobiology: Aeropalynology - Part II

This chapter deals with the study of all aspects of aeropollen studies. It is a general practice to group airborne pollen on the basis of habit of their plant sources such as trees, grasses and weeds. In temperate regions this type of grouping corresponds well with major flowering in various seasons in a year. Flowering in trees takes place in the spring, grasses in the early summer and 'weeds' in the late summer and autumn, each linked to well-defined seasons of pollinosis. In North America, the above seasonal pollen shed almost always correlates with allergic responses in the population.

PATTERNS OF OCCURRENCE OF AIRBORNE POLLEN

Regional Patterns

It is important to know not only the airborne pollen-producing plants in the general area under investigation but also their positions with respect to distance from the pollen sampler. If this is known and daily records of wind direction are kept, these data are of value in understanding the catch with respect to the sources. Some pollen may travel long distances and in large quantities (Tyldesley 1973), so knowledge of occurrence over regions beyond the local area is desirable. Range maps for many species are available.

Seasonal Patterns

The seasonal occurrence of various atmospheric pollen is primarily determined in two ways:

1) The presence and abundance of these pollen investigated by samplers and field observations of the source plants growing in the vicinity.
2) Periodic examination of plants known or suspected to shed airborne pollen and dated observations as to the maturation of the male cones

and flowers over a period of several years will accumulate the desired information. These data greatly aid in the identification of pollen from dated samples. The presence of some genera in the samples and the local observations may not always exactly correlate. Winds may bring in some pollen before the local sources release them.

Diurnal Patterns

Variations in concentration close to a source are influenced by diurnal patterns of emission, but distant from sources these variations in concentration are influenced more by meteorological variable factors. Weather plays a significant role in this process.

Among grasses, it is known that some blue grasses (*Poa*) shed pollen between 3-8 am in the morning; some fescues (*Festuca*) between 3 and 8 pm in the afternoon. Timothy (*Phleum pratense*), orchard grass Cocksfoot (*Dactylis glomerata*), and reed grass, (*Phragmites communis*) are morning shedders. Quack grass Twitch (*Agropyron repens*) sheds pollen in the afternoon.

In case of common or dwarf ragweed (*Ambrosia artemisiifolia*), giant ragweed (*Ambrosia trifida*) and probably the other species of *Ambrosia*, the peak at the source during dry weather is usually from 7 to 9 am . This two-hour period may shed 75% of the pollen in a day. Nearly all of the pollen will be shed before noon. At approximately 6 am the mature flowers change shape; at 6:30 am the anthers are exposed; if the humidity is low the flowers may open in 15 minutes; if high, it may take 2 or 3 hours; if it is raining they may not open at all. Dispersal may be divided into four phases:

1) Ejection of pollen in clusters from flowers.
2) Temporary attachment of pollen clumps to neighbouring foliage.
3) Reflotation by air currents.
4) Final distribution in the atmosphere.

Little is known about the diurnal patterns of emission of tree pollen. In *Pinus banksiana* (Jackpine) pollen, diurnal patterns appear to be correlated with humidity. The apple is primarily entomophilous, but appreciable quantities of pollen may be in the air close to the tree; the highest concentrations are in the afternoon.

Carica papaya (Papya) with heavy pollen is entomophilous. However in the vicinity of the male plants, pollen is found abundantly and, if inhaled, may induce allergic symptoms in sensitive individuals (Shivpuri 1964).

In the temperate climate of North America and Europe, there are generally sharply demarcated pollen seasons, such as spring or summer. However, in the tropics some plants will always flower and at any given time pollen will be always found in the air, though their concentration in the atmosphere may vary in different seasons.

CORRELATION OF AEROALLERGENS WITH METEOROLOGICAL FACTORS

Although the flowering nature is a pre-determined factor for a given species, it is observed that climatic factors influence the flowering pattern by either shortening or lengthening the flowering period or shifting the flowering period earlier or later. The length of the sporulation period is genotypically controlled by external factors such as temperature, wind speed, humidity and rainfall. The environmental conditions are also known to be important for the release, dispersion, transportation and removal of pollen from the atmosphere. Other than the climatic conditions, the presence of the pollen in the atmosphere is the result of numerous developmental stages which may occur interdependently or independent of many stages such as the flowering intensity, formation of flower buds, wintering of buds, ripening process of pollen, opening of the anther, dispersion of pollen in the air through wind and turbulence, transportation by air, deposition, redeposition and viability.

After a six-year survey at Cardiff, Hyde (1952) had pointed out that the magnitude of annual catches of individual pollen types varied considerably, and attributed this mainly to variation in pollen productivity. According to him, pollen productivity of the herbaceous types appeared to be affected by rainfall during the period of vegetative growth. The other reasons that could bring about variation in the magnitude of the annual catches of the individual pollen types were; weather during the preceding year, weather during the current year and intrinsic cyclic variation in flowering.

A review of aeropalynology in Britain was published by Hyde (1969). It has also been found that the allergy symptoms are more severe when there is a change in the weather. Hence, it is worth assessing the pollen count of allergenically important pollen during climatic variations like a thunderstorm and depression so that the symptoms can be correlated to the climatic changes.

AEROBIOLOGY AND POLLEN MORPHOLOGY OF SOME COMMON AND ALLERGICALLY IMPORTANT PLANTS

Parthenium hysterophorus pollen

Allerginicity of Parthenium hysterophorus

Parthenium is a genus of basically the western hemisphere belonging to the family Asteraceae. It has been included in subtribe Ambrosiinae along with the well known and notorius genus *Ambrosia* (ragweed) and *Iva* (marsh elder). *Parthenium hysterophorus* is a ubiquitous annual weed native of the Gulf of Mexico region and West Indies but now disseminated in warm and semi arid subtropical regions of the United States and almost the whole of the Indian subcontinent. It occurs abundantly in the gulf coast regions from Texas to the southernmost point of Florida. It is stated that in 1956, a *Parthenium hysterophorus* 'reserve' was introduced from the gulf regions of U.S.A. to the Poona region in India.

Subsequently it has spread like a wild fire in other regions of India resulting in serious agricultural problems and a medical hazard as a major source of allergic dermatitis. The major offender is the Sesquiterpene lacton, Parthenin found throughout the plant body including the pollen. There are different views on the pollination mechanism in *Parthenuim*. Pollination has been described as entomorphilous where insects are attracted to the pollen-containing disc flowers by glandular hair secretions. The plants produce an enormous quantity of pollen (an average 624 million /plant). The pollen are carried away at least for short distances in clusters of 600 to 800 grains. The pollen become airborne to great heights in significant amounts either as individual grains or in clumps (Sitaramaiah *et al.*, 1981 and Agashe *et al.*, 1987). Sitaramaiah *et al.*, have noted that *Parthenium hysterophorus* pollen could be a source of nasobronchial allergy. A clinical survey conducted by Sitaramaiah and co workers in Bangalore, India on 71 patients suffering from seasonal and perennial nasobronchial allergy indicated that 40% of them developed symptoms during July to December, i.e., at a time of maximum incidence of pollen of *Parthenium hysterophorus* in the atmosphere.

In another research investigation from Bangalore Rao *et al.*, (1985) and Uthyashankar *et al.* (1985) performed extensive studies on the allerginicity of *Parthenium hysterophorus* in patients. Both skin tests and radio allegro solvent test (RAST) were performed on patients with classic symptoms of allergic rhinitis during July and December. These classical studies demonstrated that 34% of patients with classic allergic rhinits and 12% of patients with asthma were sensitive to extracts of *Parthenium hysterophorus* pollen.

Similar studies were carried out by collecting sera from 18 patients selected on the basis of symptoms of allergic rhinitis or asthma related to the fall season and were shown to be reactive to *Ambrosia* species on the basis either skin testing or RAST.

The pollen of *Parthenium hysterophorus, Ambrosia* and *Quercus sp.* were suspended in 10 mm phosphate buffer, ph 7.2 and gently stirred at a ratio of 1 gm of pollen to 10 ml of buffer for 48 hours at 4°C Later, the pollen extracts were classified by a combination of filtration and centrifugation.

RAST

Rast studies were carried out with the help of sera of the patients and the pollen extracts of *Parthenium hysterophorus* and *Ambrosia*. A number of conclusions were drawn from these studies.

First, it is clear that there is significant non-reactivity between *Parthenium hysterophorus* and *Ambrosia*. *Parthenium hysterophorus* pollen may be an important environmental allergen similar to *Ambrosia* species. *Parthenium hysterophorus* represents an unique allergen where importance has largely been overlooked in U.S.A.

Agashe and Vinay (1975) described the pollen morphology of *Parthenium hysterophorus* the obnoxious weed growing abundantly in different parts of India **(Fig. 14.1a, b)**.

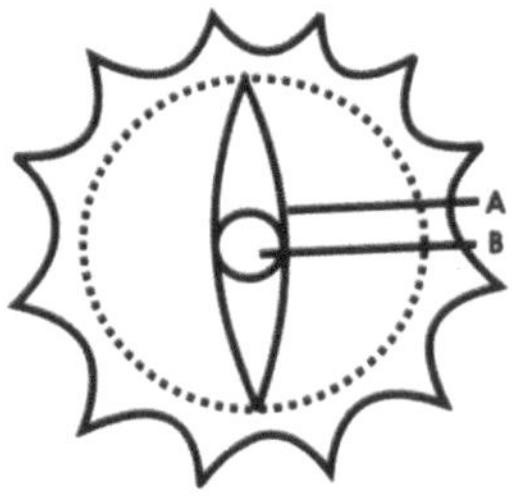

Fig. 14.1a Diagrammatic sketch of *Parthenium hysterophorus* pollen showing aperture and ornamentation (a. colpus, b. pore).

Fig. 14.1b Enlarged view of exine showing spinules.

This weed is commonly referred to as 'Congress Weed' , 'Congress Grass' or 'White Top' in India. It has been further reported that the food grains imported into India from U.S.A. in 1950s were contaminated with the seeds of this weed.

Parthenium hysterophorus Linn., a member of the family Asteraceae (Compositae), is known to produce pollen abundantly. The toxic effects of the pollen grains of this weed with reference to allergies have been reported both by Wodehouse (1965) and Shivpuri *et al.* (1963). The flowers of this weed are known to be amphiphilous.

Pollen diagnosis of *P. hysterophorus* Linn. Pollen grains 3-colporate (peritreme), oblatespheroidal (16 × 17 μm). Apocolpium diameter about 3.5 μm. Colpi (10 × 2 μm) tenuimarginate, with tapering ends, membrane smooth.

Ora circular (diameter about 2 μm). Exine (spinules included) about 4.4 μm thick. Sexine about 3 μm thick, pertectate suprategillate, provided with pointed spinules. Tegillum undulating, differentiated into supra and infrategillar layers, each less than 0.5 μm high supporting the tegillum of each layer. Spinules about 2 μm high with pointed solid apices, base about 2.2 μ wide made up of slender rod like elements. Nexine as it seems, consists of a homogeneous layer, inner margin smooth. There appears to be a thin distinct region about less than 0.5 μm wide between the baculate layer and the nexine.

Fig. 14.1c shows SEM photomicrograph of pollen of *Parthenium hysterophorus* in which spiny exine is clearly seen.

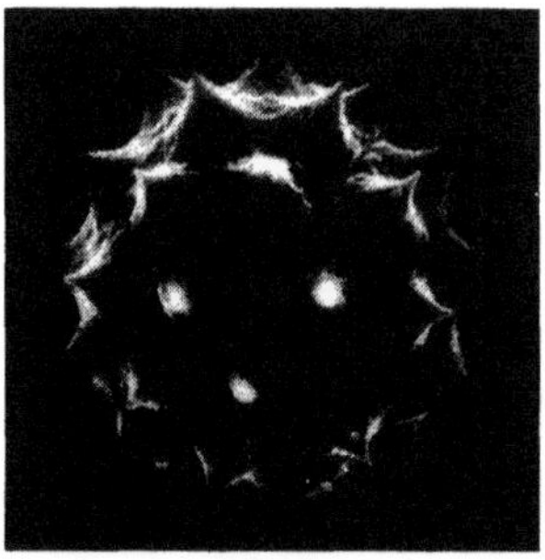

Fig. 14.1c Photomicrograph (SEM) of pollen of *Parthenium hysterophorus* showing detailed pollen morphology. Courtesy Siwert Nilsson.

The weed *P. hysterophorus*, according to Shivpuri and Kartar Singh (1971) is known to cause skin allergies. It has been clinically proved by Lonkar and others that severe skin allergies are caused (contact dermatitis) when the sensitive persons come in contact with *Parthenium* leaves and stems which are covered by unicellular hairs loaded with a chemical substance called parthenin.

Gupta and Chanda (1991) worked on various aspects of the aerobiology and chemical composition of this pollen. This investigation was done on the basis of an aeropalynological survey carried out by them in Salt Lake City (a newly developed township in the eastern fringe of the metropolitan city of Kolkata), India.

For the determination of pollen count, undehisced anthers from unopened disc florets of *Parthenium* were placed in a haemocytometer, covered with a cover glass and microscopically examined to count the number of pollen per anther. The mode of pollination in this plant was also studied.

The rate of pollen production was proved to be 9,600-pollen/staminate flower or 3,45,600 pollen per inflorescence. The maximum amount of pollen release was 31.4% at 32°C and 86% relative humidity on July 7, 1989, whereas 41.3% release was recorded on July 8, 1989 at 8 am. When the temperature was 32°C and 86 % Rh. on July 9, 1989, 100% pollen release was recorded at 8 am when the temperature was 31°C and relative humidity was 96%.

With regard to the mode of pollination, *Parthenium hysterophorus* appears to be entomophilous. Physical observations on these plants showed that honeybees were frequent visitors followed by ants, houseflies and other members of the *Diptera*. Further it was deduced that *Parthenium* is not a self-pollinated plant though normally ants induce self-pollination. At the most it may be placed in the category of amphiphilous plants where pollen is dispersed mainly by insects and partly by wind.

Chemical Composition of Pollen of Parthenium hysterophorus

Gupta and Chanda (1991) worked out the chemical nature of pollen of *Parthenium hysterophorus* with particular reference to carbohydrate, protein, and lipid contents. The results are indicated below:

Chemical analysis of the pollen of *Parthenium hysterophorus* (per cent dry weight)

Total carbohydrate	Total protein	Total lipid
10.0	17.5	15.0

The amino acid composition of the pollen of *Parthenium hysterophorus* showed the presence of free amino acids like arginine, aminocaprylic acid, proline, methionine, histidine. Whereas the bound amino acids present in the pollen were glycine, alanine, valine, proline, histidine, and tyrosine.

Stanley and Linskens (1974) had stated that most of the pollen allergens are water-soluble which enhances their degree of diffusion in mucoid tissues and increases the reactive charge groups available on dissociated molecules. In order to assess allergenicity of *Parthenium hysterophorus* pollen, clinical investigations were carried on 160 patients attending the Allergy Clinic of the Institute of Child Health, Kolkata, India. A total of 56 patients (35%) showed a positive response to *Parthenium* pollen antigen. Clinically one out of six patients showed ELISA positive reaction to *Parthenium*. Specific IgE antibodies in the patient were estimated to be 1.35 PRU ml^{-1} as against 0.35 PRU ml^{-1} in control

(PRU = pharmacia RAST uni/ ml $^{-1}$).

Though *Parthenium hysterophorus* pollen were found in a very small numbers in Salt Lake City near Kolkatta, the aeropalynological survey carried out in other centres in India such as Bangalore, Hyderabad and Pune showed much greater proportions.

Various types of air samplers have been employed by aerobiologists to assess pollen count of *Parthenium hysterophorus*. Some of these results have been shown in the following **Table 14.1.**

Table 14.1 *Comparative account of samplers and respective percentage of airborne pollen of Parthenium in India.*

Author	*Sampler and sampling altitude*	*Annual Parthenium Pollen (%)*	*Peak Month*	*Place of exposure*
Bhasale (1984)	Tilak air sampler, 1.5 m	13.63	September	Aurangabad
Agashe and Abraham (1988)	Vertical cylinder, 1.9 m	66.18	August	Bangalore
Bhat and Rajasab (1989)	Vertical cylinder 9 m	22.1	September	Gulbarga
Agashe and Chatterjee (1987)	Nutrient Petridish and glycerine coated glass slides 8 m	69.0	August	Bangalore
Agashe and Chatterjee (1987)	Air craft sampling using glycerine coated glass slides 305 m	23.0	August	Bangalore
Agashe and Chatterjee (1987)	Air craft sampling using glycerine coated glass slides 610 m	21.0	August	Bangalore
Agashe and Chatterjee (1987)	Air craft sampling using glycerine coated glass slides 915 m	31.0	August	Bangalore

OCCURRENCE OF AIRBORNE POLLEN OF *PARTHENIUM* IN THE U.S.A.

Parthenium pollen occupied a small portion of herbaceous aeropollen flora of Corpus Christi, Texas, U.S.A. (Lewis *et al.*, 1990). An aeropollen survey was carried out by them in 1988 by using a Burkard seven-day volumetric sampler. Graphic representation of ambient pollen of *Parthenium* is given in **Fig. 14.2**.

Parthenium appears to be the second major asteraceous pollen producer like *Ambrosia*, its pollen was shed mainly in the autumn with the major

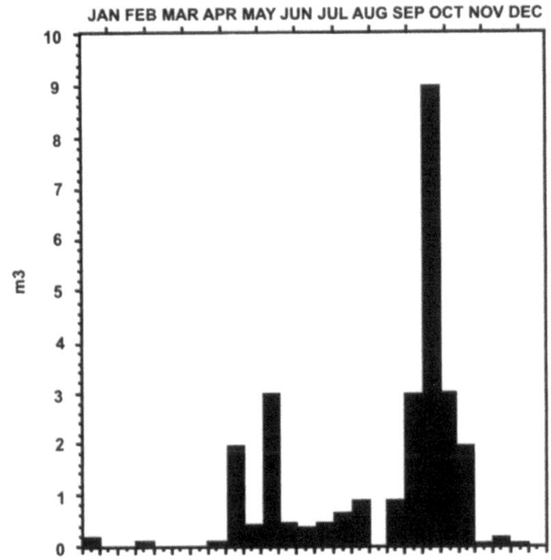

Fig. 14.2 *Parthenium* aeropollen survey at Corpus Christi, Texas, U.S.A. in 1988 (From Lewis *et al.*, 1990).

peak occurring between September 23 to October 6, though at a lower level of 8.8/m³. A secondary peak occurred during the spring (May 20 to June 2) at 2.9/m³.

Incidence of Pollen Allergy

According to Iwanami *et al.* (1988), one out of every ten Japanese suffers from pollinosis. About 90 species of plant pollen have been identified as aeroallergens. The commonest among them are: *Artemisia vulgaris* (common mugwort), *Oryza sativa* (rice) *Ambrosia artemisisaefolia* (ragweed), *Betula tauschii* (birch), *Nerium indicum* (oleander), *Castanea crenata* (Spanish chestnut). The gymnospermous pollen-causing allergy are *Ginkgo biloba* (maidenhair tree) and *Cryptomeria japonica* (Japanese red cedar).

Commonly occurring pollen aeroallergens of North America have been sufficiently listed and described along with their allergenic significance by Lewis *et al.* (1983) in a very exhaustive book. It also contains vast amount of aerobiological data of major pollen groups.

Siwert Nilsson's Pollen Flora of Europe, Fungal Spores of Europe and Tourist Guide for Allergenic Plants in Europe, Scandinavian Pollen Flora, and European Pollen Network are very useful in understanding allergenicity of common aeroallergens in Europe and the Scandinavian countries.

AEROPALYNOLOGICAL RESEARCHES IN INDIA

In India the first systematic aerobiological survey was started in Calcutta by Cunningham in the 1880s of the 19th century (1873). Kasliwal *et al.* (1959) started such studies in Jaipur, Rajasthan. In Delhi Shivpuri *et al.* (1960) conducted such studies. Later on similar studies were conducted all over India by various researchers. Nair *et al.* (1986) have summarized pollen count as reported from various cities in India in their publication entitled 'Airborne pollen, spores and other plant materials of India' which was the outcome of an all India coordinated project initiated in 1980.

AEROBIOLOGY IN THE UNITED KINGDOM

Studies on airborne pollen in the United Kingdom has a history extending over 60 years. Monitoring of pollen in the air began in Wales in the 1940s and in London in the 1950s using Hirst spore traps.

The number of sampling sites increased gradually during the ensuing decades. It was not until some 50 years late in 1990 that a collaborative network was established in the U.K. The network became known as the British Aerobiology Federation (BAF), which had its origin in a meeting held in the Harley Street consulting room of Dr. William Frankland, a leading specialist in the treatment of allergies. Some 16 pollen monitors attended, and with the enthusiastic leadership of Jean Emberlin, the network began its operations. The centre of operations for a few years was based at the North East London University where Emberlin worked. A few years later, the headquarters of the BAF moved to the university college at Worcester where Emberlin established the National Pollen Research Unit (NPRU) now the National Aerobiology Pollen Research Unit (NAPRU). The BAF network currently operates at approximately 40 sites. The number fluctuates as monitors retire for one reason or other, new sites are set up in different places. Manpower and finance are the two principal factors which determine site location and function. Sites are not evenly spread geographically. For example, of the total number of monitoring sites in operation at the turn of the century, only three are located in northern Britain: two in Scotland (Edinburgh and Invergowrie, near Dundee) and one in Northern Ireland (Belfast). The remaining sites are scattered throughout England.

Some 13 pollen monitors are members of the European Aeroallergy Network (EAN) whose database is in Vienna (Austria). Comparison between sites in the U.K. are made (Emberlin *et al.*, 1994) and between the U.K. and Spain and between the U.K. and Poland.

THE SCOTTISH CENTRE FOR POLLEN STUDIES

One of the two sampling sites in Scotland, was established in 1987 and began monitoring pollen in 1988. In addition to two volunteer scientists who helped to establish and develop the Pollen Centre's work, a succession of job seeking youngsters, opted to work in the centre to acquire some basic scientific and organizational skills to help them gain full time employment. In 1989, the first of an annual succession of French students – more than 40 – from the universities of Montpellier, Toulon and Bordeaux, requested to come and work for the six-eight weeks of their placement (stage) to gain experience of working in a scientific institute and to augment their knowledge of English language. These students were from Institute of Health (Montpellier), Institute Universite de la Sante (I.U.S.) and technology (Toulon and Bordeaux) Institute Universite de Technologie (I.U.T.) From time to time the centre has welcomed for various periods of time, pollen specialists from Greece, the Netherlands and Spain. In the first half of the Centre's existence, students from Napier University, the Centre's base, undertook degree and diploma projects on various aspects of pollen and spores. The Centre is involved in a research programme, which has included melissopalynology, but most consistently investigations into palynology of animal faeces (Caulton and Simpson 1988; Caulton *et al.*, 2005).

Current developing work involves the aerobiology of monitoring sites, both indoor and outdoor, urban and rural. Secondary school students in the fifth year of their studies are invited to apply to spend four weeks of their summer vacation working on this developing programme. The programme is generously funded by the Nuffield Foundation and is designed to encourage young students to gain work experience in science laboratories; and in the hope that science will be the basis of university and their subsequent careers. All students working in the Centre are required to assist in the pollen and spore-monitoring programme, which uses the Burkard volumetric spore sampler. The Scottish Centre for pollen studies has gained the present important status on account of the deep interest, devotion and consistent efforts of Eric Caulton, the present Director of the centre. In Scotland, the pollen monitoring season begins in March and extends to the end of September, supplying daily data to the NPRU at Worcester from April when the *Betula* (birch) season starts, and finishes at the end of August, by which time the grass season (Poaceae) is over. Monitoring of the allergenic spores of *Alternaria* continues until the end of its sporing season September/October depending mainly on the date of completion of the area's Barley(*Hordeum*) harvest.

The Scottish Centre for Pollen Studies is an independent unit based in the School of Life Sciences, Napier University. Dr. Eric Caulton, the co-author of this book, has been the Centre Director ever since its establishment in1987 (Caulton 2003). In addition, currently it is staffed by volunteer scientists.

The centre's primary aim is to provide a daily pollen count for sufferers from asthma and hay fever. For this, daily monitoring of pollen is carried out with the help of a Burkard Volumetric spore sampler.

CHAPTER 15

Aerobiology – Applications of Airborne Pollen Studies in Allergy

ALLERGY

The term 'allergy' was coined in 1906 by Dr. Clemens Freiherr von Pirquet, an Austrian physician, to describe any abnormal reaction of the immune system. The immune system is intended to protect the body against the noxious invaders. But in allergy, immunity has gone awry, and the system reacts to substances that are ordinarily harmless. By far the most familiar allergic reactions are respiratory – sneezing, runny nose and watery eyes caused by inhaled allergens from growing weeds, trees and grasses or moulds, house dusts, mites and animal danders. These symptoms are commonly referred as 'hay fever'. However, hay fever is a misnomer, since it is not necessarily caused by hay and a rise in body temperature is not one of its symptoms.

Hay fever is most commonly recognized as an allergic problem. Other allergic manifestations are skin eruptions from food ingestion for example hives from strawberries and eczema from chocolate, and anaphylactic shock from nuts are also commonly recognized. Allergy is sensitivity, which a susceptible individual develops to normally harmless substances. The tendency to become suceptible to various things in our environment is usually inherited. This is also referred as atopy. Hence, atopic individuals are more prone to develop an allergy. The same thing is true about children of both or one of the parents having an allergy.

A considerable percentage of the world's population reacts in various ways to inhaled particles. The severity of this reaction varies according to the degree of exposure over time and concentration of the particles present when inhaled. In the case of pollen and spores, being living entities the time of their release into the air and for a short time beyond, often cause reactions themselves when inhaled.

During the course of evolution, mankind has developed resistance to many bioallergens, but new ones often arise by mutation and also from human investigations into the nature and use of chemicals, such as pharmaceutical drugs, cosmetics, detergents pesticides and herbicides. Many of these chemicals may adhere to the outer wall of pollen and spores when airborne, as a result of the use of aerosol sprays. It is when many such aerosol sprays are inhaled together with allergenic pollens and spores, that atopic persons develop allergic reactions. Many airborne pollen grains on contact with fine mist droplets may absorb water, swell and burst, releasing large numbers of allergenic particles, which add to the allergen load of the air.

Air contains a bewildering range and variety of spores discharged from algae, mosses, liverworts, ferns and fungi in addition to pollen from dehiscing pollen sacs of gymnosperms and angiosperms. Many cryptogamic spores have been shown to be allergenic – spores of many fungi including *Penicillium, Aspergillus* and *Alternaria*. Spores of the fern *Pteridium aquilinum* (Bracken) are carcinogenic when inhaled in high concentration at the point of release from fertile fronds (Lacey and McCartney 1994; Caulton *et al.*, 1995). The penetrability of allergenic particles in the pulmonary system depends on particle size. The non-living micro particles designated as 2.10 or 2.50 ppm (parts per million), together with the very small spores of moulds such as *Penicillium* and *Aspergillus* can penetrate the alveoli of the lungs thereby triggering asthmatic attacks in susceptible people. Larger allergenic spores such as the fungus *Alternaria*, like smaller pollen grains, may reach the bronchii and upper regions of the trachea, initiating hay fever attacks according to the season, with varying degrees of response.

Many tree and herb pollen are allergenic. The most important of these are *Betula* (birch) and *Prosopis* trees of the temperate zone and herbs such as Poaceae (grasses) as a group, *Artemisia* sp., *Ambrosia* (Ragweed) and *Parthenium*. Herbaceous plants such as *Urtica* sp. (Nettle) and *Parietaria* are also allergenic, but to a lesser degree than the taxa mentioned above. The degree of allergenicity, or potential allergenicity, of any pollen depends to a greater or lesser extent on the abundance of a particular plant's pollen concentration as pollen grains per cubic metre of air (m^3 air) and the degree of exposure to and potential susceptibility of the person involved.

Many explanations are offered to account for the noticeable increase in the incidence of both seasonal rhinitis and asthma, especially in the ever increasing urban populations of world, such allergy susceptibility is now extending at both ends of the age range. Whatever the initial cause of these observed and recorded increases, it is obvious that the natural inborn and developed immune system is being severely weakened.

Allergic symptoms are not only seen in human beings but atopic conditions have been recorded for horses (Dixon *et al.*, 1992) and dogs (Fraser *et al.*, 2001) resulting from the inhalation of allergenic pollen. Evidence of weakening of the natural immune system in seals, thought to be due to the intake of toxic substances discharged into offshore waters from industrial and domestic sources, resulted in the deaths of a considerable number of these animals, causing great concern to conservationists. Whilst the airborne spora is not a likely culprit here, it may possibly be a part explanation of the weakening of immune systems in land fauna such as humans and other mammals, where allergenic spores and pollen with contaminated surfaces may play a major role in this phenomenon.

Aeroallergens can vary in intensity among populations, as with individuals, due to change in and fluctuations of local and regional weather patterns and geographical location, latitude and elevation. In addition, seasonal patterns, more clearly marked in higher latitudes, play an important part in the periodicity of asthma and 'hay fever' – hence the medical term 'seasonal rhinitis'.

Atopic reactions such as asthma and 'hay fever' are important. With asthma, the reaction may be so serious as to be life threatening. In the case of seasonal rhinitis, the economic effects are more noticeable, due to its widespread incidence in populations. Absenteeism for short periods and lower productivity are due to victims being so unwell as to be unable to concentrate and work efficiently. Hereditary factors can play a part in asthmatic conditions whereas in seasonal rhinitis, the condition can arise or decline throughout life. The interaction of heritable and environmental factors in allergenic responses are complex and not fully understood. The individual response can vary considerably.

Aeroallergens and allergenicity due to airborne allergens are a contemporary phenomenon of considerable importance to the health and economy of human society. Allergens are simply defined as substances that are capable of producing allergy. When an allergen enters the body, substances called antibodies are produced. The interaction of allergen and the antibody produce an irritation in the affected tissue. The swelling or inflammation of the nasal lining during the hay fever season is an example of this interaction.

Allergens usually enter the body by four routes, such as ingestion, inhalation, injection and by external contact with the skin. In view of the above, it is apparent that foods, dust, pollen, fungal spores, fumes and almost anything in the environment may cause allergic diseases.

The dangerous respiratory disorder, asthma, can be caused by almost anything that irritates the bronchial tissues, including tobacco smoke and can even be triggered by emotional stress. But about 40% of the time, asthma is related to allergy.

Immediate hypersensitivity in allergy patients is caused by the pollen proteins and other compounds perhaps recognizing molecules peculiar to the species. These are stored in the sexine and intine which get released through the apertures when the pollen contacts mucosal surfaces, giving rise in sensitized atopic individuals to allergic rhinitis, allergic sinusitis and bronchial asthma.

In temperate regions allergenic plants are traditionally grouped as trees, grasses and weeds which follow the generalized flowering sequence from spring to autumn. Trees and shrubs flower from late spring to midsummer, and weeds from mid-summer to autumn. However, there are some exceptions: *Ricinus, Albizzia, Castanea* and *Tilia* often flower in the summer months. Lewis has given an excellent summary of pollen allergenicity which includes pollen-producing plants of allergic significance in continental United States of America (Lewis 1984).

Hypersensitivity pneumonitis, an inflammation of the lungs producing breathlessness, wheezing, chills and fever is another allergic respiratory disease. Contaminated household humidifiers, dust and bird droppings may induce allergic manifestations. Many allergies strike the skin. Atopic dermatitis or eczema is an itchy rash that often appears on the crook of the arms or legs, but sometimes covers the entire body. It usually begins with redness and swelling and can progress to fluid – oozing blisters. Eczema, most common in children under the age of five, may often come from food. Hives or urticaria, are itchy welts caused by a variety of allergens, including food and sudden changes in temperature. Touching an animal, plant or chemical allergen can produce contact dermatitis – poison ivy and the Umbellifer, *Heracleum mantegazzianum* (giant Hogweed) being notable examples. In addition, the other plant and animal materials such as trichomes, fibres, insect debris, cobwebs and dust particles can also cause allergic manifestations due to their protienaceous nature (Stanley and Linskens 1974; Knox 1979).

Anaphylaxis

The direful allergic reaction is anaphylatic shock. According to Anon (1994), anaplylaxis in the allergy patient is probably one of the greatest fears of the beginner as well as the seasoned allergist.

Anaphylaxis is a systemic hypersensitivity reaction that is immunologically mediated and induced by exposure to a specific antigen in a previously sensitized individual. The word anaphylaxis, which stems from the Greek term meaning 'against protection,' was originated by Porter and Rictet in 1902 after they failed experimentally to produce prophylaxis to sea anemone toxin in dogs.

There appear to be certain features that characterize the patients most likely to suffer an anaphylactic reaction. These include: a history of atopy, a family history of anaphylaxis, multiple exposures to the same or similar antigen, and the route of administration of the antigen.

According to some allergists penicillin is the number one cause of death from anaphylaxis, with 300 cases reported per year, followed by *Hymenoptera* stings, which account for approximately 20 to 50 deaths per year. Body tissues suddenly swell, followed often by abdominal cramps and vomiting. Lung tissues go into spasm, the lining of the throat swells and the victim gasps for air. Finally the blood pressure plummets and death may follow unless prompt action is taken. The usual emergency first step is injection of epinephrine (adrenalin), which constricts blood vessels and raises the blood pressure. People unduly sensitive to shellfish or nuts, as well as insect stings or penicillin are especially prone to anaphylactic reactions.

Immediate medical management is directed toward reducing the exposure to the antigen, reducing the effects of the chemical mediators, and instituting the ABC's of life support-airway, breathing, and circulation, intramuscular or subcutaneous injection of epinephrine 0.3 to 0.5 cc of a 1:1000 dilution. Prefilled spring loaded epinephrine 'pens' are available and are prescribed for all patients who have had some type of anaphylactic reaction to stings. Epinephrine has effects on both alpha and beta receptors. The alpha properties increase blood pressure, decrease peripheral vasodilation, and decrease the cutaneous manifestations of the reaction. The beta effects include bronchodilation, positive inotropic and chronotropic changes on the heart, and increases in intracellular levels of cyclic adenosine monophosphate (AMP), which prevents further release of histamine and leukotrienes.

One of the rare instances where the patient experienced anaphylactic shock was in Lucknow, India. When the doctor administered a test dose of *Holoptelia integrifolia* antigen intradermally the patient suffered due to anaphylactic shock and had to be revived by giving emergency first aid treatment.

In response to viruses or other invaders, the immune system mounts a reaction with two kinds of white blood cells. T Lymphocytes attack the foreign substances directly. B Lymphocytes produce antibodies, which are protein molecules that attach themselves to the surface of the offender and prepare it for destruction. Once formed, antibodies or immunoglobulins continue to circulate in the blood. Two kinds turn out to be important in allergy immunoglobulin G (IgG) and immunoglobulin E (IgE).

In 1930s Drs. Mary Loveless and Robert Cooke of New York's Roosevelt Hospital, showed that IgG appeared to protect normal people from an allergic attack, just as if it were fighting an ordinary infection. When healthy volunteers were injected with ragweed (*Ambrosia*) extract, their IgG levels rose in response and they did not develop allergic symptoms. What the doctors could not explain however, was the fact that allergic people already had in their blood, elevated levels of IgG called the 'blocking antibody, yet for some reason did not benefit from it.

When a normal person was inoculated with serum from an allergy sufferer and then injected with an allergen, he/she exhibited inflammatory skin reaction. If he/she received the allergy injection alone no such response occurred. Thus, the difference could be accounted for only by something in the blood of the allergic patient.

In 1966, after performing countless skin tests, on themselves, the husband and wife doctor team of Kimishige and Teruko Ishizaka at the Children's Asthma Research Institute and Hospital in Denver, United States of America, isolated the guilty globulin: which they called IgE. It turns out that IgE skin-sensitizing or reaginic antibodies are involved in about 95% of allergic problems. These cells, part of the immune system, lie on all surfaces of the body where foreign particles might enter the skin, the lining of the respiratory and intestinal tracts and around the tiny veins or veinules. In response to infection, for example, mast cells release histamine and other chemical substances that enable disease-fighting antibodies and white cells to leave the blood vessels and enter the endangered tissues, but in allergy, IgE makes the mast cells run amok.

Allergy victims have 10 times as much IgE in their blood as compared to normal people. Usually the Y-shaped molecules embed themselves stem down on the mast cells. When allergens arrive they attach themselves to the arms of adjacent IgE molecules. This pulls them together and prompts the mast cells to behave inappropriately since there is no infection to fight and release their contents including histamine **(Figs. 15.1a, b, c, d)**. At the same time, the cell membranes of the mast cells and nearby tissue cells exude arachidonic acid. This substance through the action of enzymes produces additional potent chemicals, including prostaglandins and leukotrienes. It was once thought that the histamine was the principal chemical involved in allergic reactions, causing inflammation and contraction of the smooth muscles of the respiratory tracts, but leukotrines, researchers now know, can be far more powerful. Leukotrienes (SRS-A) induce significant bronchospasms (up to 1,000 times more potently than histamine), mucus production, and increased vascular permeability. Eosinophil chemotactic factor is responsible for the many eosinophils seen in allergic reactions. Bradykinins are also potent broncho constricting agents.

While histamine constricts the central bronchial passages, the leukotrines account for the narrowing and loss of elasticity of the smaller outlying passages that can seriously impair the passage of oxygen into the blood. IgE protects whole ravages of parasitic diseases. "Perhaps best IgE producers would best survive parasites". But now this system has turned against pollen, moulds and other substances.

Certainly heredity helps determine who suffers from allergy. Generally it can be said that one is not born allergic but with a tendency to make IgE antibodies. The first things a person makes IgE antibodies to, are the first

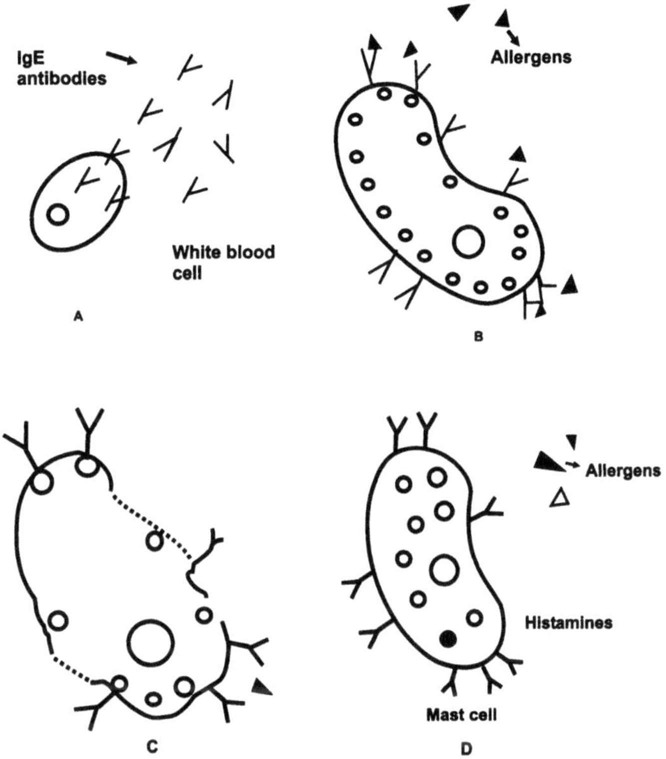

Fig. 15.1 A. White blood cell mass produces IgE antibodies after exposure to allergens; B. Allergens bridge the area between IgE molecules pulling them together; C. The bridged IgE molecules stimulate the histamines and other chemicals; D. Up to 5,00,000 IgE molecules gather on individual mast cell.

things he/she is exposed to. That is probably why babies are prone to milk and egg allergies. But it takes time and exposure to allergens to produce symptoms. Immigrants from Europe, where ragweed (*Ambrosia*) is relatively scarce, may not develop hay fever for years after moving to the U.S.A.

Allergy Tests

There are several ways to test for allergies such as scratch test, prick test, inhalation (brochoprovocation) and ingestion tests, intradermal tests and RAST (radioallergosorbent test) done on a patient's blood sample.

HISTORY OF SKIN TESTING WITH POLLEN ANTIGENS

Hippocrates was perhaps the first one to describe allergic reactions in individuals. However, the first modern description of an allergic patient was published as early as 1819 by Jonathan Bostock, a physician in

London, who had clearly described a clinical symptom known today as 'asthma' and 'allergic rhinitis'. He used the term 'summer catarrh' for this clinical condition.

Wyman at Harward University, U.S.A. in 1872, clearly identified ragweed pollen as the causative agent in hay fever. Later it was Blackley, a physician from Manchester, England in 1873 who conducted a number of experiments in which he performed the first skin test with grass pollen extract to the excoriated skin of the patient's forearm and demonstrated the development of allergic reaction in the form of a weal in the allergic individuals.

Later, in 1911, Leonard Noon, an infectious disease specialist, injected allergic patients with pollen extract in an attempt to cause the patient to develop 'pollen antitoxin' to neutralize the 'toxin' that he thought was produced by the body secondary to the original pollen exposure. He developed a protocol in which various dilutions of an allergic extract were used to drop in the conjunctiva of an allergic individual. Production of erythema and inflammation of the conjuctiva determined the strength of pollen extract. French Hansel, an orolaryngologist in 1930, began experimenting with skin testing using various dilutions rather than a single test of allergenic material.

PRECAUTIONS TO BE TAKEN PRIOR TO ALLERGY TEST

In order to evaluate the allergy test accurately, certain medications such as antihistamines and tranquilizers should not be taken 48 hours before the allergy test is performed. Testing should not be done if a patient is running a fever or has an attack of asthma or hay fever. Antihistamines suppress the weal and flare response. Decongestants, cromolyn, corticosteroids, and bronchodilators do not affect skin test results, and need not be discontinued before testing.

PROCEDURE FOR SKIN TESTING OF ALLERGENS

Antigens prepared from allergens are tested on the skin of allergy patients. The following step-by-step procedure should be adopted for an accurate response.

1. Clean the injection site with an alcohol swab, moving in a circular motion. Start from the centre and move outward. Allow the alcohol to dry.
2. Carefully pick up the syringe. Pinch a two-inch fold of skin, and with one quick motion, inject the needle into the skin. The normal injection angle is between 45 degrees and 90 degrees, or straight in.
3. Release the pinched skin. While holding the barrel with one hand, pull back slightly on the plunger. Look for blood in the barrel. If there

is blood, do not inject the allergy medication. The needle is in a vessel. Simply pull the needle out of the injection site and start again with a new syringe and a new injection site.

4. If there is no blood, proceed with the injection by pushing the plunger down and inject the allergy medication.
5. When finished, pull the needle from the skin and gently hold an alcohol swab on the injection site. Do not massage the area.
6. Dispose of syringes properly. Make the syringe unusable by carefully breaking off the needle. Drop the unusable syringe into an empty resealable container such as a coffee can bleach bottle, or liquid detergent bottle. These steps have been shown in **Figs. 15.2A to E**, modified from information supplied courtesy Becton Dickinson and Company.

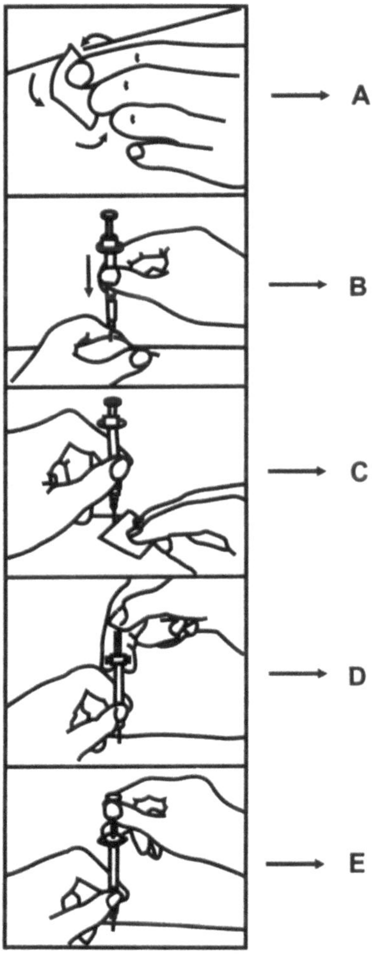

Fig. 15.2 Steps A-E involved in the procedure for skin testing of allergens.

Intradermal Skin Testing

Usually the upper half of the volar surface of the forearm is selected for intradermal skin testing. The surface selected is cleansed with spirit (70 % ethyl alcohol). The allergens with 1:500 dilution are used for testing. Negative control is maintained by using phosphate buffered saline diluent. 0.1 ml solution is taken in 1 ml glass tuberculine syringe fitted with 26-gauge needle. The selected skin is stretched taut and the syringe is placed at an angle of 45° to the arm, introducing the needle into the superficial layers of the skin. A small amount of allergen solution (approximately 0.02 ml) is gently injected that can raise a bleb of 1-3 mm in diameter.

The skin reactions are read after 15-20 minutes. The size of the weal is measured with a reaction gauge or by using the Shivpuri technique (1964). The pseudopodes, erythema are also noted. If a hypersensitivity (allergic) reaction is present, a positive weal will enlarge at least an additional 2 mm beyond the size of the 'negative' (5 mm) weal within 10 minutes.

The size of the skin reaction is influenced by a number of factors, including the volume and potency of allergen injected, the degree of sensitization of cutaneous mast cells (related to levels of circulating IgE specific for that allergen), and the reactivity of the skin to histamine and other mediators released from the mast cells.

Allergen exposure has the same effect on skin reactivity as the use of stronger testing antigen; that is, the production of a more intense skin reaction. Therefore, patients will usually be found to be more sensitive to skin testing carried out during the season when the allergens are at their peak, and endpoints obtained during coseasonal testing will often shift when that antigen goes out of season. These data are useful for preparation of prescription for antigen extracts for immunotherapy.

SCRATCH TESTING

Scratch testing is the original technique first described by Charles Blackley in 1873. This technique consists of making 2-mm superficial lacerations in the patient's skin followed by the application of a drop of concentrated antigen.

PRICK TESTS

The technique of the prick test was first described by Lewis and Grant in 1926. As currently performed, a single drop of antigen concentrate is placed on the skin. A sterile 26-gauge needle is passed through this drop and inserted into the skin in a superficial manner so that no bleeding is caused.

A variant of this test is the 'multitest', in which a sterile disposable applicator with eight puncture heads is used, which allows for the simultaneous testing of six antigens and a positive (histamine) and negative (glycerine) control. Grading of skin reactivity is done on a subjective 0 to 4+ basis.

Treatment of Allergy

Actually the treatment of allergy is a cooperative venture with both the patient and physician as participants. The goal of treatment is to produce an allergic balance so that the patient is freed of existing allergic illness and to avoid the development of new allergic problems. A basic part of the management of allergy is educating the patient to eliminate offending allergens whenever possible. Since certain allergens such as inhalants (air, along with airborne pollen and spores) cannot be eliminated, the reaction of the patient to these substances is controlled by desensitization. The process of desensitization is known as immunotherapy.

Desensitization is actually nothing but immunization. The desensitization is the technique of injecting small but increasing amounts of the antigens to control the allergic reaction so that the patient has little or no reaction when he is exposed to the offending substances.

The physician administers extracts of the allergen in gradually increasing doses. In time the patient acquires some protection against the allergen in his environment, much as he would after vaccination against an infection. During the treatment, IgE levels rise at first, and then drop. As they fall, levels of 'protectitive' IgG increase. Typically shots are given once or twice a week and are gradually reduced to once a month usually ending after three years. The treatment must be used judiciously as an accidental overdose can trigger an anaphylactic shock.

In about a third of patients, the shots relieve symptoms permanently, another third have recurrences in a year and the rest from 5 to 15 years after treatment. Sometimes, treating allergies can be as frustrating for the doctor as the suffering is for the patient (Seligmann *et al.*, 1982). The basic treatments that allergists offer include injection therapy and telling people to avoid the things they are allergic to. In addition to desensitization or immunotherapy, pharmocotherapy is also equally important.

Immunotherapy

A classical case history has been given by Lewis (1984), which is reproduced below, to prove the importance of immunotherapy to allergy patients by using pollen extracts.

Case History

A 23 year old opera student complained of severe rhinitis and opthalmitis in spring and fall. The patient's symptoms began in mid-March and lasted until the end of April. He was then symptom free until mid-August when his symptoms recurred. He continued to be symptomatic until the first frost. The patient's symptoms included profuse rhinorrhea, nasal occlusion, and intense pruritus. In addition, he had severe ocular pruitus and intense lachrymation. Due to his symptoms he was considering giving up his singing career.

The patient stated that he was much better indoors than outdoors, especially in air conditioning. He was also significantly better when he went to the West Coast of U.S.A. for a vacation in late August. The patient received some benefit from antihistamines but was loath to utilize them because of excessive somnolence.

Laboratory investigations revealed an IgE of 540 IU/ml but otherwise normal.

Skin testing using a multitest apparatus revealed a 4+ reaction to giant ragweed (*Ambrosia trifida*) and oak (*Quercus*) pollen. All other skin tests were negative. Immunotherapy was instituted with ragweed and oak pollen extracts and was carried through to 2500 PNU/injection of each extract. On this regimen he had a marked reduction in both spring and fall allergy symptoms. He was able to continue to pursue his singing career.

PREPARATION OF POLLEN ANTIGENS FOR INTRADERMAL TESTING EXTRACTS

Antigens for the diagnosis of respiratory allergy are prepared from the concentrated solutions of allergenic pollen extracts, which are usually aqueous or glycerinated solutions. Normal saline is used for preparation of aqueous pollen extracts, whereas 50% glycerine saline is used for glycerinated pollen extracts. In both cases, preparations are buffered to pH 8 and phenol is added as a preservative. Precaution is taken that antigenic extracts are prepared in a dust-free sterile place, preferably using dehumidifiers, air conditioners and laminar flow.

The following steps are involved in the procedure used for preparation of antigenic pollen extracts:

1) Properly selected mature unopened flower buds are collected in large quantities followed by drying them naturally or in an oven.
2) Anthers are separated with clean sterile forceps and crushed to extract pollen. This is followed by sieving through No 100 or 300 (180 μ) mesh sieves to get only pure pollen material.

3) Subjecting pollen to dehydration at 39°C for preventing microbial contamination.
4) Defatting is done by soaking the pollen in petroleum ether overnight. The process of defatting is repeated at least 3-4 times.
5) The defatted pollen are subjected to extraction using saline phosphate buffer of pH 8 for 8 to 24 or 72 hours.
6) The extracts are clarified by passing through Whatmann filtre paper No. 1.
7) Dialysis is done using dialyzing tubes for 24 hours to get rid of the irritating and colouring material.
8) The dialyzed pollen extracts are sterilized, sterility tested, standardized and are ready for use for testing in 1:500 w/v concentration.

For clear understanding, the following flow chart for the preparation of antigenic extracts from pollen can be pursued.

SIMPLIFIED FLOW CHART OF THE PREPARATION OF ALLERGENIC EXTRACTS (Courtesy Dr. Govind)

Dried, ground, 98% pure pollen sieved

↓

Defatted using peroxide free diethyl ether

↓

Stored in air tight bottles at +2 - +8 C

↓

Extraction
[1 gm of defatted material in 50 ml phosphate buffer saline (PBS)]

↓

Clarification
(Separation of allergenic extract from the rest using Whatmann filter paper No.1)

↓

Dialysis using dialysis tube

↓

Sterilization using millipore filtres

↓

Sterility testing of aerobic and anaerobic organisms

↓

Standardization
(w/v- clinical use Modified Lowry's method – Protein estimation)

↓

Aseptic filling into 30 ml / 100 ml sterile vials using swinex adapter millipore filtres

FUNGAL ANTIGEN PREPARATION

Fungal antigen preparation for testing purpose involves the following steps:

1) Mass culturing of isolated fungi
2) Harvesting and drying
3) Grinding
4) Defatting
5) Extraction
6) Clarification
7) Sterilization
8) Sterility test

1) Mass Culturing of Isolated Fungi

A few tube slants containing Czapek agar medium are inoculated with fungi isolated earlier and incubated at 30°C ± 1°C for 5-7 days till maximum sporulation occurs. The subculture spores are then suspended in sterile saline solution. The spore suspension is then used for inoculation of Ehrlenmyer (500 ml) flask containing 100 ml of Czapek Dox broth or Saubauracid medium and then incubated at 30°C ± 1 for 10-30 days till maximum spore production takes place.

2) Harvesting and Drying

When sporulation is maximum the fungal pellicles are harvested and kept in 95% alcohol at 4°C for 24 hours. The fungal pellicles are very gently squeezed between two filter papers and left to dry at room temperature.

3) Grinding

To obtain active antigen, it is essential to extract appropriately the fungal extract for which the dried fungi are pulverized with pestle and mortar and then filtered through No 100 mesh to obtain a fine powder.

4) Defatting

Various defatting solvents are available such as ethyl ether, petroleum ether, hexane, chloroform, and carbon tetrachloride. However, ethyl ether is most commonly used. The solvent is used in weight per volume ratio (W/V) of 1:500 that is 1 gm of fungal powder in 500 ml of solvent.

The fungal powder is sealed in an appropriate solvent and kept for sometime with periodical shaking and stirring. The oily supernatant is discarded. To ensure complete defatting, fresh changes of solvent 2-3 times are added till colourless supernatant is obtained. Finally it is filtered and the particulate matter is kept and desiccated and dried over anhydrous Calcium Chloride ($CaCl_2$).

5) Extraction

Extraction of fungal powder is generally done in Coca's solution as suggested by Sheldon *et al.* (1967).

Coca's solution consists of:

Sodium chloride	5.0 gm
Sodium bicarbonate	2.5 gm
Phenol	4.0 gm
Double distilled water to make it to	1,000 ml

Agarwal *et al.* (1968) used both buffer saline and Coca's solution for extraction of fungi. However, Sheldon suggested use of buffered saline for pollen extract.

The defatted fungal powder is added to Coca's solution and kept for 72 hours with periodical shaking 8-9 times for every half an hour. Eventually they are dried in desiccator over anhydrous calcium chloride.

6) Clarification

Before sterilization with Seitz filtre it is essential to remove particulate matter from the Coca's solution containing biologically active antigens. To separate, it is centrifuged at 4,000 rpm in refrigerated centrifuge, the supernatent is kept and the particulate matter is discarded.

7) Sterilization

As the antigens are thermobiable they are sterilized by bacterial filtre. The Seitz filtre and filtre flask are sterilized by autoclaving. The Coca's solution containing biological antigen are filtred through Seitz filtre apparatus. These are eventually stored in sterilized vaccine vials and capped with sterilized cap and stored at 4°C.

8) Sterility Test

Before using the antigen on patients its sterility both for aerobic and anaerobic microbes should be checked.

MODERN RAPID METHOD OF FUNGAL ANTIGEN PREPARATION

This was suggested by Agarwal *et al.* (1973) and Vyas (1969). These methods involve defatting of fungal pellicles in different solvent solutions usually ethyl ether and ethanol (3:1) and chloroform methanol (2:1) followed by 24 hours extraction with continuous shaking or 72 hours extraction with 8-9 shakes of half an hour each. It was found that the defatting with chloroform, methanol (2:1) was very effective and 24 hours extraction was as good as 72 hours extraction. Thus this method was found to be less time consuming.

ANTIGENS

Antigen may be defined as any substance which when introduced in an individual stimulates the production of antibodies. Any foreign substance when introduced in the individual stimulates antibody production and antigen-antibody reaction (serological reaction) which occurs in an observable manner is called antigen.

Antigenicity may be defined as the property of an antigen to lock antibody. The two antigenicity properties are:

1) To stimulate the antibody production.
2) To effectively cause the serological reaction.

Antigens are basically proteins, polysaccharides, lipids and nucleic acids. They may be simple or complex according to their building blocks. According to their nature there are three classes of antigens.

1) Synthetic and natural antigens.
2) Whole and purified antigens.
3) Recombinant and idiotype.

Synthetic and Natural Antigens

Synthetic antigens are products of laboratories while natural antigens are products of the cell. Synthetic antigens are prepared by DNA recombination. The synthetic antigens are simple, whilst natural antigens are complex. The synthetic antigens consist of one building block only, whereas natural antigens always comprises different types of building blocks.

Whole and Purified Antigens

The whole of the cell acts as an antigen, for example a bacterial cell, which is known as whole antigen. In purified antigen the biochemicals of whole antigen act as antigen.

Recombinant and Idiotype

Recombinants are products of the laboratorary and derived by recombinant DNA technique. Idiotype antigens are antibodies that themselves act as an antigen. The recombinant antigens are obtained by selection of gene recombination DNA while the idiotypes are derived by selecting an antibody in selected cells.

Determination of antigenicity, depends on the following factors, some of which are discussed below:

1) Cell size
2) Chemical nature
3) Susceptibility to tissue enzyme
4) Foreigners
5) Antigen specificity
6) Species specificity
7) Isospecificity
8) Autospecificity
9) Organspecificity
10) Heterogenic specificity

1) *Cell size*

Antigenicity is found to be related to cell size. The antigens with high mol. wt. e.g. Haemocyanin (mol. wt. 6.75 million) are found to be active antigens. But those molecules with less molecular wt. (> 10,000) are found to be non-antigenic or feebly antigenic.

2) *Chemical nature*

The antigenicity depends on the chemical nature as cell proteins and polysaccharides are found to be very antigenic. Whole nucleic acids and lipids are found to be less antigenic.

It is obvious that aerobiology has a significant role to play in allergy as convincingly reported by Agashe *et al.*, (2003). Earlier Agashe (1994) had used a cartoon on the cover page of 'Recent Trends in Aerobiology, Allergy and Immunology', which was the compilation of keynote addresses by eminent aerobiologists from all over the world during 5th International Conference on Aerobiology held in Bangalore, India, in 1994. The cartoon summarizes the role and objectives of aerobiology for allergy. The same has been reproduced here **(Fig. 15.3)**.

It summarizes that the ultimate goal of reducing the suffering of allergy patients can be achieved by the symbiotic association of aerobiologists and allergologists.

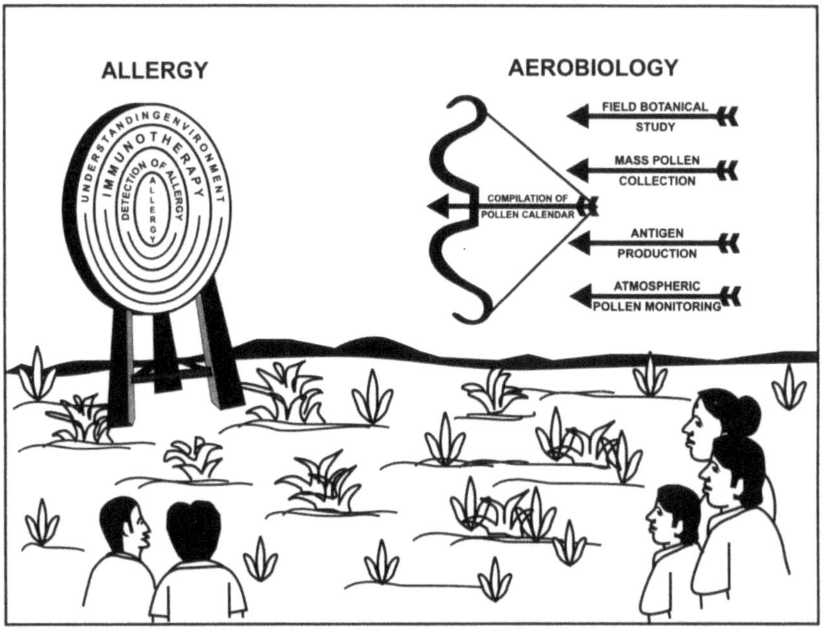

Fig. 15.3 Cartoon summarizing the role of aerobiology in allergy.

ALLERGY IN INDIA

The morbidity surveys conducted at a few places in India have revealed that more than 1% of country's population suffers from bronchial asthma and 3-4% of the population is estimated to suffer from allergic rhinitis. A rough estimate reveals that more than 10% of the population suffers from major allergic diseases.

India has a wide range of vegetation and ecozones ranging from desert to alpine tundra. This variation in vegetation contributes to enormous differences in the quality and quantity of airborne pollen from diverse ecozones of India (Singh 1987).

Wind-pollinated (anemophilous) plants are the chief source of the majority of inhalent pollen allergens. The flowers of these plants possess a set of adaptive characteristics comprising reduction in size and number of floral parts, lack of nectar, small smooth dry walled , simple buoyant pollen produced in large quantities. In contrast to the above, the plants pollinating through vector mediation (entomophilous) are less significant in respiratory allergy on account of their limited occurrence in the air to sensitize the patients. However, many entomophilous taxa such as *Carica papaya, Kigelia, Cocos,* and *Phoneix* produce abundant pollen in the atmosphere.

Pollen surveys carried out in different parts of India indicate the occurrence of plant families such as Asterceae (Compositae) Cannabinaceae,

Casuarinaceae, Chenopodiaceae, Euphorbiaceae, Fabaceae, Moraceae, Palmae, Pinaceae and Poaceae, all of which contain allergenic species.

Some of the dominant airborne pollen types reported from various parts of India are: *Ailanthus, Artemisia,* Amaranth-Chenopod, *Azadirachta, Brassica, Cannabis, Dodonaea, Eucalyptus, Holoptelea, Morus, Prosopis, Putranjiva, Ricinus, Xanthium* and pollen from members of Poaceae such as *Cenchrus ciliaris, Cynodon dactylon, Poa annua* and *Sorghum vulgare.* Airborne pollen types confined to the coastal regions of India are: *Borassus, Casuarina, Cocos, Peltophorum* and *Phoenix.* Airborne pollen confined to Himalayan and sub-Himalayan regions are: *Alnus, Aesculus, Betula, Cedrus, Corylus, Quercus, Pinus* and *Pyrus*

Predominant Tree Pollen Include

Ailanthus excelsa, Anogeissus pendula, Casuarina equistifolia, Gauzuma tomentosa, Holoptelea integrifolia, Prosopis juliflora and *Salvadora persica*

POLLEN ALLERGY

Airborne pollen of *Phoenix sylvestris* of the Palmae family commonly occurs in and around areas of Kolkata, the North East and the South eastern regions of India. The plants grow wild and are also cultivated on account of their fruit-yielding sugar and alcoholic drinks. Janaki Bai and Subba Reddi (1982) had reported its occurence in and around Visakhapatnam in Andhra Pradesh, South eastern India. Recently Gupta-Bhattacharya and Chakraborty (2006) have exploited the allergenic significance of airborne pollen of *Phoenix sylvestris.* Their pollen extracts were found to be effective in seasonal respiratory allergic subjects. This conclusion was brought out by them after subjecting 18 patients for immunotherapy at the Institute of Child Health in Kolkata. They found noticeable increase in FEV1 and specific IgG1 and IgG4 in these patients.

Predominant Weed Pollen Include

Amaranthus spinosus, Artemisia scoparia, Brassica campestris, Cannabis sativa, Chenopodium album, Mimosa pudica, Parthenium hysterophorus, Ricinus communis, Xanthium strumarium.

While working on the allergenicity of various plants belonging to the family Mimosacae, Panicker (2002) had estimated protein contents in pollen grains of some plants of this family. He had reported that the highest protein contents of 201.3 mg/gms of dry pollen of *Acacia auriculiformis* followed by *Acacia mangium* (196.4) and *Prosopis juliflora* (186.6). *Mimosa pudica* pollen grains had the least amount of proteins (68.2).

ALLERGENICITY OF POLLEN OF MIMOSACEAE

The literature abounds with reports about the allerginicity of Mimosaceae, members such as *Acacia*, *Albizia*, *Prosopis* and *Mimosa*. It is also known that proteinaceous material in the exine and intine especially at the apertural region of pollen grains acts as allergens in human beings. Out of 48 patients who underwent skin test for pollen antigens of various members of the family Mimosaceae, the pollen of *Prosopis juliflora* had proved the highest incidence of allergenicity with 66.66% of the patients showing two positive reactions. As early as 1964 Wilkinson had exhibited that *Prosopis* pollen is the main cause of the allergy in Kuwait. Latif *et al.* (1989) had reported a total 70% positive reaction to *Prosopis* pollen. The allergic nature of *Prosopis* pollen was also ascertained in India by Agnihothri and Singh (1971). 8.33% of the patients had shown two positive reactions to the pollen antigens of *Acacia auriculiformis* whereas *Mimosa invisa* var. invisa showed a total allerginicity of 58.33% with 4.17% showing two positive reactions. The skin test positivity reactions showed a higher incidence in the males (40.41%) as against (35.41%) in the females.

Present Trends in Allergy and Immunotherapy

An unusual site for pollen survey

Hugg *et al.* (2007) conducted pollen survey in an unusual site, i.e., indoor of private cars from July 14 to August 17, 2003 by using rotorod type of sampler and sticky tape method for inside car surface sampling. The aeropalynological survey was carried out both in the moving car on highway No. 6 in southeast Finland and also in stationary (parked) car. For comparison, ambient airborne pollen outside car were determined by using Burkard sampler.

Their main objective in undertaking this investigation was to assess the risk of exposure to pollen inside the car throughout the Grass pollen and *Artemisia* pollen season. After analyzing the data generated during this pollen survey they concluded that the concentration of Grass and *Artemisia* pollen inside the car during the flowering period were low, hence likely to cause reactions only in the most sensitive car passengers. In contrast, the concentration of *Betula* (birch) pollen even after their main flowering period, was on a level high enough to cause allergic reactions in individuals inside the car.

Pollen forecast models

Aeropalynology is intended as a useful tool to predict the beginning of the pollen season and thus of flowering of plants producing allergenic pollen and those with agricultural importance. In recent years many aerobiological studies have been carried out concerning the pollen seasons with an emphasis on allergenic pollen. This aim leads to two different lines of work.

1. The effect of meteorological factors at the beginning of pollination in anemophilous plants (*Phyllirea*, Michel *et al.*, 1979; *Ulmus campestris*, Richard 1985; *Alnus* and *Populus*, Frenguelli *et al.*, 1991).
2. Possible time relationships between pollen season in different taxa – known as bioindicative pollination. That is the pollination of one plant predicts the beginning of pollination in others (Driessen *et al.*, 1989; Bricchi *et al.*, 1995).

Davies and Smith first reported forecasting the start and severity of the hay fever season. Spieksma in the Netherlands conducted radio broadcasting of the expected influence of the weather on the subjective complaints of hay fever sufferers from 1980. By using several years' data collected from Stockholm, Sweden, aerobiologists have evaluated some models, to predict the starting and intensity of pollen seasons. Driessen *et al.* (1989) studied prediction of the start of the grass pollen season for the western part of Netherlands.

Frenguelli and co-workers from Italy have done lot of work in the field of pollen season forecasting. Frenguelli *et al.* (1989a) did a predictive study on the beginning of the pollen season for Gramineae and *Olea europaea*. Frenguelli *et al.* (1989b) conducted a predictive study on the beginning of pollen season and on daily variations of pollen grains. Frenguelli *et al.* (1989 c) conducted a study on pollen forecast for some taxa in central Italy.

Freguelli *et al.* (1991c) studied the influence of air temperature on the starting dates of pollen season of *Alnus* and *Populus*. Spieksma (1988) in the Netherlands conducted retrospective and predictive system of information on pollen concentrations in relation to hay fever. Spieksma *et al.* (1989) studied biometeorological and pollen flight forecast. Buch (1986) described pollen and mould spore counting methodologies for use in pollen forecasting in Denmark.

Bricchi *et al.* (1995) studied time linkages between pollination on a set of different taxa over an 11-year period in Perugia, central Italy. The model is based on the fact that several species show highly significant correlation in the dates of their onset and consequently the dates of the start of pollination of those species linked to other ones can be easily predicted. The identification of various flowering phases allows the users to forecast the first appearance of pollen. This phenological model is a purely statistical model, which does not involve meteorological parameters but utilizes instead the correlation between the occurrence of flowering phenophase in various species.

Gonzalez-Minero *et al.* (1996) worked on the prediction of the beginning of the olive pollen season in southwest Spain. A positive relationship has been found between mean temperature of months before the pollen season (February and March) and the date when season starts (April). They found that rainfall before pollination does not affect the starting date of the full pollen season but can affect total pollen production, particularly in years with prolonged drought. Dahl *et al.* (1996) in Sweden worked on the problem of predicting the intensity of the birch pollen season and pollen dispersal, which may be used in the management of birch pollinosis, and in the planning of clinical trials.

The European Journal of Aerobiology brought out a special issue in 1992 devoted entirely to the problems and prospects of net working and forecasting in different parts of the world like North America, the United Kingdom, Spain, Sweden, France, Austria, Finland, Italy, the Netherlands. Among these countries, only in the Netherlands, was a daily hay fever forecast broadcasted on the National Radio and Tele text, particularly for grass, since 1977. No such work has been done in India. However, Agashe (2001) had indicated such a possibility in view of the active aerobiological work being carried out in several Indian cities.

CHAPTER **16**

Significance of Fungi as Aeroallergens

HISTORICAL ACCOUNT: HISTORY OF ALLERGIC DISEASES CAUSED BY FUNGAL AEROALLERGENS

The possibility of fungal allergy was first mentioned by Cadham (1924) and Van Leeuwen (1924). However, the first systematic studies of fungal allergy were those of Feinberg (1935). The first systematic aerobiological work including airborne fungi in India was carried out by Cunningham (1873) in Calcutta. Agnihotri (1980) studied the fungi in the bedroom of bronchial asthma patients and found *Aspergillus niger*, *Aspergillus flavus*, *Aspergillus versicolor*, *Aspergillus fumigatus* and *Helminthosporium* sp. as dominant fungi. The percentage of fungal spores in the air is approximately 10 times higher than that of pollen grains, however by volume the pollen dominate.

Ravindran (1991) investigated the aeromycoflora of Kerala and reported the dominance of *Aspergillus, Penicillium, Cladosporium* and *Alternaria*. Agashe *et al.* (1983) studied the aeromycoflora of Bangalore and found *Cladosporium* sp. and *Alternaria* sp. to be predominant. Agashe and Anuradha (1996) isolated the dominant fungal types *Cladosporium, Penicillium, Aspergillus, Trichoderma* and *Fusarium* from the hospital ward exposures of Petri dishes.

Hyde (1972) had suggested that a fungal spore to be qualified as a potential aeroallergen has to satisfy following features:

1. The spore must be produced in large quantities.
2. The spore must be sufficiently buoyant to become airborne.
3. The spore-producing species must be widely and abundantly distributed.
4. The spore should contain an excitant of hay fever or asthma.
5. Symptoms must occur when the spores are numerous in the air.

Regional spore calendars are based on seasonal variation and circadian periodicity. This has proved significantly useful for the clinicians in proper treatment of innumerable ailing patients of allergy. There is an absolute need of trained aerobiologists and clinicians working in close cooperation and also need of an extensive network of allergy clinics in the Indian subcontinent.

Over 200 spore types have been identified from air-spora studies in different geo-climatological regions in India. Out of these only 38 have been reported as allergenic to human beings (Tilak 1989). *Aspergillus fumigatus* spores, which are 3 μm in size cause Type I or Type III allergy. Of all the biological particulate matter suspended in the atmosphere, mould spores constitute the largest percentage in both indoor and outdoor environment (Spieksma 1995).

Fungi produce large amount of spores, which get into the air stream soon after liberation from the source plants. It is estimated that *Puccinia graminis* in a wheat field can produce at least 25 million uredospores per square metre. Whereas one fruiting body of the fungus *Ganoderma applanatum*, produces 5.46×10^{12} spores. More than 20,000 to 40,000 different species of fungal spores were recorded from the atmosphere in different parts of the world.

The indoor environmental sources of fungi have been investigated and identified by several researchers. Some of these are damp walls, dustbins, mattresses, leaking pipes, humidifiers, outdoor air, organic raw materials and also indoor potted plants. Fungi Imperfecti are the most important class of fungi in causing IgE mediated allergy. *Aspergillus* species are most abundant in indoor as well as outdoor environment and are known to be the prime aetiological agent in broncho-pulmonary aspergillosis.

INDOOR AND OUTDOOR FUNGAL AEROALLERGENS

Enclosed space is referred to any place where the air is locked within, such as within a building, room etc. A certain volume of locked air is called 'stale' air. The enclosed space has minimum conditions to exchange or entry of new or fresh air through ventilation. The fresh air which enters with structures such as windows, do not bodily remove completely the entire stale air locked inside but there is a mixing up of stale air and fresh air (50%: 50%).

Usually there is a decrease in air spora in the enclosed space. This is a gradual process and this process is called 'die away of spora'. It happens mainly because of the processes of:

1. Deposition
2. Dilution along with the ventilated air
3. Loss of inability due to condition such as temperature, humidity, etc.

The main sources of the air-spora in an enclosed space are:

1) Fresh air rushing through ventilators

2) Coughing, sneezing, talking, etc. generate droplet nuclei, which become airborne gradually.
3) Other mechanical disturbances, such as vacuum cleaning, sweeping, walking, making of beds.
4) Growth of moulds on the damp walls.
5) Other wooden structures or furniture may contain fungal growth in them.

Enclosed air space shows definite air movement due to thermal convection. When the ground level heats up, a convection current is created and this pushes the particles that are sedimented down towards the upper side. This is called vertical diffusion gradient. When the particles move to the upper side and when the ceiling cools, these particles again come back to the ground level. This way it creates a profile thermostatic convection. The convection current may also be created if there is any light source like a bulb or lamp inside the enclosed space. Up draught or down draught of air and along with it airborne particles are produced in warmer or colder climatic condition. These two types of draughts create a balanced current at the centre of the room.

Microbial movement is characteristic of indoor as compared to the outdoor environment. These convection currents inside an enclosed space may also be responsible for the increase in the number of particles. The intramural spaces contain nuclei produced due to sneezing, coughing speech, etc. It has been shown (by photographic evidence) that a single sneeze can liberate 20,000 droplets and maximum of 40,000 in a stifled sneeze. A cough produces few hundred droplets. Some of the small droplets of 2 mm and 20-40% of 5 mm particles are suspended in the air.

During sneezing and coughing droplets evaporate even before they reach the ground and are called droplet nuclei. These contain solid residues with virus or bacteria coated with semi-dried mucus, which helps to retain its activity and viability.

Fungi are an important component of the ecosystem. Out of the biological particulate matter suspended in the atmosphere, fungal spores constitute the largest percentage, both in the outdoor and indoor environments. Many are highly allergenic for sensitive individuals. Fungal infections of humans originate from an exogenous source in the environment and are acquired through inhalation, ingestion or traumatic implantation.

Mould spores are ubiquitous in nature. They are found in air, water and soil and are produced by fungi, which are mostly filamentous, 'spongy' or 'cottony' in nature. Fungi can grow anywhere and on most substrates. However, warmth and humidity favour their growth. Mould spores are very tiny, light and buoyant particles. They occur abundantly in the air, both indoor and outdoor. Invariably, mould spores outnumber pollen grains in nature.

Mould spores are considered equivalent to plant seeds as far as their function is concerned. They remain viable for varying periods of time. Hence, depending on their viability, they germinate easily and give rise to new fungal filaments. Unlike plants, fungi cannot produce their own food from sunlight and air, therefore they have to depend on other ready sources of food. Basically, they are saprophytic or parasitic in nature. On account of their buoyancy mould spores are easily disseminated through air and hence can come into contact with skin, mucosa of nose, eyes and oral cavity of human beings.

Moulds live on plants or animal matter, which they decompose for their nourishment. They are among the most widespread of living organisms. Many moulds reproduce by releasing spores into the air that settle on organic matter and grow into new clusters. Mould growth is normally encouraged by warmth and high humidity and therefore, growth is most prevalent during humid seasons of the year. Moulds in the indoor environment are produced especially in areas of high humidity. They grow abundantly in dark, damp places that are poorly ventilated and in areas where water accumulates. Moisture and warmth can accelerate the growth of dormant mildew spores on most surfaces.

Moulds can accumulate in home dust either from outdoor spores that enter a home or growing within the home environment. Air conditioners and vent openings are prime locations for trapping moulds at point of entry. Vent or central furnace filters and room air cleaneres are helpful in removing airborne spores.

AEROBIOLOGICAL TECHNIQUES: FUNGAL AIR SPORA

In any aerobiological investigation it is necessary to use pollen and mould spore samplers properly. It is also essential to make a thorough study of pollen types of plants and vegetation in the vicinity of a pollen trap. Recently Allitt (2006) of Cambridge, U.K., narrated a very interesting episode with regard to identification of a fungal spore *Apiorthynchostoma curreyi* and its unusual source in England. A pollen trap was initially installed on the roof of the Botany School in Cambridge in 1994 by Allitt. The pollen trap was supported by a wooden pallet which was new. However in due course it decayed and to her surprise, she found it was heavily infected with *A. curreyi* releasing spores to the atmosphere, and naturally, quite a few of them found their way to the trap. This particular fungus lives on decorticated pine wood and takes a long time, in this case, as much as 10 years to colonize and rot new wood. It is always advisable to examine any wooden equipment near the trap, especially wooden ladders often used to reach the trap.

RAIN WATER AIR SPORA

It has been estimated since 1850s that rainwater also contains many microorganisms but the study of rainwater was difficult for researchers, as they could not collect the rainwater directly without any contamination.

At the end of 19th century it was revealed that rainwater also contains many microbes. Experiments were conducted in which rainwater was collected in a platinum crucible that was kept at about 1.7 m from the ground on a wooden frame. Then raindrops were inoculated onto a medium. Bacterial growth in the Petri dish was noticed after incubation.

Later rainwater air spora was investigated and bacteria and pink yeasts and *Penicillium* were identified. The Scripps Institute of Oceanography in California conducted rainwater analysis and they identified about 110 colonies of bacteria/m^3 of air. Gregory *et al.* (1952) analyzed rainwater air spora on a special basis by using two funnels of 20 mm diameter. One of the funnels was covered by keeping an asbestos plate board 25 cm from the funnel.

The funnel which was kept closed (covered), helped in the dry deposition and the other which was not covered, helped in the water collected as it was kept at a height in order to avoid the splash droplets from the ground to enter to the rain water collected in the flask. Then they observed that the rainwater contained bacteria and moulds, such as *Cladosporium*, *Alternaria*, *Pneumococcus* and *Micrococcus*. Later Suzuki collected rainwater in a field affected with rice blast disease and the collected water, which contained eight times to the amount of microbes than those collected from the disease-free field.

MOST COMMON FUNGAL AEROALLERGENS

A comprehensive illustrated manual of the air-spora was recently published by Lacey and West (2006). The commonest fungal allergenic species belonging to the Zygomycetes are *Rhizopus nigricans* (black bread mould) and *Mucor racemosus*. Two significant allergenic ascomycetes are *Saccharomyces cerevisiae* (baker's yeast) yeast and *Chaetomium indicum*. In addition, skin sensitivity to conidia of powdery mildew, *Microsphaera alvi* has been reported.

Among the basidiomycetes, known allergenic fungi belonging to mushrooms and bracket fungi are *Pleurotus ostreatus*, *Ganoderma lucidum*, *Geaster saccatum* and *Agaricus*. Other allergenic species are *Ustilago* and *Urocystis* (Smut fungi – Ustilaginales).

By far the commonest highly abundant airborne fungi belong to Deuteromycetes (Fungi Imperfecti). Some of these genera are: *Phoma*, *Aspergillus*, *Penicillium*, *Botrytis*, *Monilia*, *Trichoderma*, *Alternaria*, *Cladosporium*, *Helminthosporium*, *Nigrospora*, *Curvularia*, *Fusarium*, *Epicoccum* and *Sporobolomyces*.

SEASONAL VARIATION

Unlike pollen, there is no sharp seasonal variation in the atmospheric fungal air-spora. However, some variations occur due to extreme weather factors. All the dominant fungi: *Aspergillus niger, Aspergillus flavus, Cladosporium* spp., *Penicillium oxalicum* and *Alternaria* spp. were found during all the four seasons: summer, South West monsoon, North East monsoon and winter. However, the total concentrations of dominant fungal spores in the indoor and outdoor environments were maximum during the monsoon seasons in South India. In a recent study in Kerala, Govind (2004) reported the following concentrations of fungal spores:

	Indoor – 10666 CFU/m^3
	Outdoor – 7700 CFU/m^3
In winter:	Inside 3154 CFU/m^3
	Outside 3370 CFU/m^3
In summer:	Inside 3154 CFU/m^3
	Outside 2718 CFU/m^3

Thus, throughout indoor concentration was more than outdoor. Govind (2004) has also described morphology of most commonly occurring airborne fungal spores. Some of them have been scanned under SEM, which have been illustrated in **Figs. 16.1, 16.2, 16.3 and 16.4**.

Agashe *et al.* (1983) have reported high fungal concentration at high humidity. Several researchers have reported high fungal concentration at low temperatures 20-30°C. With the increase in temperature 32-35°C, concentration of air-spora decreases. Some work has been done on the seasonal variations of fungal aeroallergens in specific occupational indoor environments.

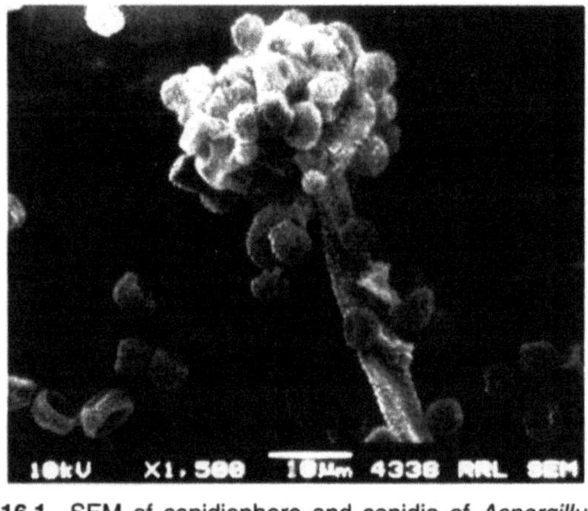

Fig. 16.1 SEM of conidiophore and conidia of *Aspergillus flavus.*

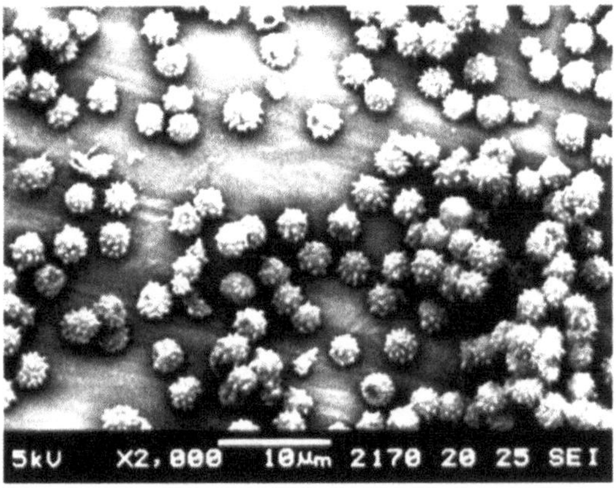

Fig. 16.2 Enlarged view of conidia in SEM of *Aspergillus fumigatus*.

Fig. 16.3 *Aspergillus niger*, SEM picture of conidiophore and conidia.

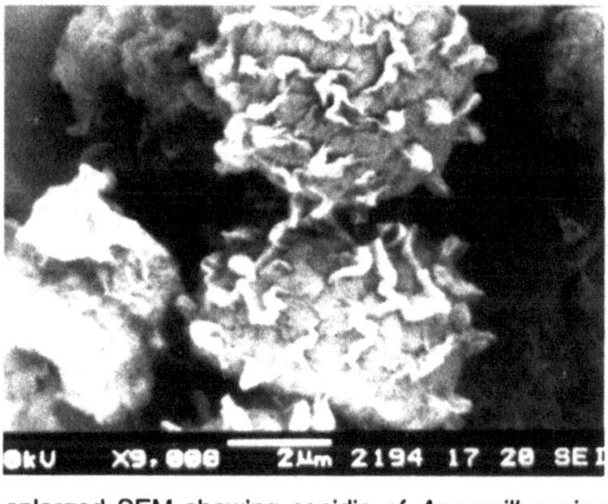

Fig. 16.4 Highly enlarged SEM showing conidia of *Aspergillus niger* (**Figs.16.1** to **16.4** from Govind, 2004).

In Bakery environment: *Aspergillus niger* dominated in a summer, South West monsoon and North East monsoon. *Cladosporium* spp. recorded the highest count in winter. Relative humidity (Rh) had a significant effect on the concentration of fungal spore deposition.

SCOPE OF FUNGAL AEROALLERGEN RESEARCH

Many asthmatic patients develop nocturnal attacks and are unable to sleep at night in their bedrooms. The identification and quantification of the fungal spores prevalent in occupational sites and in asthmatic patients' bedrooms along with their clinico-immunological studies are of immense help to the clinicians to identify the offending allergens and provide effective treatment.

The clinician can inform asthmatic patients if they have fungal allergy and the patient can be advised to take necessary measures against fungal exposure. Accurate diagnosis and treatment by an experienced and trained allergologist will help the patients to obtain relief from allergic symptoms, keep in good health and lead a normal life. It is imperative that both the diagnosis and treatment will yield better results if standardized extracts are used. Currently, the best method available is to detect antigen sensitivity by skin testing and provide allergen immunotherapy for immunomodulation in asthmatic patients.

Data on seasonal and annual prevalence of fungal species recorded in detailed aerobiological studies will be of immense importance to clinicians, and which will serve as a ready reckoner for them to select the appropriate fungal extract for the diagnosis and treatment of allergic rhinitis and asthma in patients.

This knowledge obtained from an atmospheric survey of fungal aeroallergens is useful in compiling spore calendars and prediction models of airborne fungal spore occurrence. It is necessary to carry out a survey of airborne pollen and fungal spores at three year intervals as the dynamics of environmental substances are dependent on several factors, including weather parameters which continuously change. The research will be helpful in compiling a fungal spores calendar and making it available to the clinicians for its use in allergy diagnosis. The fungal spores calendar indicates the occurrence of most predominant and common types of airborne fungal spores in the atmosphere.

Fungal Spore Survey

Aerobiologists have not lost any time in carrying out very useful indoor and outdoor airborne surveys of New Orleans, U.S.A., soon after the catastrophic Hurricane Katrina, on August 29, 2005. Following Hurricane

Katrina, several New Orleans homes remained flooded for weeks. Thus, the indoor atmosphere was ideal for microbial growth, responsible for significant risk of respiratory exposures for residents upon re-entry to their dwellings. Bioaerosol concentrations in flood-damaged homes, were extremely high and health surveys indicated an increase in allergy and upper airways infections (Muilenberg *et al.* 2006).

Dumon *et al.* (2006) carried out intensive surveys of moulds in the dwellings after they were flooded on December 2-5, 2003. The protocol in their investigation included a visual inspection of the dwellings, the computation of the mouldy area in each room and a sampling of visible growths, using the paper-gummed technique. They found that there was over representation of *Alternaria* and *Stachybotrys chararum*. They certainly cause health hazards as these species are known to produce mycotoxins. In addition, *Aspergillus* spp. and *Penicillium* spp. were also encountered in large numbers.

The aftermath of the catastrophic event of the post Katrina flooding of homes in New Orleans area resulted in increased regional total mould spore levels. The health effects of elevated levels of airborne fungi were recently discussed by Levetin *et al.* (2006). They reported mean concentrations of total mould spores as 5,762 m^3 in 2003, 4,690 m^3 in 2004 and 7,396 m^3 in 2005. During this period there was a tremendous increase in the representation of *Penicillum/Aspergillus, Cladosporium, Chaetomium* and *Eurotium* ascospores and *Stachybotrys* conidia.

Fungal succession and airborne concentrations of indoor and outdoor fungi in post hurricane Katrina in New Orleans, Louisana, U.S.A. were thoroughly investigated by Hjelmroos-Koski and Solomon (2006). Katrina, a category 4 hurricane which hit the Louisiana-Mississippi border and New Orleans on August 29, 2005 resulted into flooding of an estimated 120,000 homes (80% of the city).The homes remained flooded for several weeks.They were reflooded three weeks later when category 3 Hurricane Rita hit the city, which, enhanced the indoor airborne fungal spore population. During this period, the average daily mould spore concentration ranged from 21,000 to 102,000 spores m^3 of outdoor air and 11,000-645,000 m^3 of indoor air.

FUNGAL SPORE CALENDAR OF BANGALORE, INDIA

Agashe and Vidya (1999) compiled an excellent fungal spore calendar of Bangalore city for the year 1997 **(Fig. 16.5)**. They had used the most efficient air sampler known as rotorod sampler (model 40), which has been described in detail in Chapter 13 above.

An aerobiological survey carried out by them for the compilation of fungal spore calendar, brought to light 70 fungal spore types representing

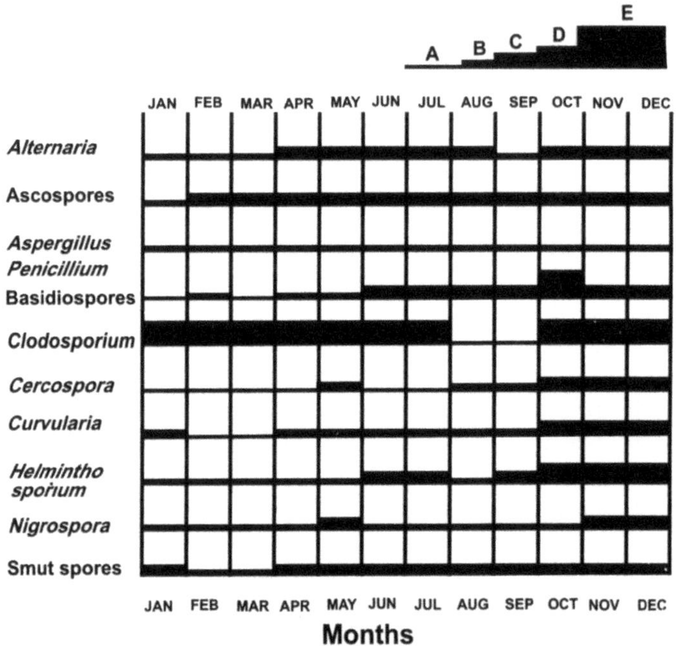

A = 1 - 49 Fungal spores
B = 50 - 99 Fungal spores
C = 100 - 499 Fungal spores
D = 500 - 999 Fungal spores
E = 1000 - 4999 Fungal spores

Fig. 16.5 Percentage contribution of predominant fungal types in Bangalore atmosphere during 1997.

three major groups of fungi: Ascomycotina, Basidiomycotina and Deuteromycotina. As expected, Deuteromycotina constituted the largest fraction of airborne spores accounting for 71% of the catch. It was represented by commonest genera such as *Cladosporium, Alternaria, Aspergillus, Penicillium, Nigrospora, Helminthosporium, Cercospora* and *Curvularia.*

The fungal spore calendar depicted in the pie chart **(Fig. 16.6)** shows that the incidence and concentration of fungal spores varied from month to month during 1997. The highest monthly concentration **(Fig. 16.7)** was recorded in October with a total of 7,698 spores m^{-3} and the lowest was recorded in the month of August with 2,639 spores m^{-3} per cubic metre. The spores of *Cladasporium, Alternaria, Cercospora, Helminthosporium, Aspergillus, Pencillium* and Smut spores were found throughout the year though their peak varied from month to month. The peak season for *Cladosporium* occurred in April and July **(Fig. 16.5)**. *Aspergillus – Penicillium*

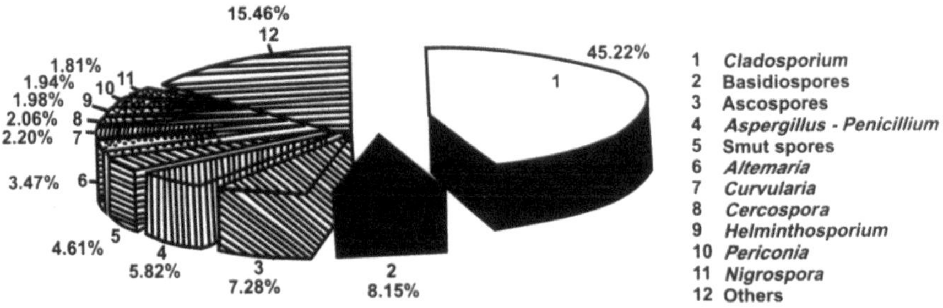

Fig. 16.6 Monthly variation of the presence of fungal spores in Bangalore atmosphere for 1997.

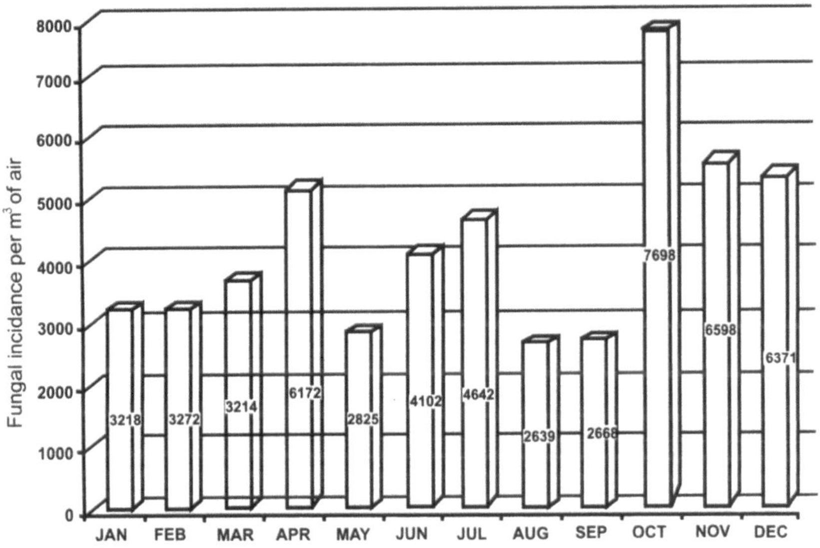

Fig. 16.7 Airborne fungal spore calendar of Bangalore for the year 1997 (**Figs.16.5 to 16.7** from Agashe and Vidya, 1999).

established their peak in April. The major types of fungal spores showed an increase during monsoons associated with high relative humidity and maximum rainfall.

Agashe and Vidya further pointed out that the fungal spore calendars should be compiled and updated every year. The annual fungal spore calendar gives a clear picture of the change in trends of the peak and concentrations over the years. Allergy practitioners should keep a close watch on the fungal spore calendar, which well help in selecting antigenic extracts of fungal aeroallergens for skin testing to assess allergenicity. This will help in a better correlation of allergy symptoms of patients with atmospheric fungal spores and the proper diagnosis leading to treatment of allergy.

The atmospheric fungal spore monitoring was continued by Agashe and Mamlakatoi Khyderova who prepared fungal spore calendar of Bangalore for the year 2000-2002 (**Fig. 16.8**, Khyderova, M. 2003).

FUNGAL SPORE CALENDAR FOR BANGALORE 2000-2002

Taxon	JAN	FEB	MAR	APR	MAY	JUN	JUL	AUG	SEP	OCT	NOV	DEC
Cladosporium	29.80	20.82	15.35	51.12	78.55	28.56	36.06	8.93	39.14	27.12	33.74	35.64
Periconia	19.19	13.11	9.05	10.13	8.68	8.24	7.84	8.00	13.24	16.36	17.80	24.56
Smuts	22.35	14.63	8.55	7.31	4.05	6.96	7.56	6.42	4.85	7.11	18.32	22.30
Nigrospora	11.82	10.21	6.84	9.05	5.40	6.89	4.90	3.34	4.59	6.56	12.88	14.57
Alternaria	6.63	5.49	6.12	10.94	7.80	6.51	6.99	4.12	4.11	6.17	7.10	8.20
Ascospores	1.59	1.79	2.79	8.86	3.76	4.14	7.27	10.74	7.18	14.25	5.84	2.59
Basidiospores	1.95	1.24	2.16	2.12	3.93	3.04	5.23	7.17	6.16	17.52	10.40	6.61
Aspergillus	4.44	3.43	3.62	5.65	3.71	3.39	3.52	1.47	3.17	2.10	4.97	9.06

Few	1-25
Optimal	25-50
High	50-75
Very High	75-100

Fig. 16.8 Fungal spore calendar of Bangalore, India for the years 2000-2002 (From Agashe and Khyderova, 2003).

STUDIES ON FUNGAL AEROALLERGENS IN BANGALORE

While investigating the scope of aerobiological studies in immunotherapy in Bangalore, Agashe *et al.* (1983) had proved allergenicity to these fungal spores. The extracts of these fungal spores were tested on 172 nasobronchial allergy patients in Bangalore. The results of which are presented in the following table **(Table 16.1)**. *Helminthosporium* showed 3+ reactions in five patients, *Alternaria* and *Nigrospora* showed 3+ reactions in three patients, whereas *Cladosporium* showed 3+ reactions in two patients.

Table 16.1

Sl. No.	*Name of Antigen*	*Total No. of Positive reactions*	*Further grading of Positive reaction*			
			1+	*2+*	*3+*	*4+*
1	*Alternaria*	7	_	4	3	_
2	*Cladosporium*	5	_	3	2	_
3	*Helminthosporium*	8	_	3	5	_
4	*Nigrospora*	9	2	4	3	_

CIRCADIAN PERIODICITY OF FUNGAL AEROALLERGENS IN BANGALORE, INDIA

Circadian periodicity of predominant fungal spores in the atmosphere in Bangalore, was carried out by operating Volumetric – seven-day automatic Burkard spore trap from April 1988 to June 1988 at the Indian Satellite Research Organization's ISEC centre in Bangalore by Agashe *et al.* (2000).

The Burkard spore trap works on the suction principle. It has a suction capacity of 10 l/min. The pollen grains and fungal spores, which are sucked through a 2 mm orifice impact on the vaseline coated-tape wound around the rotating drum. After operating the trap for a week, the tape was removed from the drum and mounted on scale with seven equal parts. The exposed tape was cut into seven parts each representing one-days' catch. Each day's exposed tape was further divided into 12 equal parts to indicate the catch of pollen grains and fungal spores for every 2 hours interval. Scanned pollen and fungal spores were converted to number/m^3. For the sake of convenience and proper understanding, the day was divided into six zones of 4 hours each, that is 2200-0200, 0200-0600, 0600-1000, 1000-1400, 1400-1800, 1800-2200 hours.

Among the 40 fungal spores recorded regularly the most dominant spores such as *Alternaria, Cladasporium, Helminthosporium, Nigrospora* and Smut were investigated for their circadian periodicity. The results were expressed as percentage of peak geometric mean concentration **(Fig. 16.9)**.

The peak incidence of fungal spores was recorded between 1000 to 1800 hours, thus falling into a midday pattern. In April, *Helminthosporium, Nigrospora, Cladosporium* showed peak incidence between 10 am to 2 pm. In May and June, these fungal spores showed peak incidence between 2 pm to 6 pm. However, the maximum concentration of these fungal spores in the atmosphere was recorded at 4 pm.

FUNGAL SPORE CALENDAR FROM LEIDEN, THE NETHERLANDS

Nikkels *et al.* (1996) exhaustively studied microscopically identifiable airborne fungal spores from he Netherlands. They concentrated on non-viable fungal spores trapped by Burkard continuous volumetric sampler for a period of 10 years. They found that the spores of Deuteromycetes dominate the fungal air spora. Among them, in the descending order of predominance are: *Cladosporium* being the highest airborne contributor with an average annual total of the daily average of over 700,000. *Botrytis, Ustilago* and *Alternaria* are represented by annual total spore count between 20,000 to 30,000. The other spore types encountered in the atmosphere were *Epicoccum, Erysiphe, Entomophthora, Torula, Stemphylium* and *Polythrinicum,* all of these were represented by annual spore count of less than 10,000.

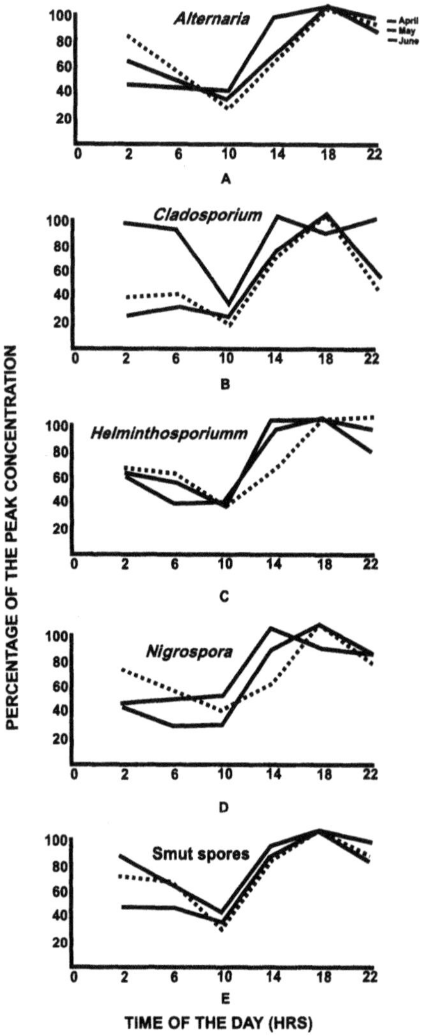

Fig. 16.9 Circadian periodicity of fungal aeroallergens in Bangalore, India. Percentages of the peak concentrations shown in A to E (From Agashe *et al.*, 2000).

On the basis of a 10-year survey from 1980-1989, the fungal spore calendar depicted in **Fig. 16.10**, clearly shows the seasonal course in the airborne concentration of 10 selected spore types. The numerical key to the spore calendar is given in **Fig. 16.11**, while compiling the spore calendar they have followed the method i.e, presentation technique; calculated for 10 m^3 of air, first introduced by Stix (1971) for pollen grains and later employed by Spieksma (1984).

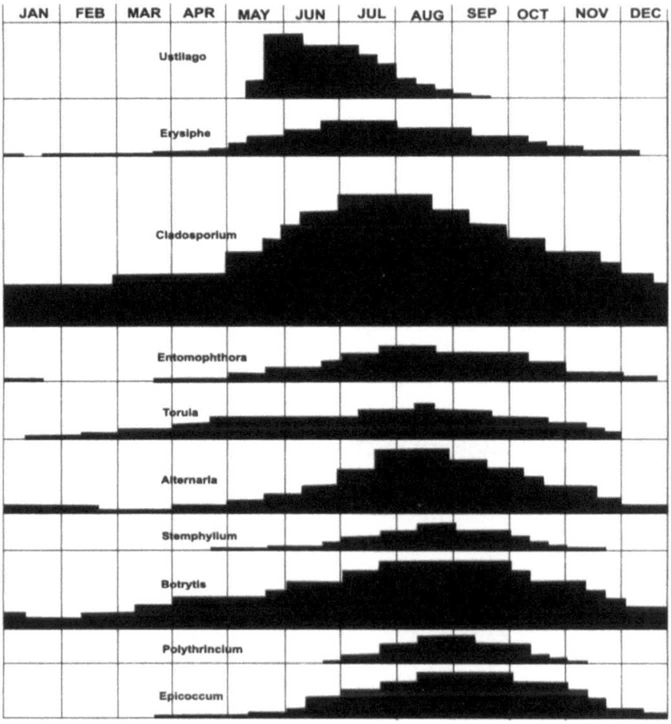

Fig. 16.10 Fungal spore calendar of 10 types of microscopically identifiable fungal spores of Leiden, The Netherlands for the period 1980-1989 (From Nikkels *et al.*, 1996).

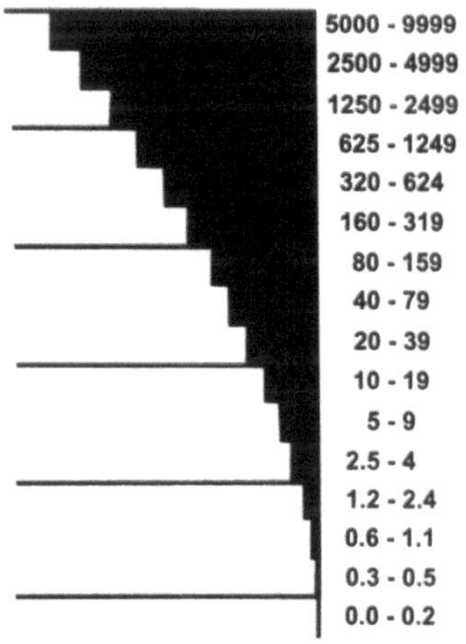

Fig. 16.11 Key to the pollen calendar shown in Fig.16.10 to show the relation between column heights and experimental classes of average daily airborne concentration of fungal spores (From Nikkels *et al.*, 1996).

Essentially the spore calendar represented in **Fig. 16.10** shows the annual appearance of the 10 spore types over the 36, 10-day periods. In this presentation, the logarithmic means for the 10 years of the average daily spore counts for the 10-day periods are calculated. Subsequently, these means are placed in exponential classes, which are plotted as columns in the graphically presented spore calendar.

AIRBORNE FUNGI OF SCOTLAND, U.K.

In addition to pollen, Eric Caulton carried out airborne fungal spores monitoring in Scotland during 2004. The peak of *Alternaria* spores was found to be in mid-August **(Fig. 16.12)**.

Application of Aerobiology in Disease Forecasting

In fungi, the liberation of spores, their dissemination and dispersal and deposition are to a large extent dependent on meteorological factors. In this connection, the following aspects need consideration:

1. Plant pathogen dissemination
2. Plant pathogen dispersal
3. Plant pathogen deposition
4. Plant pathogen concentration in air
5. Plant pathogen that is primary inoculum

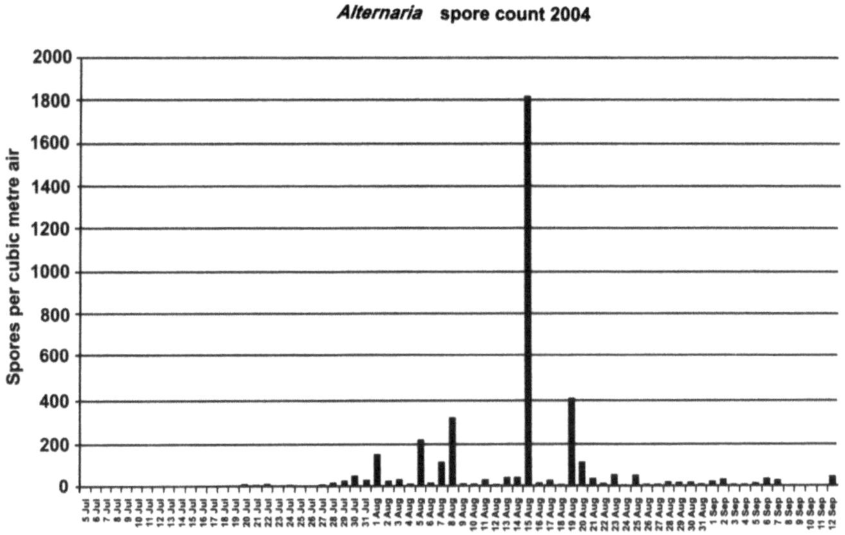

Fig. 16.12 *Alternaria* spore count in 2004 at Scottish Centre for Pollen Studies, Scotland, U.K. (From Eric Caulton, 2004).

Based on the above factors, one can derive the disease gradient and can take preventive measures.

Plant disease forecasting can be of two types: 1. Short period forecasting, 2. Long period forecasting.

DISEASES CAUSED BY FUNGAL AEROALLERGENS

Medical palynologists are concerned with interaction of pollen and spores with the human respiratory tract. Medical palynology is highly interdisciplinary. The medical community needs to know how much allergen is required to precipitate a disease response among subjects in variable health conditions. Such limits could be useful in the examination of the efficacy of drugs. Depending on their aerodynamic size, inhaled particles are deposited in three sections on the respiratory tract:

Mouth and nose – oropharyngeal deposition – most pollen and few mould spores >15 μm.

Throat – extra thoracic deposition

Lungs – thoracic deposition – most mould spores and few pollen < 15 μm. Large particles caught in the nose and mouth can result in allergic rhinitis and other allergic symptoms. The smaller penetrating particles may be involved in Airway Obstructive Disease (AOD) such as asthma.

Fungi mutate readily, filling any available niche. The allergen content often depends on growth factors such as temperature, time, availability of oxygen, size of inoculum and carbon source. Moulds are capable of producing allergies because they exhibit more varied numbers of proteins. For example, 56 proteins in *Alternaria alternata* are capable of inducing IgE antibodies. Sixty proteins have been identified in *Cladosporium herbarum*. This species contains 1 major allergen, 10 intermediate allergens and 25 minor allergens.

Frequent cross-reactivity among moulds is suspected because of multiple mould sensitivity among patients, but specific cross-reactivity studies are few. There is a great variation in the skin test results, mainly because of insufficient extract standardization, poor characterization, and possible lack of skin reactivity to some moulds.

Exposure to pollen and spores may cause disease reaction in different ways: when inhaled by immuno-compromised individuals, airborne spores such as *Aspergillus fumigatus* and *Coccidioides immatus* act as pathogens by colonizing lung tissue and reducing its function (Frankland 1977). Recovered individuals are then susceptible to secondary diseases, such as, chronic bronchitis and bronchiolitis. Colonizing fungal diseases have become increasingly prevalent with the advent of AIDS.

Fungi produce toxins and the best known is aflotoxin, a carcinogen associated with peanuts (Baxter *et al.*, 1985). Exposure to toxinogenic fungi can cause a serious health risk including liver cancer to allergic individuals affecting the liver in particular (common among African population). Pollen and spores carry proteins, polysaccharides and lipids. When inhaled, some of these foreign materials, acting as antigens may stimulate the production of IgE antibodies. These antibodies cause the release of chemical mediators, histamines being the most important, from mast cells. Histamine triggers vasodilatation, sneezing and rhinitis, which are classical symptoms of hay fever.

FUNGAL SPORE ALLERGY

Airborne mould spores in indoor and outdoor atmosphere are known to cause mild to severe allergy, including asthma and rhinitis. Based on the indoor concentration of airborne mould spores , clinical and immunological studies were carried out in 85 children suffering from bronchial asthma and allergic rhinitis in Delhi, India by Sharma *et al.* (2006).

Antigenic extracts of predominant indoor fungal spores of *Aspergillus spp, Alternaria, Cladosporium, Penicillum, Rhizopus* and *Cladosporium* were used for skin testing and immunotherapy. In their study, they observed that the highest sensitization was to *Alternaria alternata* followed by *Aspergillus fumigatus, Penicillum citrinum* and *Cladosporium cladosporites.* Raised specific IgE was observed in cases with 2+ and above skin reactivity against fungi.

INDOOR FUNGI AND THEIR ALLERGY

Indoor mould spores and their occupational health hazards in the potato cold storage facility in West Bengal, India were investigated recently by Majumdar and Barui (2006). In a two-year (2002-2004) survey of potato cold storage houses, they found a preponderance of fungal spores of *Aspergillus niger, Curvularia lunata, Echinobotryum, Alternaria alternata, Fusarium solani* and *Rhizopus nigricans.*

Skin prick tests on 126 patients showed a high positive reaction (2+ or more intensity) to antigenic extracts of the above moulds. They also found that allergenic symptoms were pronounced during the monsoon and the summer season or whenever there was prolonged disruption of power supply.

FUNGAL ALLERGY AND EPIDEMIOLOGY IN ST. LOUIS AREA, U.S.A.

An extensive survey of airborne fungal spores was carried out by Lewis *et al.* (1975) from the Missouri area. This study was also backed by skin prick tests for evaluating fungal allergenicity. As many as 1,720 patients were prick tested for predominant fungal aeroallergens comprising *Alternaria tenuis*, *Aspergillus* spp. (a mixture of *A. flavus*, *A. fumigatus*, *A. glaucus* and *A. terreus*, *Fusarium roseum*, *Helminthosporium sativum*, *Hormodendrum hordei* (= *Cladosporium*), *Penicillium chrysogenum*, *P. notanum*, *Phoma hybernica*, *Alternaria alternata* and *Drechslera*.

Skin tests involving primarily the standard prick technique for immediate skin reactivity were recorded using the 0 to 4+ scale for allergens tested:

0 : no reaction or equivalent to control
1+ : erythema smaller than 10 mm in diameter
2+ : erythema larger than 10 mm in diameter
3+ : erythema larger than 10 mm in diameter plus weal larger than 10 mm in diameter
4+ : erythema larger than 10 mm in diameter and weal with pseudopods

It should be noted that the fungi from which extracts are administered are classified in four families in the fungi imperfecti: Dematiaceae including *Alternaria*, *Helminthosporium*, *Harmondendrum* and *Spondylocladium*; Moniliaceae including *Aspergillus* and *Penicillium*; Tuberculariaceae with *Fusarium* and Sphaeropsidales with *Phoma*. Many moulds represent asexual (or imperfect) faces of Ascomycetous fungi.

Several allergists have stated that the clinical impression within this group indicates the spores of *Alternaria* are the most potent aeroallergens.

Lewis *et al* concluded that of all the eight genera representing a diverse group of common fungal aeroallergens, only *Alternaria* elicited a distinctly high allergic reaction and frequency in the population sampled. This may be due to the fact that *Alternaria* may be a common spore indoors. They also supported that other commonly occurring fungal aeroallergens such as *Geotrichum*, *Aureobasidium (Pullularia)*, *Trichoderma* and *Ustilago* (smut) be incorporated into routine regimens so that additional aeroallergens may be recorded for subsequent treatment based on more diagnosis to date. They also pointed out that prick test in paediatric patients responded at a higher level than the adults to all fungal extracts probably due to early exposure of the children to the fungi.

SCOPE OF MEDICAL PALYNOLOGY

Threshold Values for Allergy Forecasts

In the past, several attempts have been made to define thresholds for pollen counts in connection with allergic symptoms. Most of them are based on personal experience and most of them are close to meeting the target. Unfortunately they are valid only for a given regional or local situation, dependent on several circumstances such as location of the pollen sampler or sampling height. As such, these threshold values cannot be generalized.

Jäger (2006) has described this important aspect of aerobiology for allergy very well. According to him, recent investigations have shown that there is a dependency from average regional pollen load in the air with sensitization intensity of the respective human population.It is observed that in areas with low pollen load, the threshold is low, and high again in areas with very high pollen loads, possibly due to 'natural hyposensitazation' effects.

Thus, it is seen that there is great variation in the threshold values in different regions. To some extent this variation also exists due to variation in the density of plant population in the regional vegetation.This has been worked out thoroughly in ragweed (*Ambrosia*) in European countries. Scientists from France and Switzerland have reported threshold values of 5-6 pollen per cubic metre of air (daily mean) for *Ambrosia* in their regions. In contrast, for the same plant pollen threshold value of 16-20 for eastern Austria and 30-40 pollen per cubic metre of air in Hungary have been reported.

Earlier the threshold for grass pollen was calculated as 20.51 on EA (European average) with minimum of 20.63 ± 11.67 for Edinburgh, U.K. and maximum of 61.86 ± 22.6 for Aurillac in France. In case of *Betula* pollen, threshold value reported for Pavia, Italy was 21.41 ± 9.42 and maximum of 568.21 ± 700.05 for Kangasala, Finland.

In one of the clinical studies for the evaluation of thresholds, subjects are exposed to controlled allergen concentrations in a test chamber. The symptoms and serum responses of the subject are evaluated for each of the groups. This type of study measures direct response to a controlled dose of allergen. So causal relationships can be demonstrated directly. However, getting a human or animal subject approval is difficult, moreover, such studies are very expensive and subject numbers are usually low. However, such controlled studies failed to examine the cumulative effect of multiple allergens and the interaction among cofactors. Evaluation of exposure through epidemiological studies appears to be more appropriate.

O'Rourke *et al.* (1990) carried out the characterization of the pollen and moulds inside homes where people spent 8-10 hours per day in Tucson, Arizona, U.S.A. They did not find much qualitative difference indoors and outdoors as the same types of pollen are found in both environments. In 75% of the homes, pollen infiltration is 10% of outdoor values. Only in 5% of the tested homes, indoor concentrations exceed the outdoors.

Mould taxa infiltration varies seasonally. In good weather, open windows and doors promote infiltration, and about 40% of the homes have low mould concentrations (10% of outdoor values) with similar taxonomic compositions.

On the other hand, during seasons when houses are closed, concentrations of indoor moulds are elevated, as they are in northern temperate regions during the winter. In Tucson, only 24% of the homes have mould concentrations exceeding outdoor values.

DIAGNOSIS AND TREATMENT

For proper diagnosis and management of fungal allergy a detailed knowledge of the different fungal types with daily, seasonal and annual variations both inside and outside environment is of great importance. It has been proved experimentally that a strong relationship exists between airborne fungal spore concentrations and the occurrence of allergic respiratory disease. Fungal allergens are used for diagnosis and immunotherapy of mould allergy patients.

Considerable variation has been observed between allergens produced by different manufacturers and batch-to-batch variations within the same production unit. Allergen standardization and characterization based on clinico – environmental and immunological relationship only would solve this problem.

Allergy-suffering patients can be treated by proper medication for suppression of symptoms and allergen specific immunotherapy. IgE levels in healthy human subjects are known to vary from 1 to 1,000 i.u/ml. Saha *et al.* (1976) reported IgE levels as high as 820 i.u/ml in normal Indian subjects. Patients with atopic diseases including allergic asthma generally have moderately elevated IgE levels. The IgE antibodies play a significant role in the pathogenesis of atopic respiratory diseases and suggest that quantitation of IgE antibodies may be useful in the assessment of IgE mediated respiratory symptoms. Immunotherapy with fungal antigens has been successful in many countries (Gaur and Gupta 1996; Shivpuri 1982). There are certain variations in the presence of different fungi in different geographical locations based on climatic conditions. Hence, it is necessary to know the fungal concentration of the inside and outside environments of every geographical region.

CONTROL MEASURES

Fungal allergic diseases can be controlled by avoiding heavy exposure during work hours, keeping the indoors clean, regular monitoring of air and maintaining low level bioaerosols. Certain 'mildewcides' and Fungicides are available to kill the mould spores. Increased ventilation and proper drainage should be used to discourage mould growth. A tight-fitting facemask can be used for preventing the inhalation of mould spores when gardening, especially lawn mowing or house cleaning.

CHAPTER 17

Comprehensive Account of Most Common Aeroallergens and their Source Plants

AMBROSIA (RAGWEED)

Ambrosia artemisiifolia is known as one of the most dangerous aeroallergen yielding plants. It is also a very obnoxious weed in the agricultural fields. The common ragweed is an annual species, which reproduces entirely by seed. It is helpful to have the knowledge of this plant's annual life cycle, which was studied in detail by Béres and Biró (1993) in Hungary and reported by Juhász (1998) **(Fig. 17.1)**. According to them, the most important weather parameter is the temperature, which influences not only the pollen release from the anthers but also germination of seeds.

It is clear from the above figure that after primary and secondary dormancy of seeds, the germination starts at the end of March or at the beginning of April. Intensive growth of the young plants is observed in June. The main flowering period when 65-70% of the plants are blooming is in the second part of August. This is the time when maximum concentration of airborne pollen of *Ambrosia* is recorded. However, its pollen continuously remains in the atmosphere even in September. Seed production takes place in October-November.

Ragweeds are also known to produce large quantities of seeds from a few thousand up to sixty thousand seeds per plant every year. In addition, it has a long survival period in the ground (from 5 years up to more than 20 years). Thus, there appears to be continuous contribution to a very extensive ragweed seed bank.

Ragweed is one of the most allergenic plants in the flora. Its pollen production is prolific giving rise to high percentages of seasonal rhinitis (hay fever) symptoms (Comtois, 1988). The plant is anemophilous with regard to pollen dispersal, herbaceous, having both annual and perennial species. Ragweed has no economic value *per se*, but constitutes a serious health problem, affecting millions of people in the temperate zone. It is

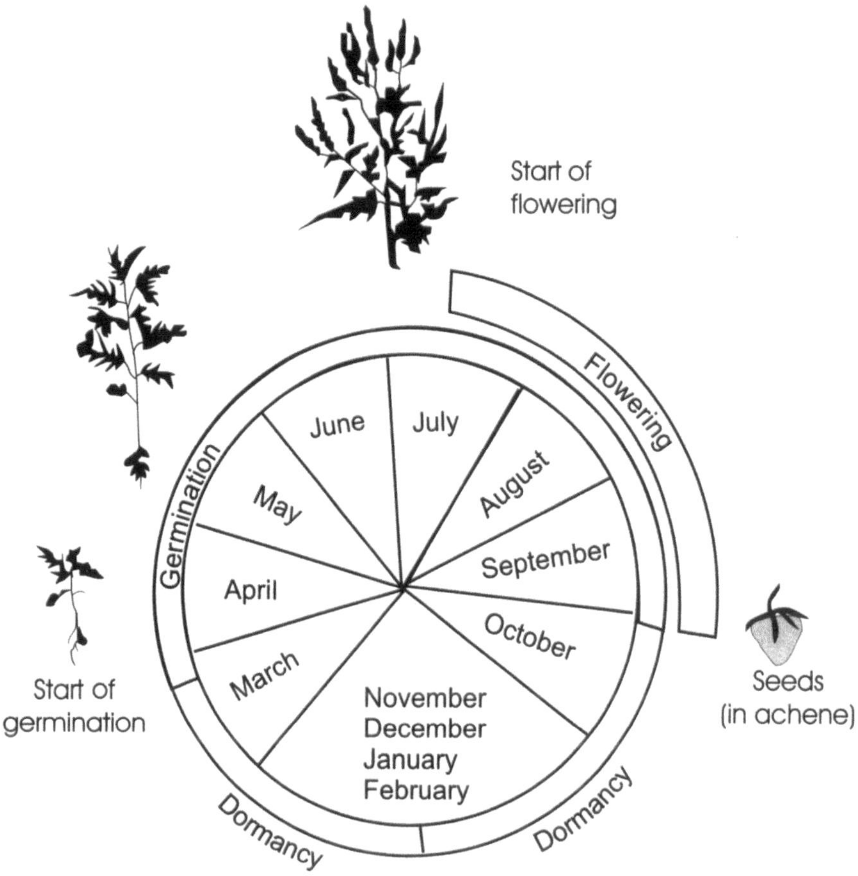

Fig. 17.1 Annual life cycle of *Ambrosia artemissifolia*, common Ragweed in Hungary (From Béres and Biró, 1993).

highly adaptable to a range of environmental conditions, which causes it to be regarded as a noxious weed and consequently has become the subject of legal requirement for its eradication in some areas where it is abundant. As with many highly competitive 'weed species', it takes advantage of human habitat disturbance, such as clearance leading to open spaces, which suit its growth habit.

Ambrosia – Ragweed in Europe

It is known since a long time as the main cause of hay fever or pollinosis in North America. It was unknown in Europe. However, since 1960s *Ambrosia* has been reported growing in the Rhine Valley, northern Italy, south of European Russia, Balkan states and in southern Austria.

Ambrosia is a weed belonging to Asteraceae same as *Parthenium*. The plant can develop into a real nuisance as a pioneer plant in disturbed soils. The plants flower in late summer, in August and September.

Botanically ragweed is known as *Ambrosia*. Its probable origin was in southern North America. Until recently European allergologists and aerobiologists thought that Europe had no ragweed pollinosis problem. They used to state that "this plant simply does not grow here". This misconception was proved to be totally wrong as there were a number of records of the existence of *Ambrosia* in different European countries.

One view indicates that ragweed seeds came to Europe as contaminated American wheat which was imported to Europe. This plant had extraordinary adaptation characteristics which helped the establishment and profuse growth. It is said that ragweed first invaded agricultural land in 1930s and within a matter of 4-5 decades it was found abundantly along the roadsides. It appears that ragweed seeds are transported contaminants among wheat grains as a tyre-dirt contaminant from cars and trucks. As reported by Jäger from Austria, ragweed introduction around Vienna is related to sunflower seeds dispersed by bird lovers. The main source of these seeds is Hungary, where seed contamination by ragweed is found in sunflower fields.

Ragweed has attracted the attention of allergologists on account of the growing allergy-patient population suffering from rhinitis or asthma. It is a strange coincidence that a similar situation is found with respect to *Parthenium hysterophorus* in India. Histrorically, the story of ragweed invasion in Europe can be duplicated for invasion of *Parthenium* in India. The source of both these plants comes from southern North America or the southern states of U.S.A. A similar history of *Parthenium* is narrated in Chapter 14 in this book.

The original centre from which ragweed has spread is thought to have been the south-west of the United States of America. From this region it gradually spread north and eastwards developing the annual habit as opposed to the perennial. Historically, ragweed followed human colonization over the past four centuries, reaching Montreal (Canada) in recent times (Comtois, 1988). The distribution of ragweed in Europe, like *Artemisia* spp. (Mugwort) is confined mainly to hedge-rows and wayside spaces and margins of agricultural land.

There are many species of *Ambrosia*, but the commonest and wide spread species of *Ambrosia* are *Ambrosia artemisiifolia* (= *A. elatior*) which is the dwarf species of ragweed **(Fig. 17.2a)** and *Ambrosia trifida* commonly known as the giant ragweed. The plant body is characterized by opposite or alternate soft-green leaves, a hairy stem, generally less than 1 m in height bearing green to yellow infloroscences **(Figs. 17.2b, c, d, e)** composed of tiny inverted capsules **(Fig. 17.2f, g)**. The plant is a prolific-pollen producer.

(a) *Ambrosia* growing at roadside.

(b) Sketch of *Ambrosia trifida* with male inflorescences.

(c) *Ambrosia trifida* in the field with erect male inflorescences.

(d) *Ambrosia trifida* near residential area.

(e) Close up of male inflorescence of *Ambrosia trifida*.

(f) *Ambrosia* with drooping capsules.

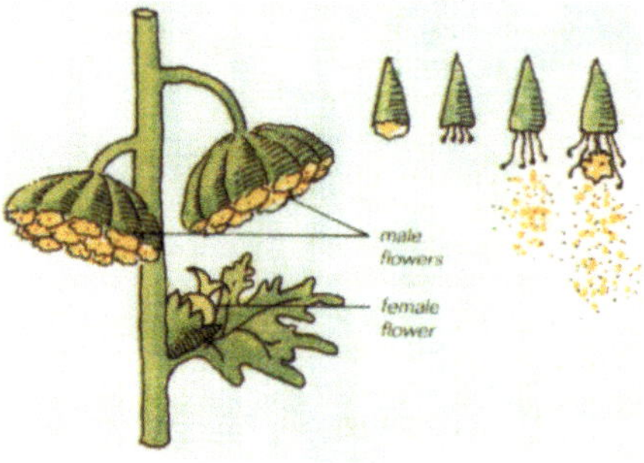

(g) Diagrammatic sketch of *Ambrosia* male inflorescence with anthers shedding pollen.

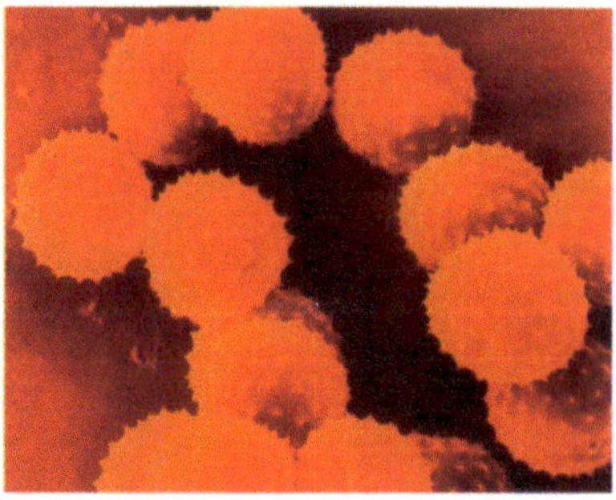

(h) LM of spiny (echinate) pollen of *Ambrosia*.

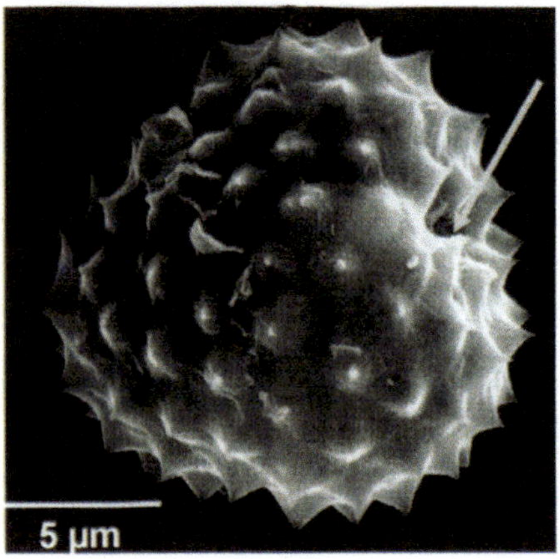

(i) SEM of *Ambrosia* pollen with spiny exine.

(j) Map of North America showing area occupied by *Ambrosia* (Ragweed)

Fig. 17.2 Colour photographic plate of *Ambrosia* (Ragweed) a to i and j-map which indicates the distribution of Ragweed (*Ambrosia artemissifolia*) in North America **Figs. 17.2c, d, f:** courtesy Déchamp, Chantal. **e, g, h** from Allergy Plants by Mary Jelks **b** and **i** from Eric Caulton) [**Fig. 17.2j** Courtesy Déchamp and Méon]

Each plant produces tens to hundreds of millions of pollen grains which get easily airborne, (one mature plant can produce as much as eight billlion pollen grains). The pollen is typically characterized by echinate or a spiny outer wall **(Fig. 17.2h)**. The pollen contains one of the most potent allergens known. It appears that ragweed is one of the most exhaustively investigated non economic plants. The plant causes a numerous health problems and hence, laws have been enforced and eradication campaigns carried out in different countries. This plant has evolved as a reaction to dry climate and open environment and from an entomophilous pollination.

The pollen grains are highly allergenic causing severe pollinosis in sensitive individuals. The pollen grains are spheroidal to oblate spheroidal, with a spiny wall (sub-echinate), tricolporate measuring around 20 μm **(Fig. 17.2i)**. *Ambrosia artemisiifolia* is the commonest and widespread species.

It would not be an exaggeration if one refers to ragweed as the "Number 1 enemy" for allergy sufferers in North America and Europe and like an enemy, the plant may change strategies to establish supremacy over man. It is difficult to say who will have a bright future for either ragweed or allergy sufferers? It seems that ragweed is a more potent enemy than thought of and even if it can be controlled and eradicated, it will always be able to reconquer present or past occupied lands. In other words one can say that battles can be won but the war will never end. The same thing can be said about its obnoxious weed cousin *Parthenium hysterophorus.*

Ambrosia or Ragweed Pollen Occurrence and their Allergy in North America

Lewis (1991) while investigating the distribution and incidence of pollen aeroallergens in various locations in North America, gave particular emphasis to *Ambrosia* (Ragweed). It was reported as the most significant airborne and allergenic herbaceous weed in eastern North America. **Fig. 17.2j** indicates the distribution of Ragweed (*Ambrosia artemissifolia*) in North America which is predominant in Central, Southern and Northern U.S.A. and Canada as reported by Déchamp and Méon (2002). They have also reported the occurrence of *Ambrosia* in different European countries and U.K. (**Fig. 17.11a**). [Courtesey Déchamp and Méon].

Weeds are known to dominate not only the ground vegetation but also the pollen flora in the atmosphere. Airborne pollen of weeds replace those of grasses at the end of summer and autumn in the north temperate zone of the U.S.A. Grass pollen are known to cause rhinitis and hay fever. Among the weeds the most prominent plants are ragweeds (*Ambrosia*) belonging to the *Asteraceae* family. Ragweed pollen cause intensive allergic symptoms. Both giant and short ragweeds are found growing profusely throughout

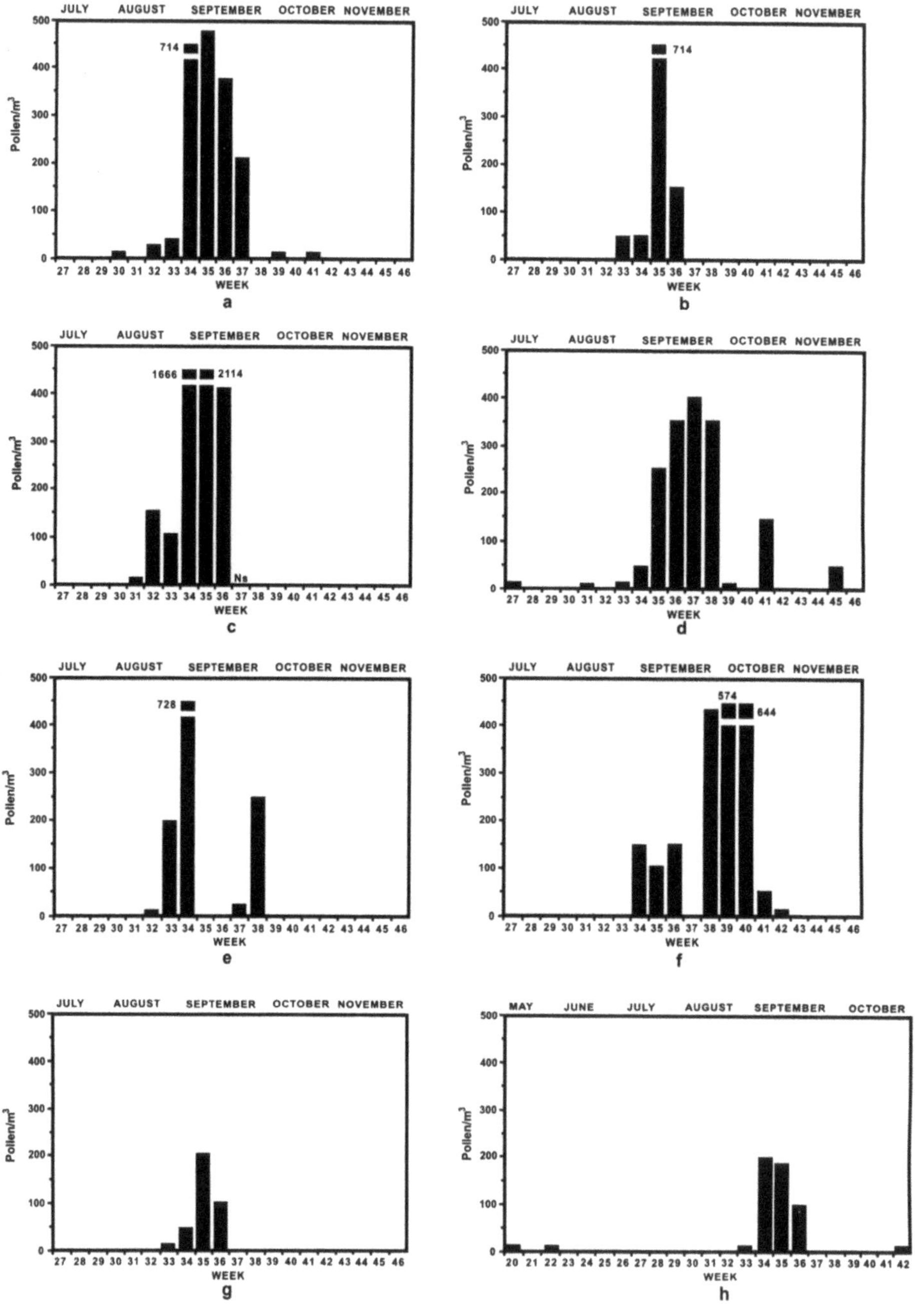

Fig. 17.3a to h Weekly pollen count of *Ambrosia* at different sites in U.S.A. a. Pittsburgh, PA. b. Jackson, MI. c. Winston-salem, NC. d. Venice, FL. e. Yankton, SD. F. St.Louis, MO. g. Dallas, TX. h. Denver, CO. (Modified from Lewis *et al.*, 1991).

Missouri and eastern North America. It was observed that about 78% of weed pollen in the atmosphere of St. Louis area comprised *Ambrosia* pollen.

Low frequencies of atmospheric ragweed pollen reported by them included: 28/m^3 in September, 3-16 in Hyannis, MA, and 112/m^3 from September, 10-16 in Atlanta, G.A., 140/m^3 at Hobbs N M from October, 1-7 and at Pendleton, OR 42/ m^3 during August, 27 to September, 2. In British Columbia (Vancouver) and California (Los Angeles, Templeton) no ambient *Ambrosia* pollen was found.

In contrast to the above reports, *Ambrosia* plants are abundant in North America, east of the Rocky Mountains. As illustrated in the **Figs.17.3a to h** fairly high amounts of *Ambrosia* pollen were encountered in the air of Pittsburg, PA (714/m^3 during August 27 to September 2), Winston,-Salem, NC (445/m^3 from August 20 to September 16). In the southern states of the North America, ragweed pollen concentration was reported in the air of Venice, FL (364/m^3 from September 17 to October 7). It appears that airborne pollen was staggered at different times of the year.

In the southern states, the incidence of airborne pollen of ragweed was approximately one month later than that in the central and northern states. However, it should be borne in mind that this scenario may slightly differ from year to year influenced by various environmental factors. It is for this reason air monitoring for aeropollen has to be a recurring activity (Agashe 2000). With reference to diurnal impact of pollen, Lewis (1991) reported that more pollen of ragweed (67.5%) was captured in the Burkard pollen trap from 3.00 am to 3.00 pm than during the other 12 hours.

The *Ambrosia* pollen count seems to be equal or higher in Europe as compared to one in the U.S.A. as indicated by the Clinical Trials Group in the University of Iowa, U.S.A. It is reported that the maximum pollen count of *Ambrosia* was 327 on August 28, 1997.

Figures 17.4 and 17.5 give a comparative account of pollen counts of trees , grasses and weeds in Baltimore, Northwest U.S.A. **(Fig. 17.4)** and Kalamazoo, Midwest, U.S.A. **(Fig. 17.5)**.

It has been reported from Montreal that the *Ambrosi*a pollen alone accounts for about a third of the annual pollen captured every year, even though its main pollination period accounts for only less than 12% of a year and is known to cause, depending on the yearly climatic conditions, from 50 to 75% of seasonal rhinitis symptoms.

Of the five recognized species of ragweed present in Europe, only the short or common ragweed, *A. artemisiifolia* (syn *A. elatior*) is the principal species concerned with its pollinosis effect (Jäger, 1988). *Ambrosia* pollen was first recorded in Basel (Switzerland) in 1970 (Leuschner 1974, 1978), but is irregular in its occurrence, apparently being the highest during periods when the prevailing winds are strong coming from the west or south-west – The Rhone Valley (France) and the open flat country in the Po Valley (Italy) (Peeters 1988). The highest and therefore the most important

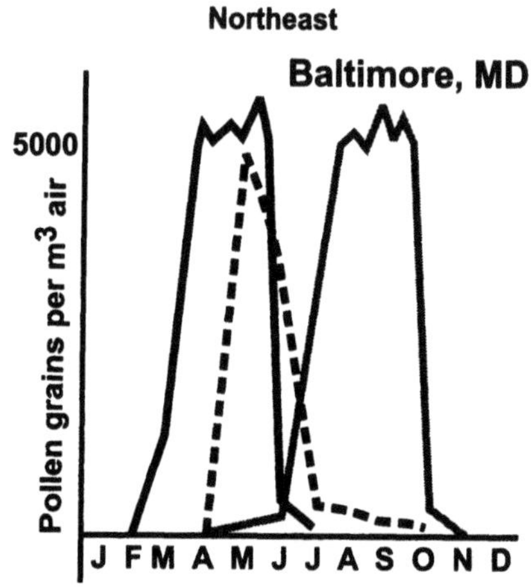

Fig. 17.4 Pollen counts in Baltimore, Northeast U.S.A. The weeds, (August-October) include Ragweed. Concentrations gain to over 5,000 or about grains/m³ respectively.

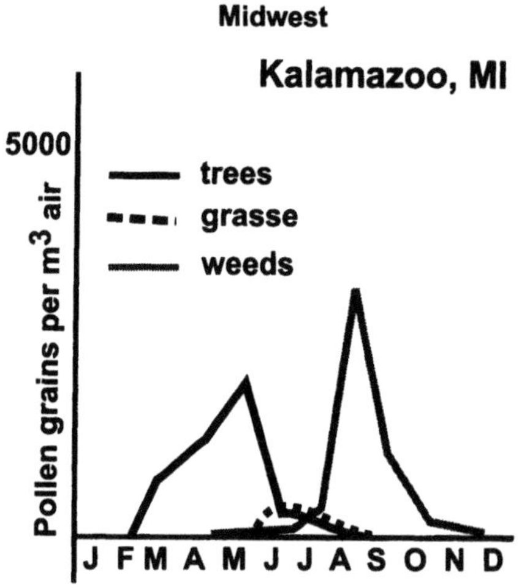

Fig. 17.5 Pollen count in Kalamazoo, Midwest, U.S.A. The weeds, (August-October) include Ragweed. Concentrations gain to over 5,000 or about grains/m³ respectively (**Figs.17.4 and 17.5** from Jäger, 1998).

ragweed pollen at the various pollen-monitoring stations in Switzerland has been ongoing since 1969 (Basel) with the remainder since 1993 at 14 stations (Peeters 1998).

Of the four remaining recognized species of *Ambrosia* occurring in Europe, only one, *A. maritima*, is native, the other species, including *A. artemisiifolia*, have been introduced from the U.S.A.

Both France and Hungary have for a long time pursued active research programmes into the status of ragweed in their respective countries. Hungary has reported the highest annual counts from 50 published in Hungary, Austria, France, The Czech Republic and Slovakia. European counts appear to be less than those recorded from the American stations (Jäger 1988). In France an organization, I' Association Francaise d' Etudes des Ambroisies (AFEDA) has been established to research all aspects of ragweed (Déchamp 1995; Déchamp and Meon 2002).

The spread of *Ambrosia* in Italy (first recorded in 1902) has been monitored in the Lombardy region, where, at the Busto Arsizio monitoring site, annual records obtained over the five year period 1993-1997, showed an increase in the pollen count, which more than doubled. Evidence indicates a movement eastwards and southwards in Italy (Zaron *et al.*, 1988). Total pollen counts of *Ambrosia* recorded in Vienna (Austria) between 1970 and 1997 rose from 200 to almost 1500. Such high and possible increased concentrations of ragweed pollen predict increasing health problems in Italy (Jäger 1988). *A. elatior* is now the dominant 'weed' species in Hungary, having spread gradually in agricultural land habitats. The average daily pollen count is over 50, which is the threshold for some 60-80% of ragweed-sensitive patients to experience serious symptoms of pollinosis (Jarai-Komlodi 1998).

Ruth Leuschner studied the daily pollen count of *Ambrosia* by using the Burkard sampler in 1970, but published her results only in 1978 (Leuschner 1970). Jäger (1991) started investigating pollen concentration of *Ambrosia* in Austria in 1976. He indicated that *Ambrosia* pollen were perhaps transported from Hungary by air to Vienna, Austria **(Fig. 17.6)** with annual totals of *Ambrosia* pollen in Vienna dating from 1976-1997.

Belli and Tosco (1962) initiated and maintained aerobiological records of ragweed pollen in Turin, in Italy. Subsequent reports indicated the abundant occurrence of ragweed in northern Italy. Zanon *et al.* (1998) gave a very good account of pollinosis in various centres in Italy. According to them, *Ambrosia* pollen-sensitized patients accounted for 50-60% of all allergic patients. In 1993, 20% of all pollinosis patients in Milan (urban area) were sensitized to ragweed. Surprisingly this figure gradually increased to above 60% in 1997. It is noticed that very few patients were allergic to ragweed alone or to ragweed and perennial allergens, most were sensitized to other pollen. Among these patients, 30-40% were affected by asthma and the remaining patients had oculo-rhinitis. In an industrial area called Busto-arsizio situated north west of Milan ragweed-induced allergic symptoms were recorded at about 70% during 1993-1997. Asthma was more frequent among patients sensitized to ragweed, than among those sensitized to grass pollen, which have a similar airborne concentration and pollination period.

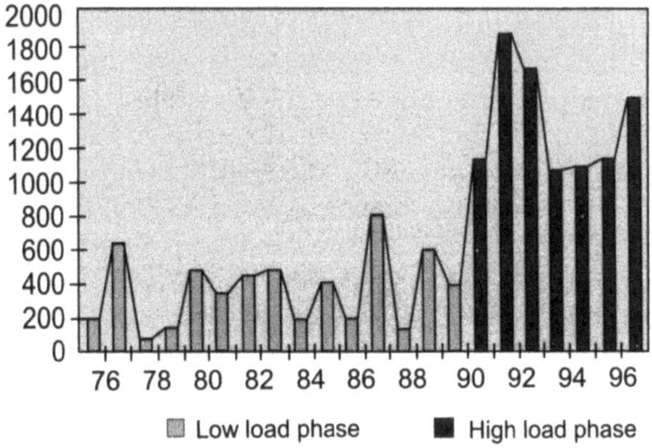

Fig. 17.6 Atmospheric *Ambrosia* pollen in Vienna, Austria: Annual Total during 1976-1997 (From Jäger, 1998).

Juhász and Komlodi (1989) thoroughly investigated the occurrence of ragweed in different parts of Hungary. Yermekova (1991) surveyed ragweed invasion in Russia. He reported the predominance of aeroallergenic plants of *Ambrosia* in major Russian cities and stated that these pollen cause hay fever.

Though ragweed pollen and ragweed pollen allergy are commomly thought of as a North American problem. Short species of ragweed i.e., *Ambrosia artemisiifolia* (*A. elatior*) occurs abundantly in Europe. Other species *of Ambrosia* namely *A. psilostachya* and *A. coronopifolia* occur on the North Sea coast and in the Mediterranean region. In some areas *Ambrosia* is found with other closely-resembling pollen-producing plants such as *Iva* and *Xanthium.*

On account of the health hazards of ragweed pollen, the Ragweed Society was established in Lyon, France and laws for prevention of ragweed pollinosis were made in Montreal, Canada, Lyon in France and Hungary. European aerobiologists have assembeled a joint pollen count data base since 1988. Jäger *et al.*, have subsequently drawn maps in order to visualize the yearly alterations of the ragweed pollen concentrations in Europe. It is observed that the threshold value for clinical symptoms for the majority of the sensitized patients is considered to be below 20 pollen grains per cubic metre of air (20 m^3).

On account of highly allergenic properties of *Ambrosia* pollen, its aerobiology has been studied extensively in different European countries. Some of these studies are briefly described below.

Ragweed in Switzerland

Ruth Leuschner observed *Ambrosia* pollen in the Basel area in 1970. Later she also made critical observations after collecting data on *Ambrosia* pollen from other centres in Switzerland, such as, Neuchâtel, Zurich and Lugano. She stated that as there were very few plants of *Ambroisa* , their pollen was perhaps transported by wind from the south and south-west of France and south of Italy.

Ambrosia Pollen Allergy in Geneva

Experimental work carried out by Davet *et al.* (2006) in the Geneva area of Switzerland has indicated that every year there is an increasing number of ragweed-sensitized patients, who are newly diagnosed. However, sensitivity in many patients may represent merely a cross reactivity between *Artemisia* (mugwort) and *Ambrosia* (ragweed) allergens. This observation served as a warning to take more urgent measures to control the spreading of the ragweed in Switzerland.

Ragweed in Austria

According to Jäger (1998) ragweed pollen are known as very potent aeroallergens in Austria. A pollination calendar shows that the pollination period of *Ambrosia* in Austria is from mid-August to end of September with a peak in early September. He also tried to correlate the production of IgE antibodies in allergic patients in the Viennese population with the atmospheric ragweed pollen count.

As shown in **Fig. 17.6** pollen count of ragweed was low during 1976-1990. However, there was remarkable increase in pollen count from 1991-1997 which was statistically significant.

It was noted that there was considerable change of the daytime distribution of ragweed pollen in Vienna from the late 1980s, ragweed pollen was reported to have peaked at midnight whereas, during 1990s, the peak time shifted to late afternoon or early evening hours. This indicates that the time for the transport of ragweed pollen containing air masses has been reduced and the main pollen source must have been closer to the pollen trap.

Ragweed in Slovak Republic

Makovcova *et al.* (1998) reported the occurrence of ragweed pollen of *Ambrosia artemisiifolia* in Slovakia. In the area around the river Danube, it is interesting to note, critical pollen seasons were when there are more than 50 pollen grains per m^3, which appeared annually almost at the same time from the years 1994 to 1997 as indicated below:

1994 = August 14 to September 21
1995 = August 19 to September 14

1996 = August 19 to September 14
1997 = August 17 to September 12

The above investigation has exhibited that the amounts of pollen of *Ambrosia* are dependent on the atmospheric conditions (such as rains) during the bloom of flowers and directions of the wind. The highest concentration of ragweed pollen was not more than 250/m³ of air in eastern Slovakia. It was probably introduced with cereals from the erstwhile Soviet Union.

Ragweed in the Czech Republic

Rybnicek (1988) reported the occurrence of ragweed pollen in the Czech Republic, although there was no major threat from ragweed pollinosis to the allergenic population there. However, he recommended that it is necessary to continue the botanical, aerobiological and allergological monitoring of ragweed in the Czech Republic.

Ragweed in Slovenia

The common ragweed or *Ambrosia artemisiifolia* has been introduced to Slovenia since the end of World War II. Seliger (1998) has recorded the daytime distribution of ragweed pollen in 1997 which indicates that the highest pollen reached between 10 to 16 hours which is attributed to local sources of pollen **(Fig. 17.7)**.

Ragweed in Serbia

Sikoparija *et al.* (2006) carried out exhaustive studies on the occurrence of airborne pollen of *Ambrosia* during 2003-2005, with special emphasis on pollen production, pollen dispersal and pollen count in rural and urban

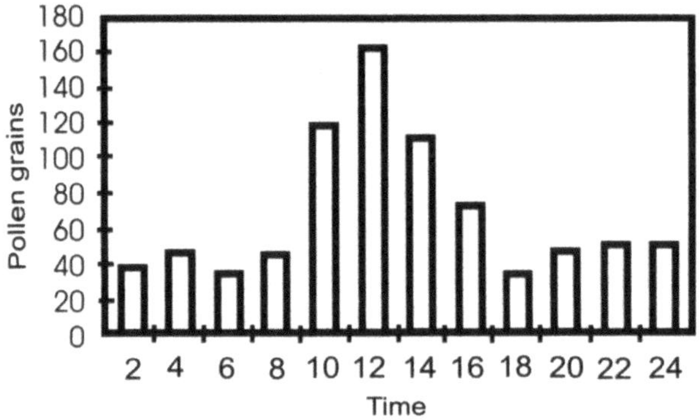

Fig. 17.7 Daytime distribution of *Ambrosia* pollen in Ljubljana, Slovenia in 1997 (From Andreja Kofol Seliger, 1998).

regions of the southern Panonnian valley of Serbia. As expected, they recorded higher pollen count (16.84% with peak concentration of 674 m^3) compared to the urban region, (14.76% with peak concentration of 439 m^3).

Ragweed in Bulgaria

The aerobiological and allergological monitoring of ragweed pollen in Bulgaria carried out since 1991, shows an increasing tendency in both of these aspects (Yankova 1998).

Ragweed in Hungary

Jarai-Komlodi, (1998) has thoroughly investigated the airborne pollen of Hungary since 1989 using the Hirst Type Volumetric Pollen Samplers at Eorvos Lorend University in Budapest and at Jozseph Attile University at Szeged. The results of pollen monitoring from these two monitoring stations have been given to the European Aeroallergen Network (EAN) and since 1990 have been published weekly by the Austrian Pollen Information Service. In southern Hungary it was noticed that the number of cases of allergic asthma is four times higher (392%) than it was 10 years earlier. The nation wide elimination of the weed and a reasonable continuous control of the airborne content of ragweed pollen are very important in presenting the autumn pollinosis.

Jarai-Komlodi and Juhasz (1993) carried out a two year (1989-1990) aeropalynological survey at Budapest in central Hungary and Paks, Szeged in southern Hungary and concentrating on airborne pollen of *Ambrosia elatior*. They found a perfect correlation between the highest percentage of airborne pollen of *Ambrosia elatior* from mid August to mid September and clinical data on prevailing pollinosis.

The weekly ragweed pollen count in Hungary during 1989-1990 has been depicted in **Fig. 17.8** whereas, comparison of annual total of daily ragweed pollen in three sites in Hungary, i.e., Budapest, Paks and Szeged has been shown in **Fig. 17.9**. About 10% of the Hungarian population suffers from pollinosis, of which the majority is sensitive to the pollen of *Ambrosia*. Ragweed grows very commonly in Hungary and its pollen plays a significant role in triggering allergic reactions (pollinosis) from August to October. The nationwide elimination programme of this obnoxious weed is continuing.

Farkas *et al.* (1998) have given a detailed account of the anti-ragweed campaign being carried out in Hungary. This step was necessary on account of the growing menace and health hazards of the ragweed pollen. The number of allergic patients has increased four fold as confirmed by

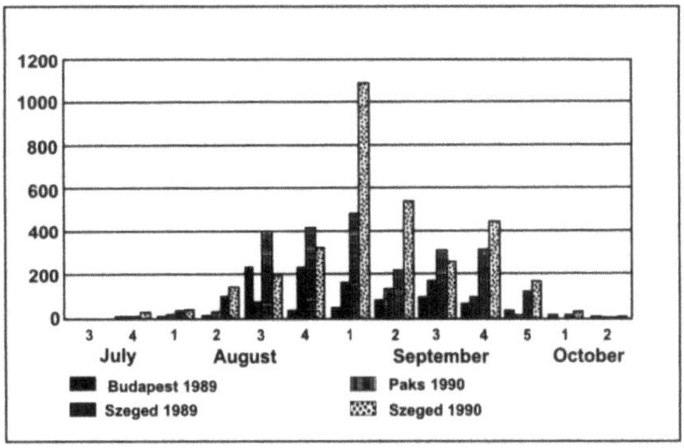

Fig. 17.8 Weekly Ragweed pollen count in Hungary:1989-1990.

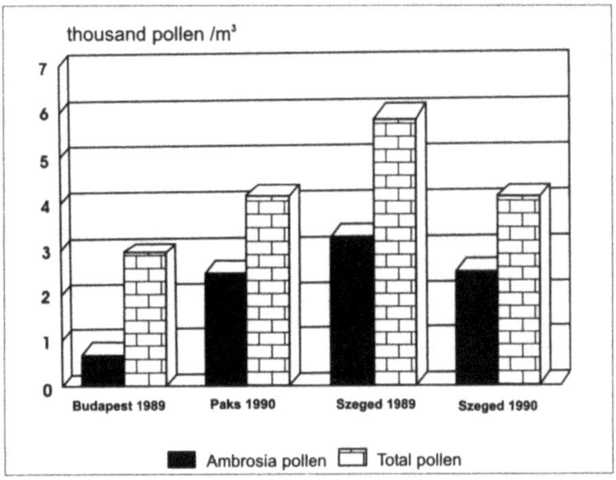

Fig. 17.9 Annual sums of daily Ragweed pollen in Budapest from Parks and Szeged (**Figs.17.8** and **17.9** from Komlodi and Juhasez, 1993).

skin prick test (SPT) positivity. Realizing the serious repercussions of the problems created by ragweed pollen allergy, an anti-ragweed campaign was launched under the National Environmental Health Action Programme. The coordinators of the Hungarian Aeroallergen Network canvassed among the general public about the danger of the growing number of allergic diseases and need for the importance of good agricultural practice. During this campaign, ten thousands of ragweed plants were uprooted in 1997, for which, help was sought from school teachers, children and non-governmental organizations (NGOs).

Ragweed in France

According to Thibaudon (1998) the pollen count of *Ambrosia* in Lyon, (France) indicated that the pollen count of more than 20-30/m³ is dangerous for allergic patients. Déchamp and Meon (1998) while elaborating on the legal aspects and directions with regard to ragweed in France have indicated that the prevalence of pollinosis rises with pollen counts, in the same way as with chemical pollutants. Infact, chemical pollution reinforces the pathogenic power of pollen. The pollen count in the French Basque county, (France) during 1990 was determined by Montouroy *et al.* (1992).

The following two maps of Europe **(Figs. 17.10 and 17.11)** show how aggressively ragweed has invaded vast areas in Europe within a matter of 8 years, i.e., from 1989 to 1997. These maps were used for the back cover of the Satellite Symposium Proceedings of Ragweed in Europe. In addition, another map of Europe and England showing countries occupied by *Ambrosia* has been incorporated **(Fig. 17.11-a)**. D'Echamp and Méon (2002) have also shown the geographical distribution of *Ambrosia artemissifolia* in the world. (**Fig. 17.11b**).

Bees foraging on pollen were critically analyzed in the Drome valley of France by Penel and de Clercq (2006). In September, flowering plants, which are the source of pollen for bees, becomes scarce and therefore, bees are forced to feed on pollen of ragweed (*Ambrosia artemisiifolia*). This was also confirmed by physical observation of a large number of bees among or on these weeds. Eight among nine bees randomly captured had full pollen baskets. Sixteen pollen sacs were analyzed, the ninth bee was washed in a vortex and its pellet analyzed. About 500 pollen grains of ragweed were found in each pellet sample.

Frequent allergic reactions, some of them of a serious nature, have been observed for patients who have consumed honey or bee pollen. It is interesting to note that similar observation were made in India in respect to honey samples contaminated with pollen of *Parthenium hysterophorus* a member (cousin of *Ambrosia*) of the Asteraceae family.

GRASS POLLEN

The grass family Poaceae (or earlier known as Gramineae) is one of the largest, comprising more than 10,000 species out of which 1,200 are known to occur in North America. Economically as well as ecologically, grass species are most important. Grasses have a worldwide distribution, the majority of which produce anemophilous or windborne pollen most of them are known to produce allergic symptoms. The other closely related grass- like family is the Cyperaceae comprising sedges, some of which produce allergically significant windborne pollen.

Pollen Morphology of Grasses

Pollen morphology is relatively similar in almost all species of grasses. Pollen are usually spheroidal to ovoid, 22-122 μm in diameter with a thin (3 μm) outer wall or exine and a thicker (up to 5 μm) intine. The aperture

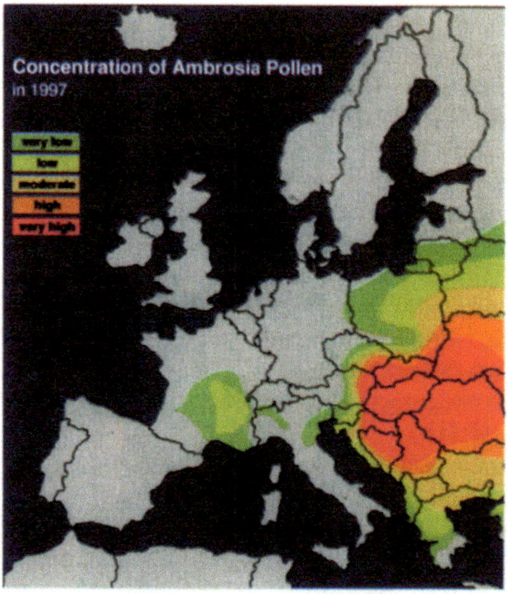

Fig. 17.10 Map of Europe showing extent of further invasion of countries by *Ambrosia* (Ragweed) in 1997.

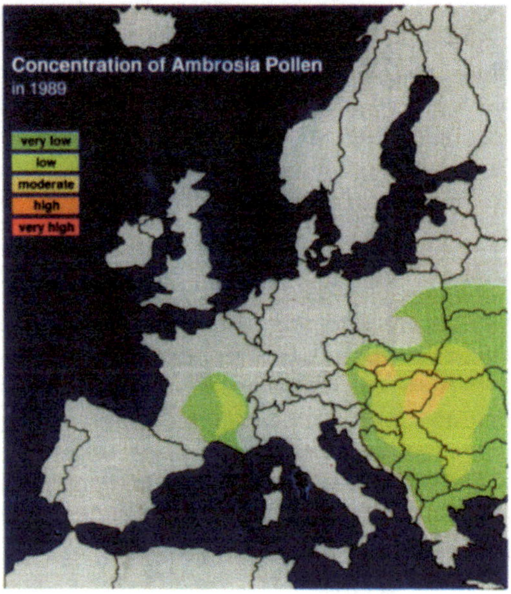

Fig. 17.11 Map of Europe showing extent of invasion of countries by *Ambrosia* (Ragweed) in 1989. (**Figs. 17.10 and 17.11** from proceedings of Satellite Symposium of Ragweed in Europe, 1998).

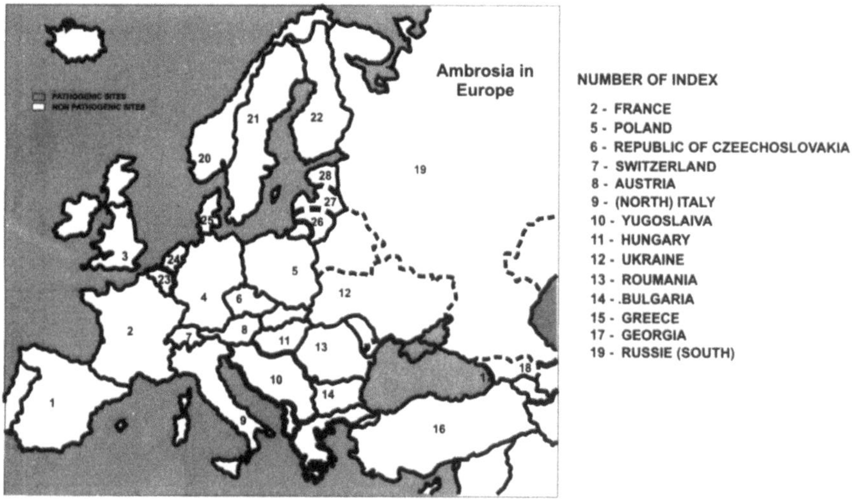

Fig. 17.11a Map of Europe and England showing countries occupied by *Ambrosia* (Courtesy Déchamp and Méon, 2002).

is a single circular pore, 2-8 μm in diameter **(Fig. 17.12)**. The portion of intine below the aperture may be granular or finely reticuloid. Invariably the pore is surrounded by a thickened protruding ring referred to as the annulus.

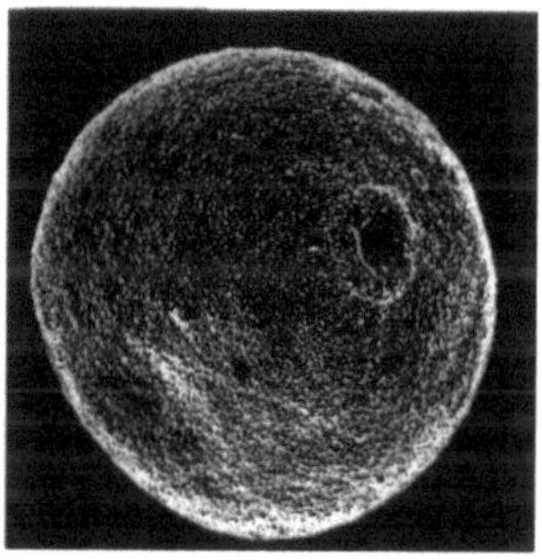

Fig. 17.12 SEM of a typical Grass pollen with a single pore surrounded by annulus.

Basically pollen of small size are noted in commonly occurring grasses such as *Agrostis* (Bentgrass); *Cynodon* (Bermuda grass); *Poa* (Blue grass); *Koeleria* (Hair grass); *Dactylis* (Cocksfoot grass); *Holcus* (Soft grass). In contrast large sized pollen occur in cultivated cereals such as *Zea mays* (Corn) ranging in size between 86-122 μm.

Airborne pollen grains of various grasses cannot be distinguished from one another morphologically. Their shape is spheroidal or ovoid, with typically a single pore surrounded by an annulus and covered by an operculum. The wild grass pollen varies from 20-35 μm in size compared to cultivated Poaceae (cereals) members where pollen though monoporate are large in size. The morphology of sedge pollen differs from grass pollen in having a pyriform shape, 24-65 μm long and 21-40 μm wide with a thin intine and very thick exine reaching up to 20 μm in thickness with 1-4 poorly defined poroid apertures. There appear to be two types of grasses; late spring and early summer flowering grasses, typical of cooler regions of North America and Europe. Most of these grasses are long-day plants,

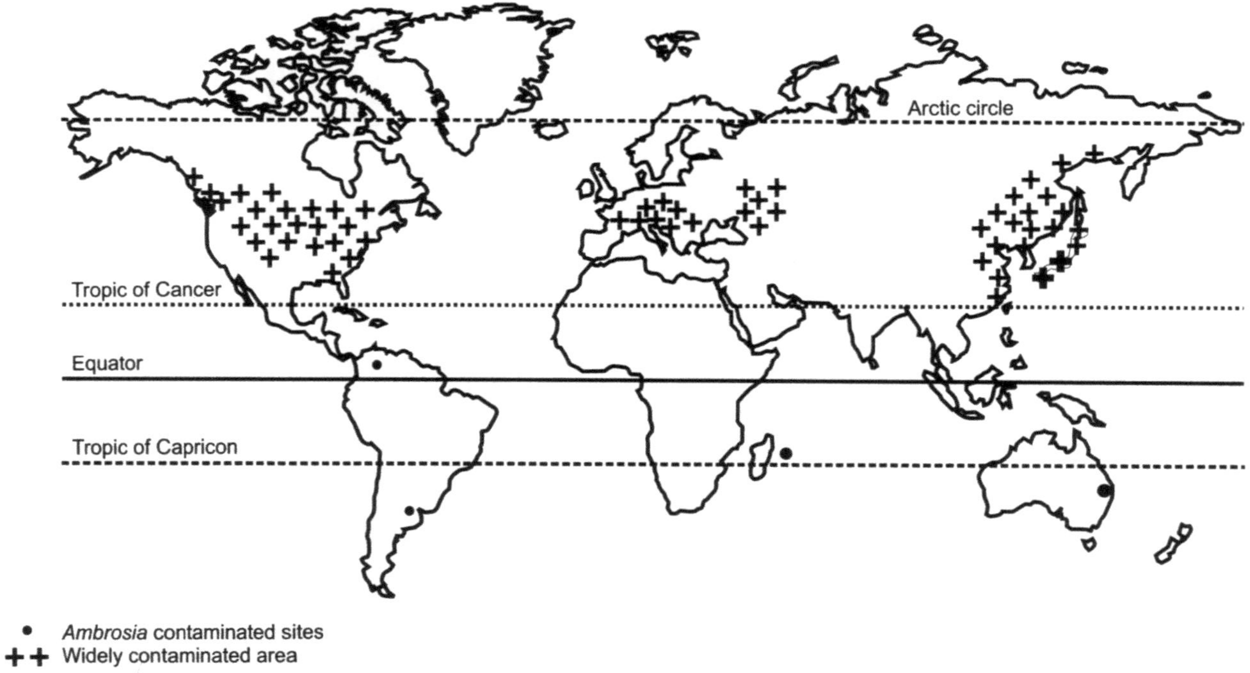

Fig. 17.11b Map of the world showing different sites known for occurrence of *Ambrosia artemissifolia* (From D'Echamp and Méon, 2002).

i.e.', they only flower or flower most profusely, with fewer than a certain number of hours of darkness in each 24 hour period as daylight hours increase through spring to summer). This photoperiodic response is found characteristically in the following cool temperate grasses earlier referred to Gramineae.

Agrostis (Bent grass), *Avena* (Oats), *Bromus* (Bromograss), *Festuca* (Fescue), *Hordeum* (Barley), *Lolium* (Rye grass), *Tritcum* (Wheat), *Dactylis* (Cocksfoot grass), and *Phleum* (Timothy).

The other types of grasses occurring in warm temperate to subtropical regions are classified as short-day plants in which floral initiation and development, and pollen emission occurs when day lengths are shorter and dark periods increase to a prescribed length in late summer and autumn.

Grass Pollen of the Western United States: Gulf Coast

Lewis *et al.* (1992) investigated airborne pollen of grasses at Corpus Christi,Texas, (U.S.A) during 1987-1989. They found that grass aeropollen accounted for 15.3% of the total pollen captured. These pollen showed two peaks one in May and another from September to October. The remaining grass pollen, about one quarter of the total, was found throughout the year with low levels in the winter and mid-summer.

Sometimes the long distance dispersal of grass pollen adds to the high quantity of grass pollen triggering allergic symptoms causing grass pollinosis out of season in acutely sensitive individuals. They also confirmed the positive correlation between drought conditions and reduced pollen shed in 1998. During normal precipitation much higher grass pollen was shed during 1987 and 1989.

These observations support the important role of environmental factors such as moisture on annual pollen frequency and levels of pollinosis, which can be variable and annually unpredictable. Thus the genetically controlled timing of pollen maturation and release which will be generally the same year by year.

Recent studies done by Emberlin and Newnham (2001), 1994 on grass pollen monitoring indicate a close relation between the pollen production and dispersal which is controlled by temperature and hence climate change. Pollen monitoring studies done during the summer of 1988-89 which is the grass pollen season in New Zealand shows the potential for predicting pollen seasons and also throw light on global warming.

In these studies, Newnham selected different geographical locations varying in latitude and hence temperature. It was reported that the grass pollen season in the southernmost localities in New Zealand commenced 8-9 days later then in northernmost localities.

Grass pollen is one of the most important causes of pollinosis during summer throughout Europe **(Figs. 17.12a and 17.13)**, where there are more than 200 species of wild grasses. The most widely distributed grasses belong to *Alopecurus* (Fox tail), *Dactylis* (Cock's foot) *Lolium* (Rye grass) and *Phleum* (Timothy). Grass pollen is one of the most potent allergen, responsible for hay fever and asthma in Spring at least in northern Europe and the U.K.

The pollen season is long on account of the number of species which grow widely. Typically grass pollen season varies in different parts of Europe and Nordic countries as indicated below:

Nordic countries – Second half of June to the first half of August

Northwestern and central Europe – Second half of May to the first half of July

Southern and Mediterranean European region – peak in May

Grass pollen is by far the most common and important cause of pollinosis in early summer all over Europe.

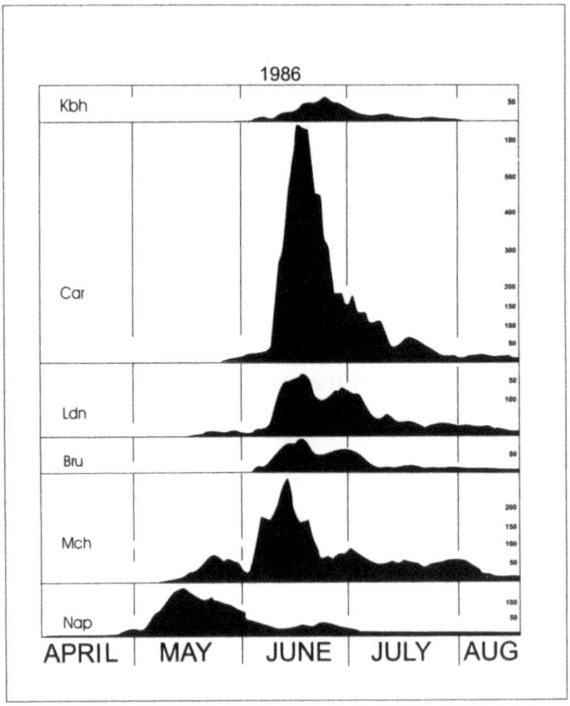

Fig. 17.12a European Grass pollen season in 1986. (Index for **Figs.17.12 and 17.13:** Kbh – Copenhagen; Car - Cardiff Ldn – London; Bru – Brussels; Mch – Markt Schwaben; Nap – Naples.

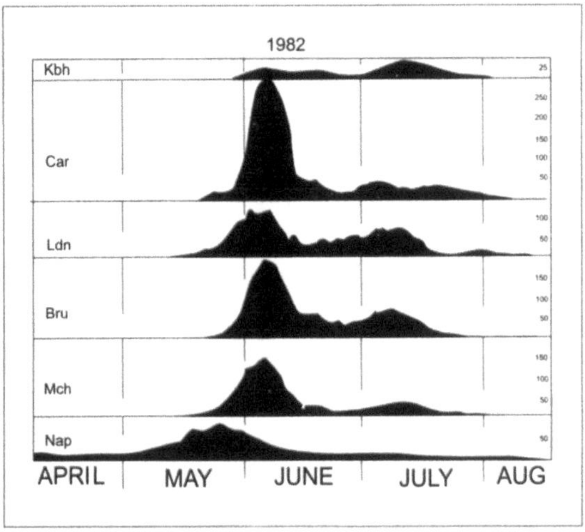

Fig. 17.13 European Grass pollen season of 1982. Ten days running of daily average of airborne grass pollen concentrations during 1982 at six European pollen trapping stations. (**Figs. 17.12 and 17.13** from Spieksma *et al.*, 1989).

Grass pollen allergy is also very common in Scotland, (U.K.). Eric Caulton, at the Scottish Centre for Pollen Studies, carried out grass pollen (Poaceae) air monitoring and reported peaks in June 2003 **(Fig. 17.14)**.

Grass pollen was also determined by Green *et al.* (2004) during five sampling years (June 1994 to May 1999) in Brisbane, (Australia) **(Fig. 17.15)**. He found that the peak of Poaceae pollen count was in December when the temperature was fairly high.

BETULA (BIRCH)

Birch trees are common in both Europe and North America. The plants produce drooping male catkins and bright red female inflorescence. **(Fig. 17.16)**. Birch pollen **(Fig. 5.9d)** is one of the most potent allergens responsible for hay fever and asthma in Spring, at least in northern Europe and the U.K. Allergenic cross reactivity exists with *Alnus* and *Corylus* pollen, all belonging to the same family Betulaceae. Emberlin *et al.* (1993, 1997) are credited for doing pioneering work on the airborne pollen surveys of *Betula* pollen in London, Derby and Cardiff.

Later, Corden *et al.* (2000) carried out a detailed aeropollen survey of *Betula* for a period of five years (1993-1997) at seven sites within the European Aeroallegen Network (EAN). These sites include the ones in London, Taunton (in the south west), Cardiff (Wales), Derby, Leicester,

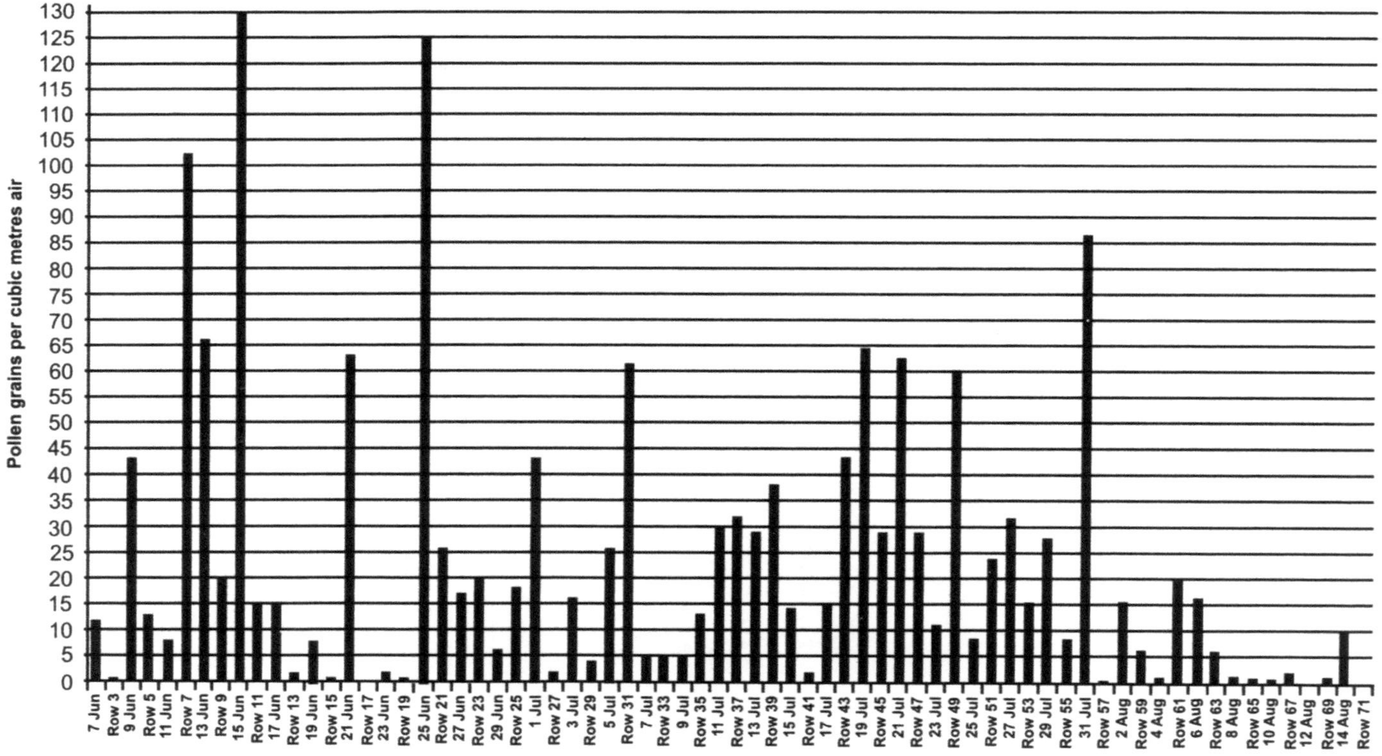

Fig. 17.14 Poaceae pollen count at Scottish Centre of Pollen Studies, Edinburgh U.K. from June-August 2003 (From Eric Caulton, 2003).

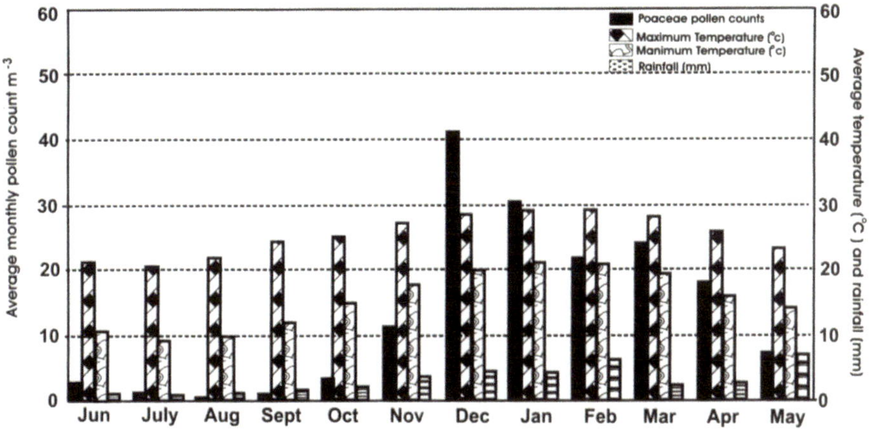

Fig. 17.15 Average monthly Poaceae (Grass) pollen count (± SD) based on six sampling years i.e. June 1994 to May 1999 of daily pollen grains / m³ in Brisbane, Australia (From Green *et al.*, 2004).

Fig. 17.16 Showing a bright red small female inflorescence at the top and yellow long male inflorescence (catkin) loaded with pollen in male flowers in *Betula* (Birch).

Central London, Edinburgh (Scotland) and Belfast (Northern Ireland) **(Fig. 17.17)**. They found significant variation not only in the total pollen count of *Betula* but also variation in the beginning and duration of the pollen season. It was observed by them, that *Betula* is very dependent on meteorological factors for its pollen release. They concluded that a period

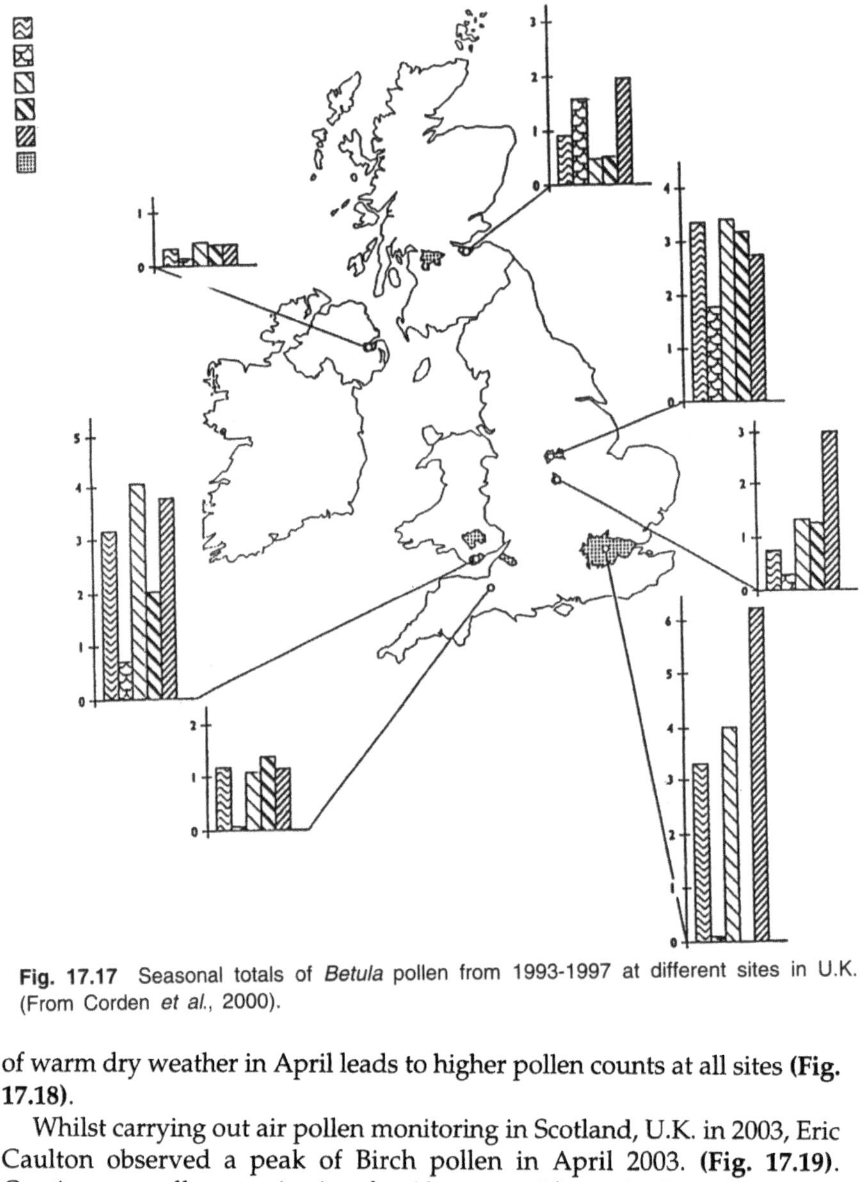

Fig. 17.17 Seasonal totals of *Betula* pollen from 1993-1997 at different sites in U.K. (From Corden *et al.*, 2000).

of warm dry weather in April leads to higher pollen counts at all sites **(Fig. 17.18)**.

Whilst carrying out air pollen monitoring in Scotland, U.K. in 2003, Eric Caulton observed a peak of Birch pollen in April 2003. **(Fig. 17.19)**. Continuous pollen monitoring for 10 years with particular reference to *Betula* (Birch) pollen was undertaken from 1988 to 1997 at Reykjavik, Iceland. Their peak was noted mostly in May **(Fig. 17.20)**.

Dahl and Strandhede (1996) have developed models to predict the magnitude of *Betula* pollen season. Some of the phenological and morphological observations of *Betula* inflorescences of the previous year are significant in this connection.

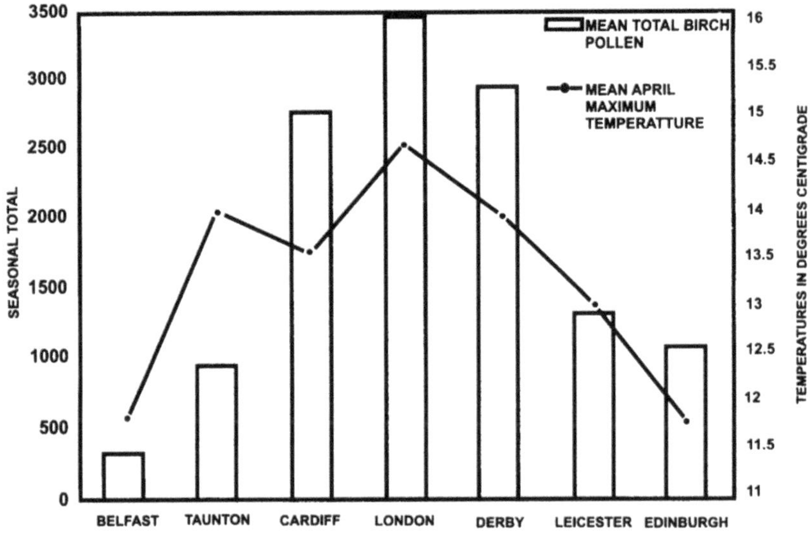

Fig. 17.18 Five year (1993-1997) mean *Betula* pollen and maximum temperature (From Corden *et al.*, 2000).

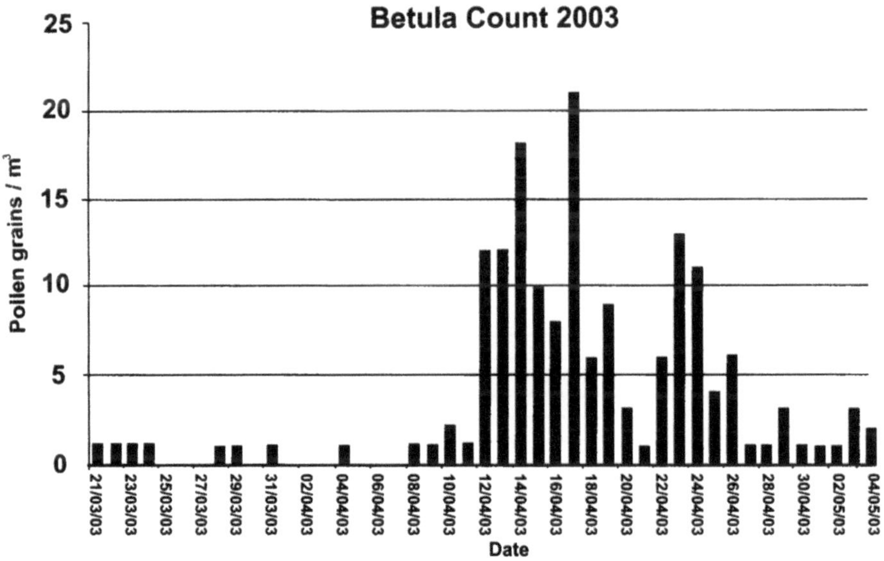

Fig. 17.19 Total *Betula* (Birch) pollen count (March-May) in 2003 at Edinburgh, U.K. (From Eric Caulton, 2003).

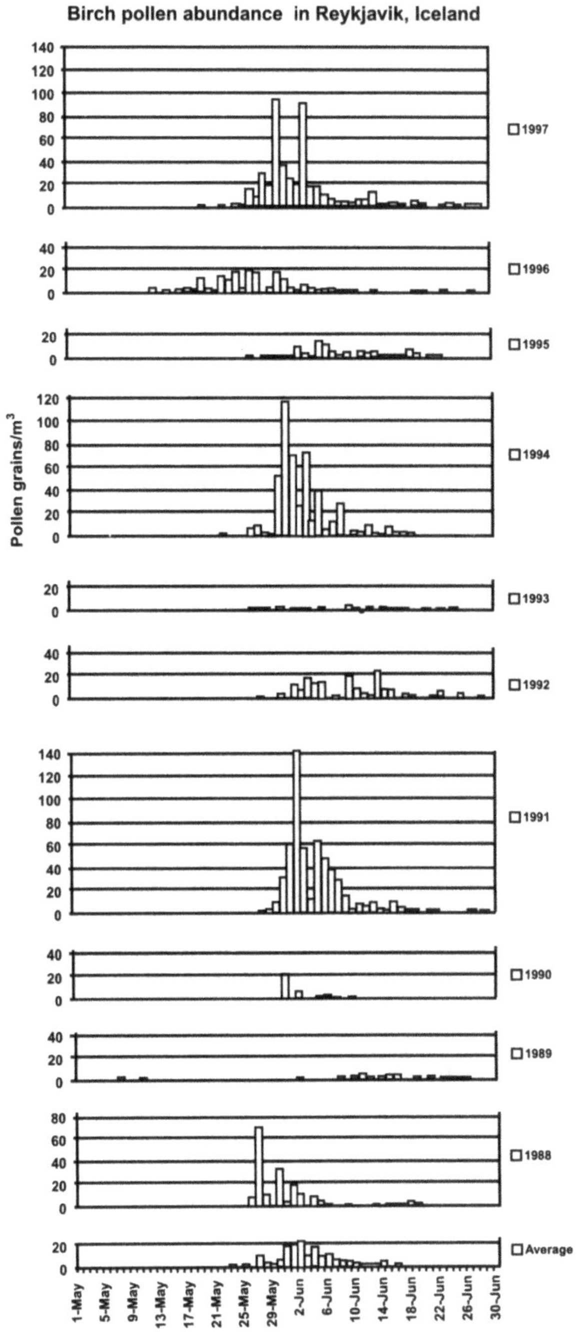

Fig. 17.20 Daily Birch pollen count in Reykjavik, Iceland, plotted for the period May. June each year, and the 10 year average (From Hallsdóttir, 1999).

CHAPTER **18**

Pollen Calendars–Global Scenario

Pollen calendars are very useful for practising allergologists for the diagnosis and treatment of allergy in patients. The compilation of pollen calendars is the ultimate aim of aerobiologists. A pollen calendar, once compiled, has to be updated year after year. In this chapter pollen calendars compiled for different regions of the world have been incorporated.

POLLEN CALENDAR OF SOUTH TAIWAN, CHINA

Two most outstanding Chinese palynologists Huang and Lin (2001) have undertaken a two year study of air spora at Pingtung station in Taiwan from 1993-1995. These studies preceded a comprehensive account of an aeropalynological survey in order to compile a pollen calendar and to establish the relationship between weather parameters and airborne pollen in Taiwan.

The aeropalynological survey was carried out using the Burkard spore trap. As a first step, field botanical studies were carried out in the area around the sampling site, which indicated the occurrence of 118 families, 379 genera and 539 species of vascular plants. Of these, pteridophytes contributed 7 families and 10 species, Gymnosperms, 92 families and 441 species and Monocoytyledons 14 families and 76 species, total pollen count with the relative representation of various taxa are given in tabular form **(Table 18.1)** and pie diagram **(Fig. 18.1)**.

The study supports the statement of Agashe (2001) that pollen monitoring of the atmosphere should be carried out continuously, year after year, as variation occurs in different years. The present study showed a predominance of airborne pollen of *Ardisia squamulosa* in May 1993 whereas in 1994 various predominant airborne pollen taxa consisted of *Trema* (May), grass pollen (June and November), *Fraxinus formosanus* (July), *Alnus* (September), *Alnus formosana* (October), *Mallotus paniculatus* (February), *Broussonettia* (March), and *Trema* (April).

Table 18.1 *Total numbers and percentages of the major aeropollen bi-yearly at Pingtung Station during May, 1993.*

Taxa	*Amount*	*Per cent (%)*
Broussonetia	23,235	15.30
Trema	20,018	13.18
Mimosa pudica	16,114	10.61
Gramineae	15,484	10.19
Alnus	8,327	5.48
Mallotus	7,101	4.68
Macaranga tanarius	5,767	3.80
Lauraceae	3,084	2.03
Cruciferae	2,849	1.88
Areca catechu	2,749	1.81
Ardisia squamulosa	2,725	1.79
Phoenix hanceana var. *formosana*	2,025	1.33
Fagaceae	1,948	1.28
Humulus	1,914	1.26
Syzygium samarangense	1,746	1.15
Acalypha	1,626	1.07
Others	35,180	23.16

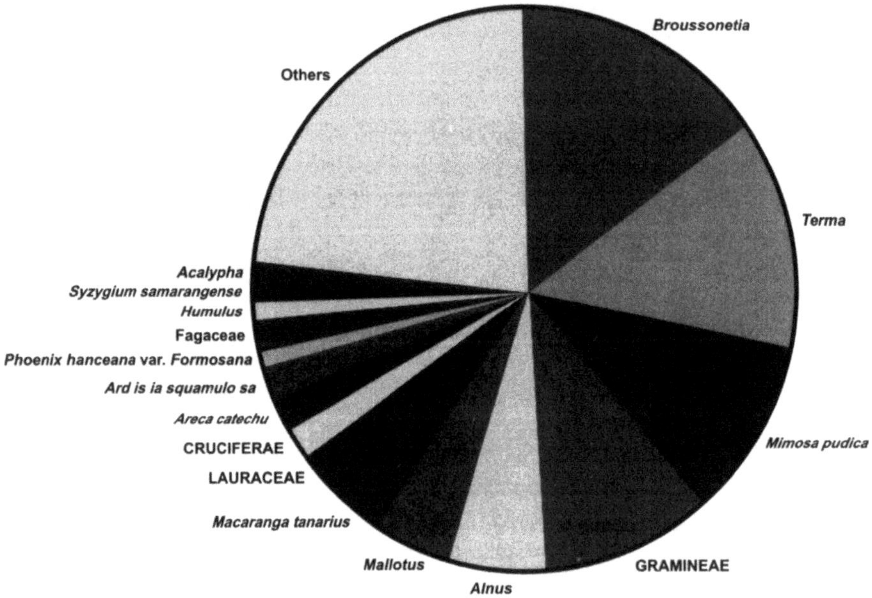

Fig. 18.1 Pie diagram representing relative percentages of the minor aeropollen at Pingtung Station, South Taiwan during May 1993 and April 1995. (From Huang and Liu, 2001).

POLLEN CALENDAR OF JAPAN

Like the U.S.A., in Japan there are basically three pollen seasons based on the habit of their plant sources: the tree pollen season from March-May; the grass pollen season from May-July and the weed pollen season from August-October. However, except in the Hokkaido region where orchard grass and Timothy grass are cultivated as pastures, the weed season is not so pronounced. The grass pollen season appears to be an extended one, i.e., from May to June and August to September **(Fig. 18.1a)**.

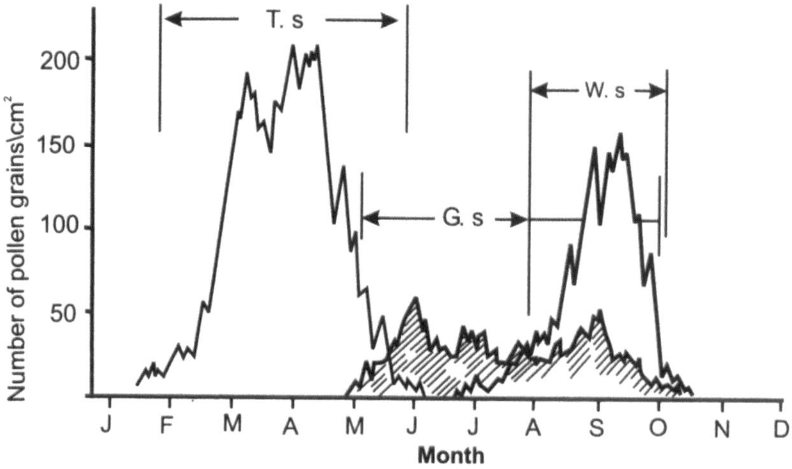

Fig. 18.1a Seasonal changes of anemophilous pollen in Japan (From Iwanami *et al.*, 1988). T.S: Tree season; G.S: Grass season; W.S: Weed season.

A general pollen calendar of Japan was compiled in the Kanagawa Prefecture which was compiled by Shida and the same is presented here in **Fig. 18.1b**. Though anemophilous plant pollen dominate the atmosphere, the pollen of entomophilous plants such as *Solidago altissima* (Goldenrod), *Pyrus serotina* (Pear) and *Nicotiana tabacum* (Tobacco) were also trapped in the exposed slides.

POLLEN CALENDAR OF MOSCOW

Severova and Polevova (2001) compiled a pollen calendar of Moscow by carrying out an aeropalynological surveys during 1993-1995. They used the Burkard seven-day recording volumetric spore trap. Prior to this only a few studies were done by using non-volumetric gravity samplers. Earlier volumetric sampling of airborne pollen of Moscow was initiated by Siwert Nilsson of the Palynological Laboratory, Swedish Museum of Natural History, Stockholm, Sweden (1992).

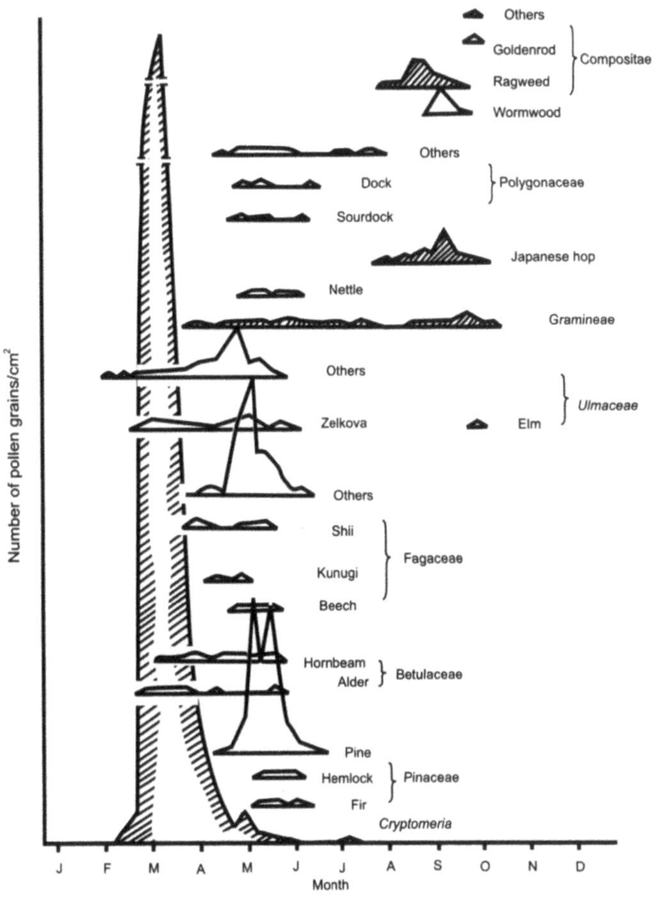

Fig. 18.1b Types and amounts of pollen observed over one year in Kanagawa prefecture, Japan (Recorded by Dr. T. Shida).

Moscow, the capital of erstwhile USSR and now of Russia, is situated within a temperate continental zone characterized by long, cold winters (the coldest month is January with –10°C temperature) and warm summer periods (July being the warmest month with temperature of 18°C). A snow cover exists for about 115 days a year.

The pollen season lasts from mid-March to mid-September. The annual total pollen count fluctuates from year to year and season to season. The daily pollen catch exceeds $1000/m^3$ in the first pollination season for mid-March to mid-May **(Fig. 18.2a, b, c,).** The airborne pollen spectrum during this period consists of *Alnus, Corylus, Salix, Populus, Fraxinus, Ulmus,* and *Acer*. The duration of the main pollen periods of early spring taxa is considerably short which rarely exceeds 25 days. For *Betula* pollen, it varies from 12-23 days whereas for *Alnus* it is from 19-24 days. The pollen spectrum

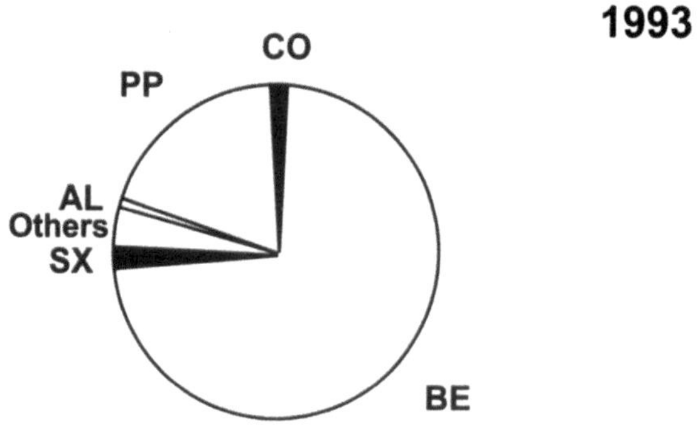

(a)

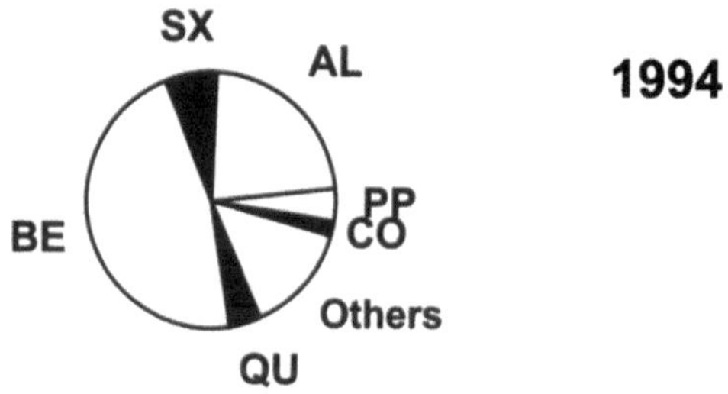

(b)

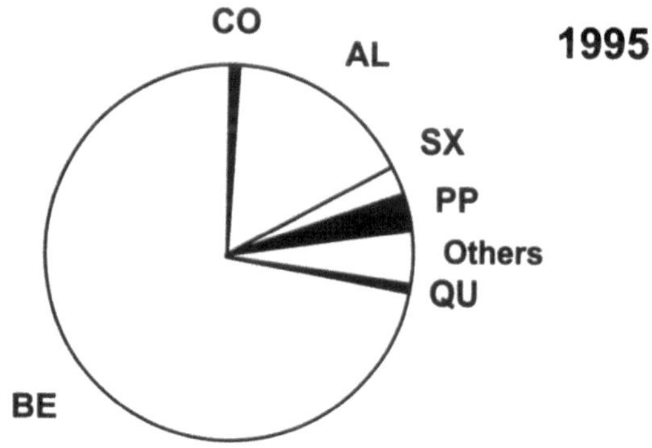

(c)

Fig. 18.2a, b, c Pollen spectrum in the first period of pollination in Moscow. Diameters of pie diagrams are proportional to the number of pollen grains. Index for **Figs.18.2, 18.3** and **18.4:** a – 1993; b – 1994; c – 1995. CO – *Corylus*, BE – *Betula*, PP – *Populus*, AL – *Alnus*, SX – *Salix*, QU – *Quercus*, PO – Poaceae, AR – *Artemisia*, PL – *Plantago*, CH – Chenopodiaceae. (**Figs. 18.2, 18.3, 18.4** adapted from Severoa and Polevova, 2001).

in the second pollination period end of May till to the beginning of July) and the third period of pollination (July to mid September) are depicted in **Figs. 18.3a, b, c** and **Figs. 18.4a, b, c.**

Though 40 different pollen taxa were identified from the air spora, only 14 of them found their place in the pollen calendar on account of their abundance and allergological significance. The total yearly sum of mean daily concentration of airborne pollen have been indicated in the **Table 18.2.**

They also reported perfect correlation of aerobiological and phenological data.

POLLEN CALENDAR OF SYDNEY, AUSTRALIA

Extensive aeropalynological work was carried on by Katellaris (2000) in Sydney, (Australia) with the objective of correlating allergy symptoms with predominant airborne pollen. A Burkard seven-day pollen trap was used by her for surveying airborne pollen.

While compiling the pollen calendar of Sydney, she concentrated on eight predominant angiospermous and two gymnospermous pollen. They are Poaceae (grasses) members, *Casuarina* (Australian pine), *Plantago (Plantain), Platanus (Sycamore), Eucalyptus, Callistemon, Populus* (Poplar, Cottonwood, Aspen), *Alnus* (Alder), *Cupressus*(Juniper family) and *Pinus* (Pine).

Table 18.2 *Total yearly sum of mean daily concentration of airborne in Moscow, 1993-1995, (pollen grains/m³).*

Taxa	*1993*	*1994*	*1995*
Acer	**692**	**118**	
Alnus	*173*	**3493**	**5216**
Artemisia	*1833*	**2636**	**1555**
Betula	*19919*	**7040**	*23965*
Chenopodiaceae	**127**	**136**	**272**
Corylus	*300*	**116**	**277**
Fraxinus	**445**	**569**	
Larix	*9*	**8**	**16**
Picea	**239**	**6**	**77**
Pinus	*897*	**2165**	**2704**
Plantago	**177**	**126**	
Poaceae	**723**	**2175**	**1622**
Populus	*5165*	**700**	**922**
Quercus	*20*	*550*	*270*
Rosaceae	**168**	**125**	
Salix	*591*	**942**	**909**
Tilia	*177*	**34**	**5**
Ulmus		*165*	*288*
Urtica	7387	5430	4185

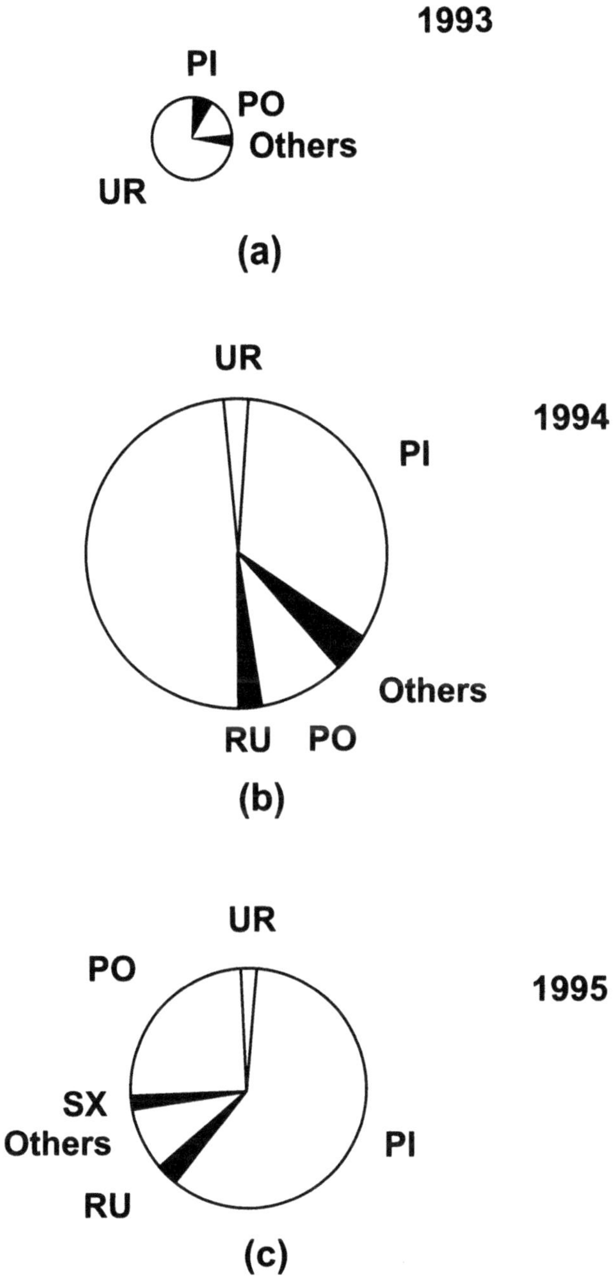

Fig. 18.3a, b, c Pollen spectrum in the second period of pollination in Moscow. Diameters of pie diagrams are proportional to the number of pollen grains.

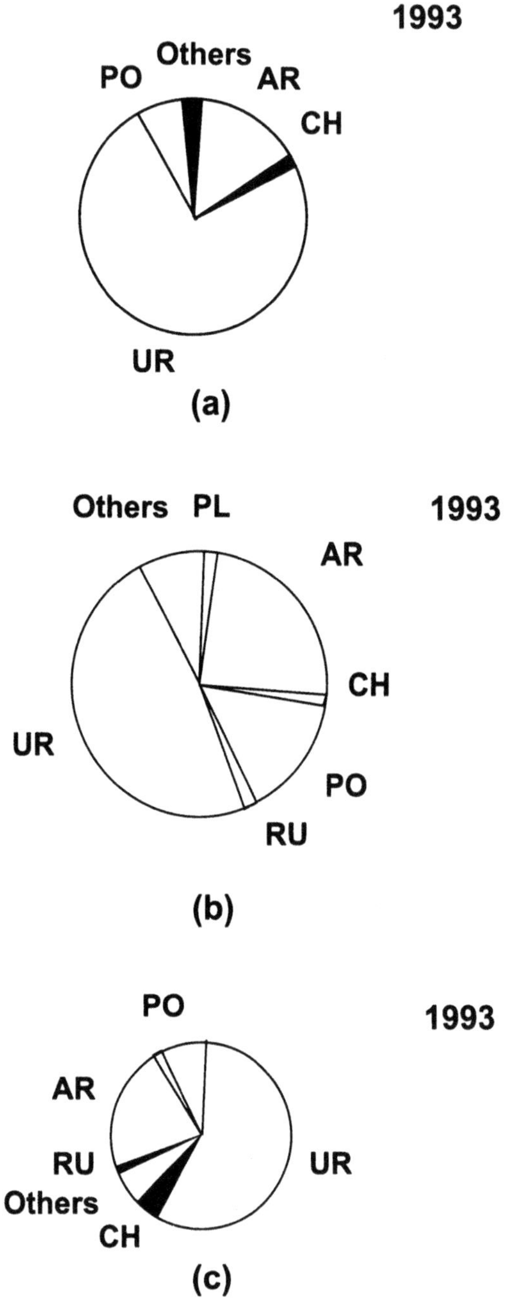

Fig.18.4a, b, c Pollen spectrum in the third period of pollination in Moscow. Diameters of pie diagrams are proportional to the number of pollen grains.

The pollen calendars depicted in **Figs. 18.5** and **18.6** indicate the occurrence of the above mentioned airborne pollen from August to December of 2000. Most of these pollen showed peak occurrences in October.

Ong *et al.* (1995) conducted an aeropalynological survey from 1991 to 1993 in Melbourne, Australia. The survey was undertaken using a Burkard seven-day recording volumetric trap installed on the roof, 14 m above the

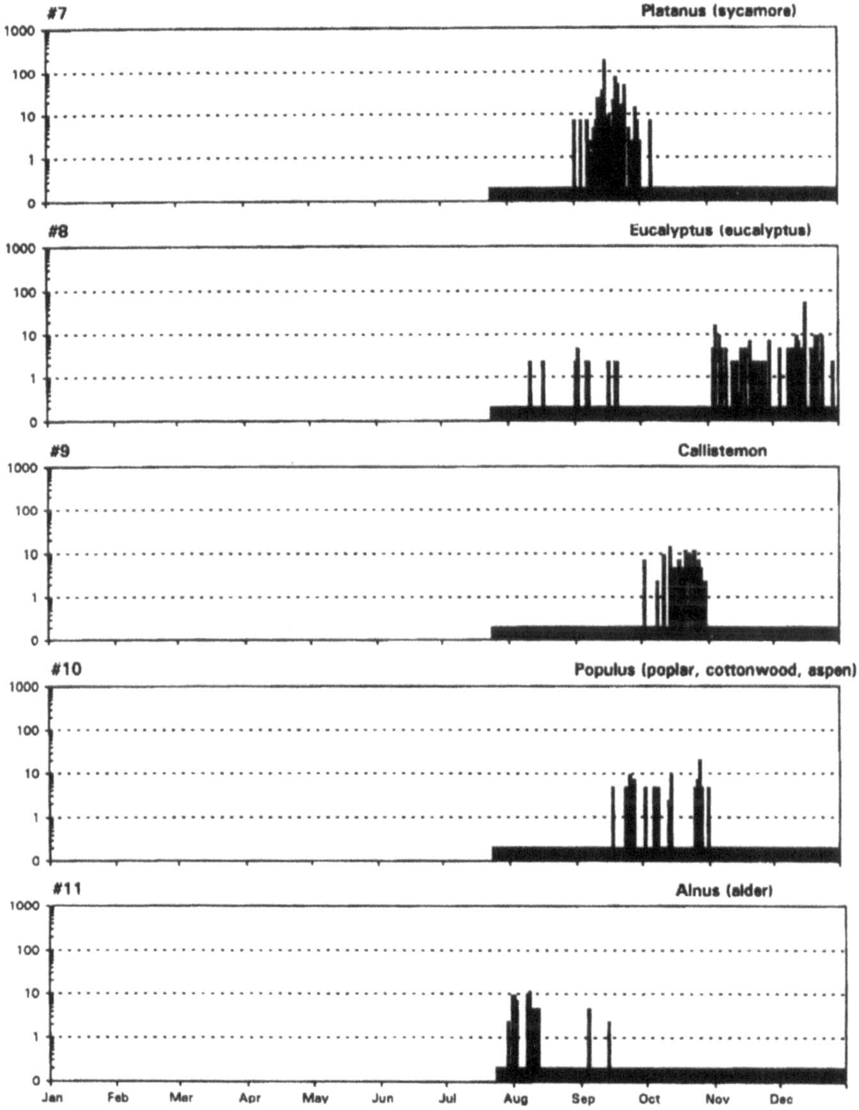

Fig. 18.5 Shortest bars indicate days sampled with a zero count; blanck spaces indicate days that were not sampled.

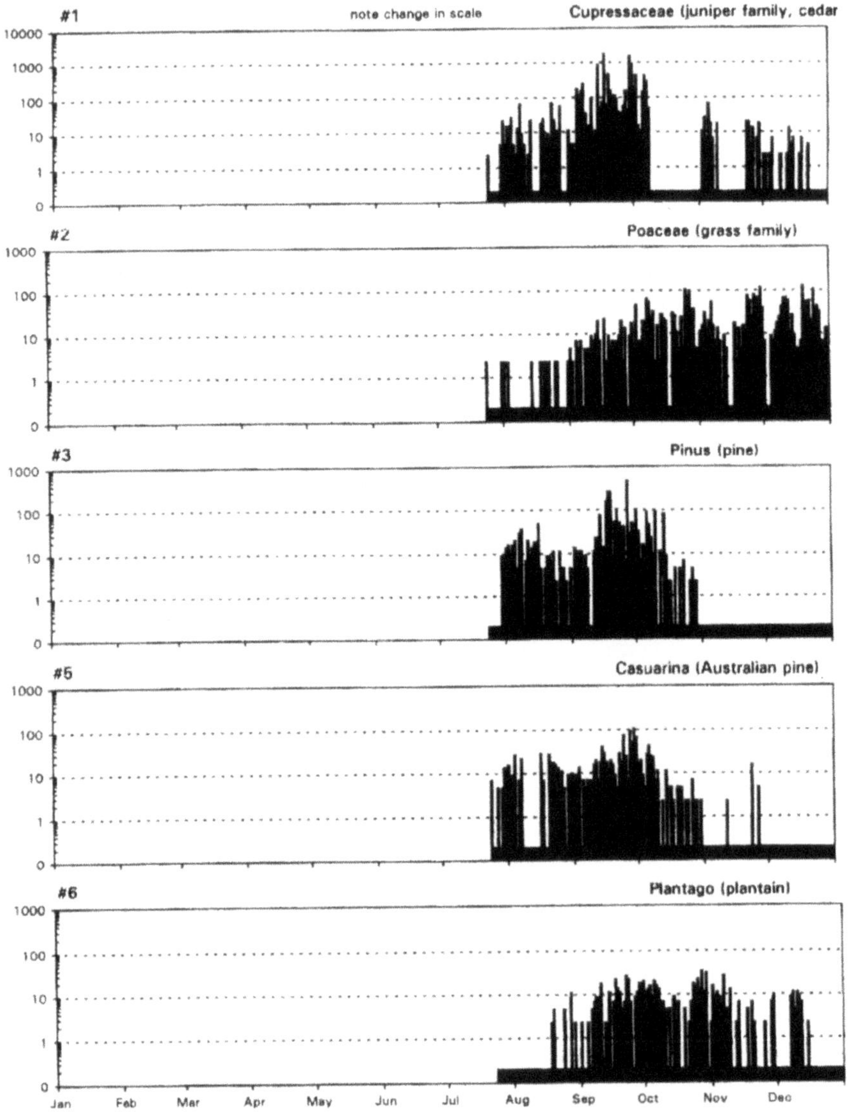

Fig. 18.6 Pollen calendar of Sydney, Australia showing representation of 11 types of pollen during 2000 (**Figs.18.5 and 18.6** Courtesy Katelaris, 2000).

ground level of the National Philosophy Building at the University of Melbourne.

The city of Melbourne is situated in Australia, the climate of which is generally temperate with warm and dry summers and cool and wet winters. The mean daily maximum temperature ranges from 26°C in January, the hottest month and 13°C in July which is the coldest month. In this cool temperate climate many introduced plants exist which are known to release allergenic pollen grains. In addition, there are some native plants which are responsible for causing hay fever. Ong *et al.* have reported the daily seasonal and monthly varieties of different types of airborne pollen which are included in the pollen calendar of Melbourne.

The pollen sampled in the atmosphere of Melbourne were classified into 22 families of flowering plants and conifers. The high pollen incidence in the atmosphere found in winter (July to August); in spring (September to November), and summer (December to January) are depicted in **Fig. 18.7**.

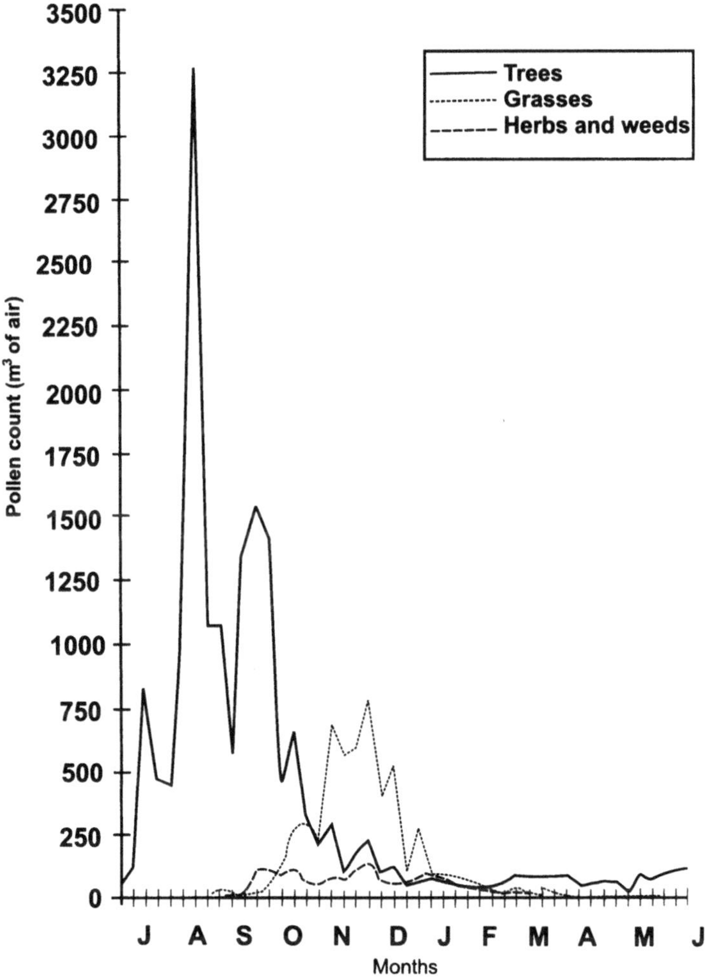

Fig. 18.7 Pollen counts of plants classified into three groups: trees, grasses and herbs and weeds. The graphs shows the average for the two years (1992-1993) of weekly pollen counts from July to June the following year in Melbourne, Australia (From Ong *et al.*, 1995).

Most of the trees flower in the late winter and spring. The summer peak was lower due to grasses, some herbs and weeds. About 62% of these pollen belong to trees. Grasses accounted for 20%, herbs and weeds 9% respectively. **(Fig. 18.7)**.

Among the weeds and grasses, predominant families were CYPERACEAE, TYPHACEAE and POACEAE. The most significant pollen grains during autumn came from the families MYRTACEAE and OLEACEAE with an average of 43% of the total pollen grains trapped where as in winter, *Ulmus, Cupressus*, OLEACEAE, *Pinus*, and *Alnus* represented most of the airborne pollen grains with an average of 93% of the total pollen. The pollen of *Ulmus, Cupressus, Betula* and grassess dominated in the spring with an average of more than 60% of the total pollen sampled. Summer was dominated by grass pollen (52%).

A seasonal incidence of pollen in the form of a pollen calendar is presented in **Fig. 18.8**. The pollen calendar of Melbourne shows the precise seasonal progression of pollen types from tree pollen in winter, grasses in spring and herbs and weeds in summer. Pollen calendars are generally useful in identifying allergies against particular airborne pollen types.

POLLEN CALENDAR OF STOCKHOLM, SWEDEN

El-Ghazaly *et al.* (1993) carried out excellent work on the comparison of airborne pollen from two localities in the Stockholm area. One pollen monitoring site where a Burkard volumetric spore trap was installed was right in the city of Stockholm designated as 'A' and the other site was 15 kms apart at Huddinge, a suburb of Stockholm, which was designated as 'B'. Pollen monitoring was carried out at A from 1973-1989, whereas at B, from 1977-1982. They recorded that pollen counts were higher in Stockholm (A) than at Huddinge (B). The results, particularly the total annual sums of most predominant airborne pollen in both sampling sites did not show any periodicity that is altogether regular and reliable. These results are depicted in the diagrammatic (Pie diagram) **Fig. 18.9**.

ATMOSPHERIC POLLEN OF THE EASTERN HIMALAYAN REGION, INDIA

Sharma (2001) carried out aeropalynological work for a very short duration, in the eastern Himalayan regions such as Darjeeling and Gangtok, Sikkim **(Fig. 18.10)** in northern India. She established a correlation in recent pollen spectra and airborne pollen. The results of aeropalynological work carried out at Darjeeling revealed high representation of *Alnus* followed by *Pinus, Carpinus* and *Betula* pollen. A slightly lesser percentage representation was indicated by pollen of *Larix, Cryptomeria* and *Syzygium* followed by sporadic

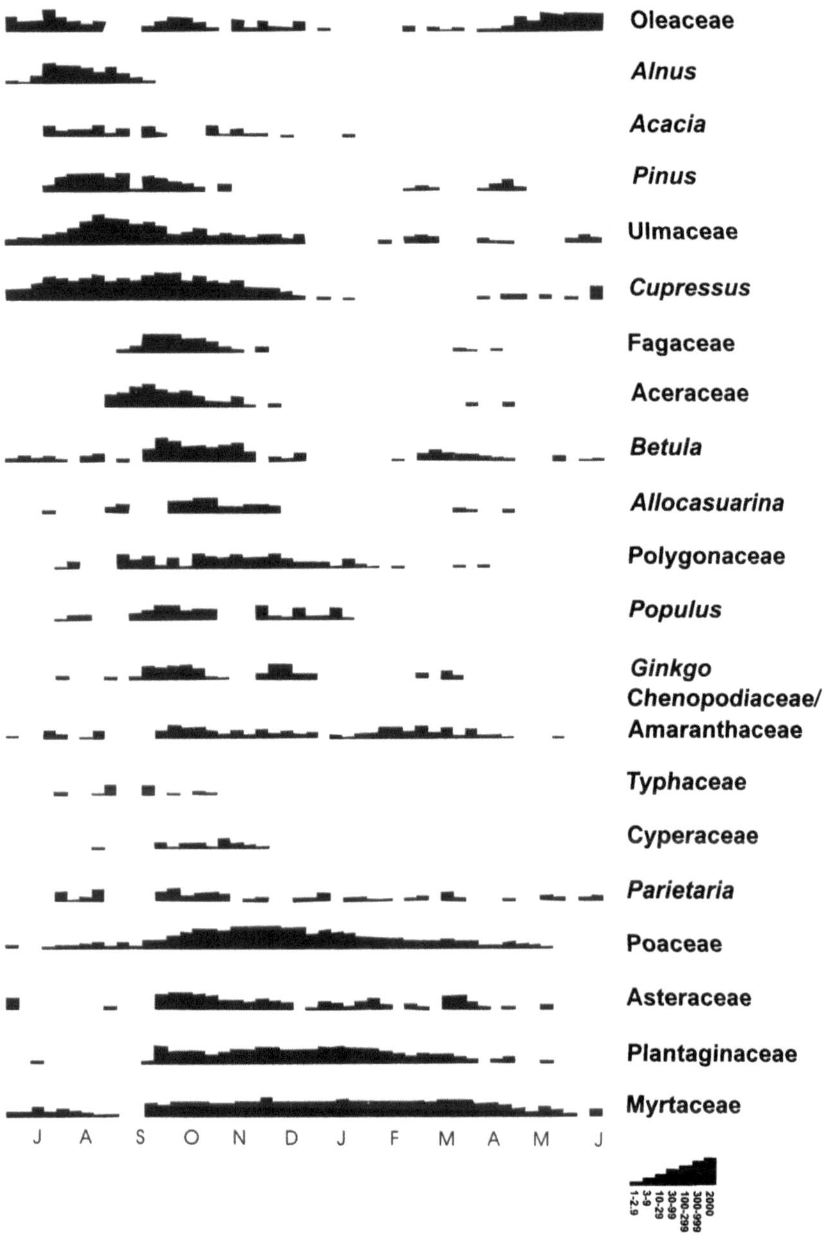

Fig. 18.8 Seasonal incidence (pollen calendar) of airborne pollen of 22 taxa collected from the atmosphere of Melbourne, Australia (1991-1993). The figure shows the average for two years of weekly pollen counts from July to June the following year, in exceptional classes as indicated at lower right figure, following the method of Spieksma *et al.* (From Ong *et al.*, 1995).

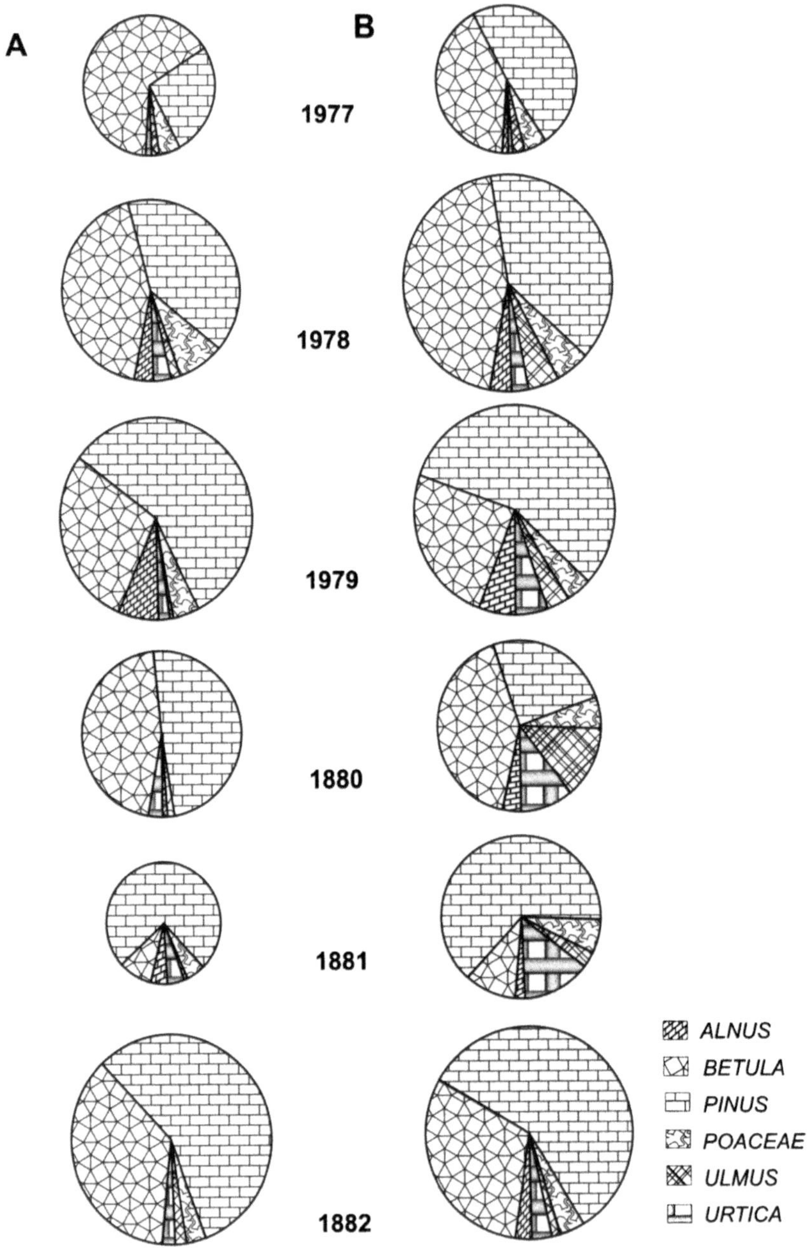

Fig. 18.9 Pollen calendar of Stockholm, Sweden depicted in pie diagrams showing relative distribution of total annual sums for common taxa. Numbers indicate the sum (pollen/ m³) for the illustrated pollen types. A: Pollen monitoring site A – Stockholm city B: Pollen monitoring site B – at Huddinge which is 15 kms from site A (From Gamal El-Gazaly *et al.*, 1993).

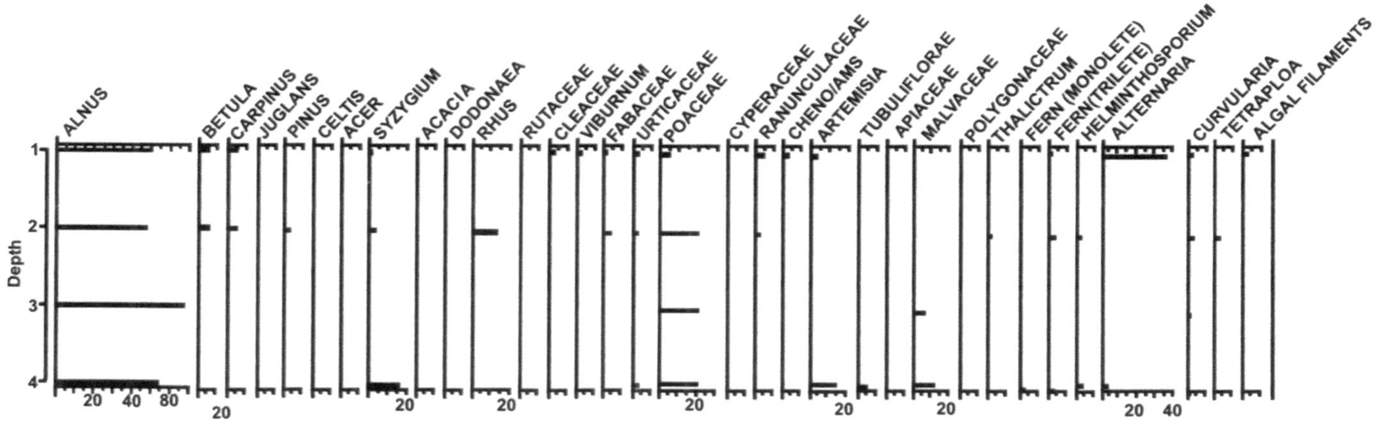

Fig. 18.10 Air catches at Gangtok in Sikkim, India during November, 8-11,1989 (From Chhaya Sharma, 2001).

occurrence of FABACEAE and URTICACEAE. Herbaceous pollen consisted of the predominance of POACEAE followed by Artemisia, Tubuliflorae, CHENOPODIACEAE/AMARANTHACEAE and RANUNCULACEAE. Fern spores were also encountered in the air sampling. Moderate representation of fungal spores was made by *Alternaria, Tetraploa* and *Curvularia.*

Aerospora studies of Gangtok in Sikkim revealed high representation of *Alnus* and low representation of *Betula, Carpinus, Syzygium, Rhus,* FABACEAE, *Juglans, Pinus, Celtis, Acer, Dodonaea,* RUTACEAE and *Acacia* **(Fig. 18.10)**. Pollen of shrubs were represented by *Viburnum* and OLEACEAE. POACEAE pollen occurred in high frequencies followed by pollen of URTICACEAE, RANUNCULACEAE, CHENOPODIACEAE, AMARANTHACEAE, ASTERACEAE and MALVACEAE. Fern spores and mould spores were poorly represented in the air.

ATMOSPHERIC POLLEN OF ISTANBUL, TURKEY

Celenk *et al.* (2006) investigated atmospheric pollen of Istanbul, Turkey in 2005. Though pollen were recorded all round the year, their peak was reported in April. The predominant pollen types in the atmosphere included CUPRESSACEAE, TAXACEAE, *Platanus, Fraxinus, Pinus,* MORACEAE and GRAMINEAE (POACEAE).

ATMOSPHERIC POLLEN OF NUEVO LEON, MEXICO

Rocha-Estrada *et al.* (2006) carried out an aeropalynological survey in the metropolitan area of Monterrey, Nuevo Leon, Mexico in 2004. They used the standard method of air sampling with the help of a Burkard volumetric pollen trap. A total of 26,831.52 pollen grains/m^3 was reported. Significant atmospheric pollen included *Fraxinus, Parietaria pensylvanica, Cupressus, Morus, Celtis,* and POACEAE members.

AIRBORNE POLLEN OF ALBANIA

Pollen calendar for Vlora, Albania was compiled by Gjebrea and Hyso (2006) after accumulating atmospheric pollen data for three years from 2002-2005. The most abundant pollen and the total annual pollen count listed by them include: *Quercus* 18%, CUPRESSACEAE 19%, URTICACEAE 8%, POACEAE 10%, *Plantago* 3%, *Alnus* 1%, *Rumex* 1%, CHENO/ AMARANTHACEAE 2%, and *Corylus* 2%.

ATMOSPHERIC POLLEN IN PORTUGAL

A four-year (2002-2005) study of airborne pollen of various cities in Portugal as part of the Portuguese Aerobiology Network (RPA) was undertaken by Caeiro *et al.* (2006) has resulted in a compilation of a pollen calendar. The chief components of the airborne pollen comprised: POACEAE, URTICACEAE, PINACEAE, CUPRESSACEAE, PALMAE, *Olea europea, Quercus* and *Platanus hispanica.*

ATMOSPHERIC POLLEN OF MUNSTER, GERMANY

Trigo *et al.* (2006) carried out an aeropalynological survey of Munster situated in the northwestern part of Germany. It was a two year study during 2004-2005. A volumetric sampler (Hirst type) was used for trapping airborne pollen. There was a significant quantitative variation in the airborne pollen in 2004 (total annual values of 42,672 pollen grains/m^3) and 2005 (27,650/m^3). The predominant pollen in both years were represented by *Betula,* URTICACEAE, POACEAE, CUPRESSACEAE, *Alnus, Quercus, Pinus, Corylus, Fraxinus* and *Platanus.*

ALLERGENIC POLLEN OF NORTH AMERICA

The Map of North America **(Fig. 18.12)** shows that there are four clear-cut zones which are characterized by typical vegetation. The plants producing allergenic pollen in the vegetation are classified in three main categories such as grasses, weeds and trees. The **Table 18.3** shows the prominent plant component in the four floristic zones namely the northern region, estern region, central region and western region.

POLLEN CALENDAR OF SCOTLAND, U.K.

In 1988, the Scottish centre for pollen studies compiled a pollen calendar **(Fig. 18.11)** based on 10-year review of pollen data. The calendar is published in the form of a colourful leaflet. On the reverse side a short explanatory paragraph in seven languages is given. This provides helpful information for those residents in (and to tourists) Edinburgh and south east Scotland who are susceptible to hay fever and pollen-induced asthma attacks.

Table 18.3 *Allergenic pollen in North American Floristic Zones*

Zone	*Grasses*	*Weeds*	*Trees*
Northern region			
1. Tundra	*Phleum, Poa, Agrostis*		
2. Boreal Forests	*Phleum, Lolium, Poa, Festuca, Dactylis, Agrostis*	*Rumex, Artemisia*	*Pinus, Betula, Alnus, Populus, Salix*
3. Pacific coast Maritime Forest	*Phleum, Lolium, Poa*	*Chenopodium, Amaranthus, Ambrosia, Rumex, Artemisia*	*Betula, Alnus, Ulmus, Acer, Quercus, Populus, Salix*
Western region			
4. California	*Phleum, Lolium, Poa, Festuca, Cynodon, Sorghum*	*Chenopodium, Amaranthus, Ambrosia, Rumex, Artemisia, Salsola*	*Alnus, Ulmus, Acer, Quercus, Olea, Populus, Fraxinus, Salix, Platanus*
5. California Prairie	*Phleum, Lolium, Poa, Festuca, Cynodon, Sorghum*	*Chenopodium, Amaranthus, Atriplex, Ambrosia, Rumex, Salsola*	*Ulmus, Acer, Quercus, Olea, Populus, Fraxinus*
6. Great Basin/ Columbia Plateau	*Phleum, Lolium, Poa, Dactylis, Cynodon, Sorghum*	*Chenopodium, Amaranthus, Atriplex, Ambrosia,* Rumex, *Artemisia, Salsola, Xanthium*	*Ulmus, Populus, Juniperus*
7. Hot Desert region	*Cynodon, Sorghum*	*Chenopodium, Amaranthus, Atriplex, Ambrosia, Baccharis, Rumex, Artemisia, Salsola, Xanthium*	*Ulmus, Quercus, Populus, Juniperus, Acacia, Fraxinus, Tamarix*
8. Western Cordillera	*Festuca, Phleum, Dactylis, Lolium, Cynodon, Sorghum*	*Chenopodium, Amaranthus, Atriplex, Ambrosia, Plantago, Rumex, Artemisia, Salsola, Xanthium*	*Pinus, Ulmus, Quercus, Populus, Acer, Juniperus, Alnus*
Central region			
9. Short-Grass Prairie	*Festuca, Phleum,* Poa, Dactylis, *Lolium, Cynodon, Sorghum*	*Chenopodium, Kochia, Amaranthus, Ambrosia, Rumex, Plantago, Artemisia, Salsola, Xanthium*	*Quercus, Populus, Ulmus, Fraxinus*
10. Tall-Grass Prairie	*Festuca, Phleum, Poa, Dactylis, Lolium, Cynodon, Sorghum*	*Chenopodium, Kochia, Amaranthus, Ambrosia, Rumex, Plantago, Artemisia, Salsola, Xanthium*	*Quercus, Populus, Acer, Ulmus, Carya, Fraxinus*

Contd.

Table 18.3 *Contd.*

Zone	*Grasses*	*Weeds*	*Trees*
11. Lower Mississippi Valley Flood Plain	*Phleum, Poa, Dactylis, Lolium, Cynodon, Sorghum, Paspalum*	*Chenopodium, Kochia, Amaranthus, Ambrosia, Rumex, Plantago, Artemisia, Xanthium*	*Quercus, Populus, Acer, Ulmus, Salix, Taxodium, Carya*
12. Central Hardwoods	*Phleum, Poa, Dactylis, Lolium, Cynodon, Sorghum*	*Chenopodium, Kochia, Amaranthus, Ambrosia, Rumex, Plantago, Artemisia, Xanthium*	*Quercus, Populus, Acer, Ulmus, Carya, Celtis*
Eastern Region			
13. Upland Southeastern Forests	*Phleum, Poa, Dactylis, Lolium, Cynodon, Sorghum*	*Chenopodium, Amaranthus, Ambrosia, Rumex, Plantago, Artemisia, Xanthium*	*Quercus, Populus, Acer, Ulmus, Carya, Pinus*
14. Lowland Southeastern Forests	*Phleum, Poa, Dactylis, Lolium, Cynodon, Sorghum, Paspalum*	*Chenopodium, Kochia, Amaranthus, Ambrosia, Rumex, Plantago, Artemisia, Salsola, Xanthium, Baccharis*	*Quercus, Populus, Acer, Ulmus, Carya, Myrica, Pinus, Liquidambar*
15. Northern Hardwood Forests	*Phleum, Poa, Dactylis, Lolium, Cynodon, Sorghum*	*Chenopodium, Amaranthus, Ambrosia, Rumex, Plantago, Artemisia, Xanthium*	*Quercus, Populus, Acer, Ulmus, Betula, Pinus*
16. Southern Florida	*Cynodon, Sorghum, Paspalam*	*Chenopodium, Amaranthus, Ambrosia*	*Quercus, Casuarina, Myrica, Schinus, Taxodium*

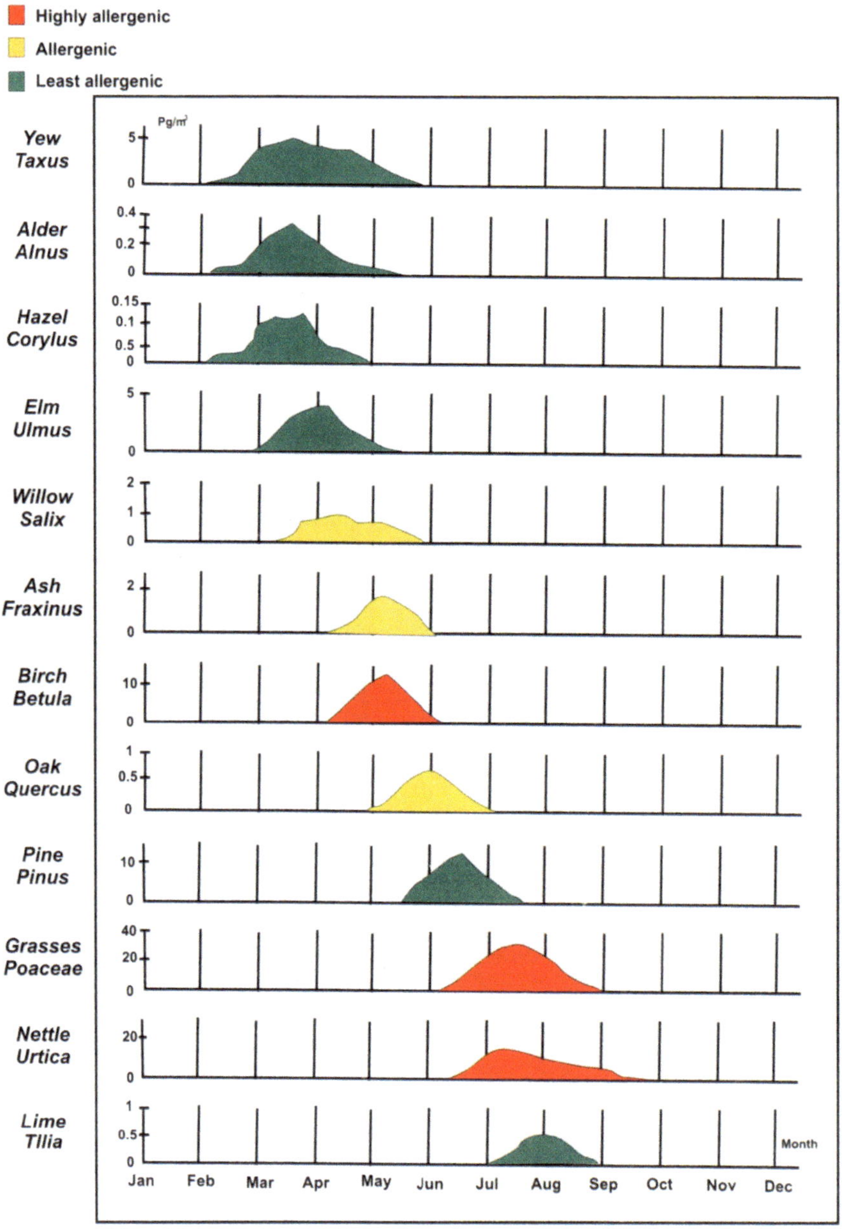

Fig. 18.11 Pollen calendar of Scotland for 1988 compiled by Eric Caulton.

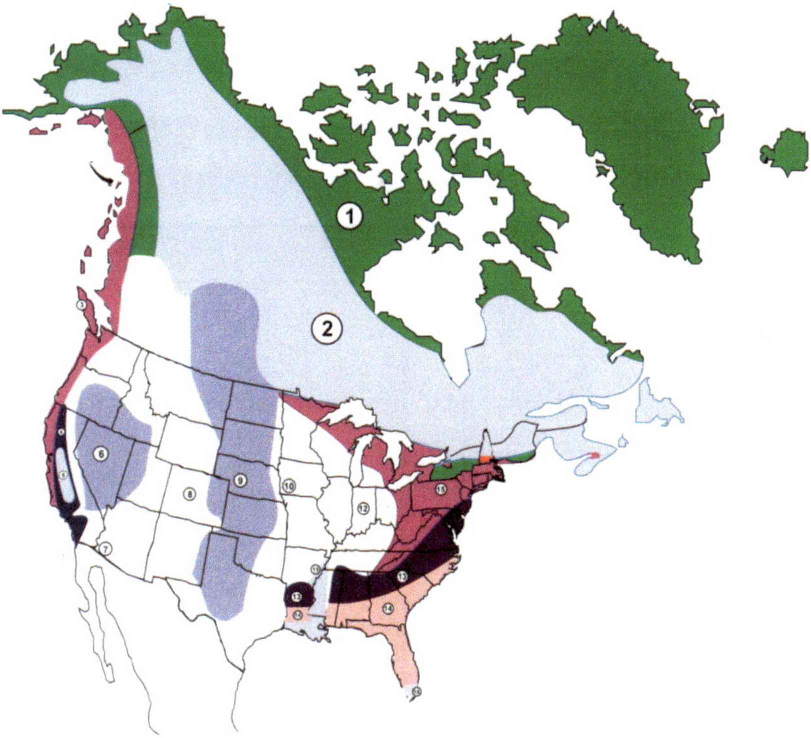

Fig. 18.12 Map of North America showing different zones of airborne pollen sources.

CHAPTER 19

Applications of Aerobiology: Pollen Analysis and Meteorology

By 1990, awareness of the impact of asthma and hay fever (seasonal rhinitis) on the medical services and loss of working hours had reached a point where the need and advantage of setting up of bodies to coordinate the growing number of pollen monitoring sites became an urgent necessity.

The driving force behind this endeavour was among others, Professor Paolo Mandrioli of Bologna University, Italy. Italian aerobiologists were prominent during the 1980s in developing a national network of pollen and spore sampling stations. Similar networks were subsequently developed in Spain, France. Finland and the U.K. Each national network receives and collects pollen data by the pollen counters throughout the season. Grass pollen data from selected sites throughout Europe are forwarded to the computer data base held in Vienna (Austria). These centres/networks constitute the European Aeroallergy Network (EAN).

In 1985, the Bologna group launched a journal entitled AEROBIOLOGIA, whose aims and scope include "the interdisciplinary fields of aerobiology: bioaerosols, transport mechanisms, biometeorology, climatology, microbiology, aeromycology, aeropalynology with (its) links to respiratory allergology, plant pathology, biological weathering, indoor air quality, industrial aerobiology and cultural heritage".

One of the first major collaborative efforts resulted in the publication of a series of maps showing the progressive spread of grass pollen during the 1990 season throughout Europe (Jäger and Mandrioli 1992). Some 171 contributors from 18 countries supplied data. The above collaborative exercise in mapping the grass season, as it progressed in time across Europe, was repeated two years later (Jäger and Mandrioli 1994). Maps relating to birch (*Betula* spp.) were also included in the project.

MEDIA APPLICATION

Perhaps the most important application of the monitoring programme is the provision of allergenic pollen and spore data to the media: press, radio and television networks.

The need to develop a reasonably reliable pollen forecasting service has been recognized as a necessary and advantageous development of the monitoring programme. Research into data analysis, by means of using data of selective taxa e.g., *Quercus* (oak) and *Fraxinus* (ash), can help in predicting the onset of birch pollen release. The application of pollen curves from different regions in the temperate zones, is being employed to produce predictive computer models (Jones *et al.*, 1994).

MEDICAL APPLICATION

The increasing incidence of both pollinosis and asthma in the population at large has involved atmospheric pollen and spore data. More recently, similar studies have been undertaken (Mackay *et al.*, 1992).

The ever increasing attention to and research into the application of aerobiology to medicine is exemplified by the publications involving the relationship between aerobiology and allergology (Morrow-Brown, 1994) and the airborne fungal populations in houses and the health implications. Between 10 and 20% of the world's population is considered to be city dwellers, and this population is increasing. The changing health patterns reflect this shift from the rural environment. City dwellers spend a major part of their lives in the indoor environment-working, at leisure, eating and sleeping, the need to investigate the aerobiological environment for causes of allergies is both paramount and urgent. Public transport, libraries, schools, hospitals, restaurants, and community/leisure centres as well as houses, can harbour pollen, fungal spores, house/dust mites, danders and other biological allergenic agents.

Three highly allergenic airborne pollen allergens are those of mugwort (*Artemisia vulgaris*), Ragweed (*Ambrosia artemisifolia*) and 'congress weed' (*Parthenium hysterophorus)*, plants characteristic of marginal land, hedgerow and waste land. Ragweed, a native of North America, has been introduced into Europe and is well known in Mediterranean countries. It is gradually spreading northwards and is currently a subject of much concern due to its severely allergenic pollen (Déchamp 1995). The pollen of 'congress weed' is highly allergenic and is widely distributed in Southeast Asia. If the predicted climatic warming occurs, then the spread of ragweed northwards in Europe might well encompass the entire continent including the British Isles. A recent report from the Baltic states refers to small pollen counts recorded from some 20 local sources of ragweed over a three-year period (Saar *et al.*, 2000). Concern has also been expressed as to an allergy risk posed by ragweed in Sweden in the event of climatic warming (Dahl *et al.*, 1999). A 14-year study in Vienna recently reported a correlation between sensitization rates with the amount of inhaled ragweed pollen (Jäger 2000).

VETERINARY APPLICATION

Animals are not immune to allergenic pollen. Pollinosis in horses in Scotland has been reported (Dixon *et al.*, 1992). The cause of the condition, apparently common in the U.S.A. but erstwhile not reported in the U.K., was feeding adjacent to a profusely flowering tree of willow (*Salix* sp.). Normally pollinated by bees, willow catkins are borne erect and exposed and their pollen are readily dislodged during anthesis and become airborne. A second investigation involved young German Shepherd dogs undergoing training for blind-guide duties (Fraser *et al.*, 2001). Pollinosis exhibited in this study was suspected to have been caused by pollen of both pine and grass (PINACEAE and POACEAE respectively). The method used in both of these investigations involved analysis of faeces containing ingested grains. (Caulton 1988).

CULTURAL APPLICATION

The role of bioaerosols, moulds, lichens and bacteria in the deterioration of buildings, archival materials and works of art are well documented in the literature. The effect of expired moist air with the relatively high carbon dioxide concentration attained during visits to subterranean limestone caves containing prehistoric cave paintings such as at Lascaux and Altamira, in France and Spain respectively, has necessitated either complete closure or very strictly controlled public access. A potential impact of pollen in subterranean conditions would be as a minor surface contaminant; pollen attached to visitors' clothing being dislodged by movement and carried by air currents to adjacent walls. Recent Italian studies on biodeterioration caused by bacteria and moulds have clearly demonstrated the deleterious effect of airbone microorganisms.

METEOROLOGICAL APPLICATION

The role of climate on a large scale and weather in the more localized situations have a major impact on aerobiology, especially relating to transport locally or over long distances of pollen and spores (Tyldesley 1973). Studies undertaken on the presence of pollen grains in the upper reaches of the atmosphere over the Atlantic produced interesting results (Erdtman, 1937; Lacey and McCartney 1991). The application of airborne pollen distribution and concentration over time to meteorological parameters of rainfall, humidity, temperature, wind direction and speed, enable correlations to be made and an explanation of observed results (Pedgley 1980). In coastal areas, sampling sites of airborne pollen and spores are much affected by wind speed and direction, probably more so than those whose sites are inland.

ROLE OF FUNGAL SPORES IN THUNDERSTORM ASTHMA

Corden and Stepalska (2006) concentrated on the occurrence of airborne fungal spores of *Alternaria* and *Didymella* captured by a Burkard Volumetric Spore Trap in Derby, (U.K.) and in Cracow (Poland). Further, they tried to correlate higher concentration of these spores with asthma admissions to hospitals in a retrospective study for the period 2000-2004 in Derby and 1997-1999 in Cracow. Richordson (1966) studied airborne spores of *Didymella* in EDINBURGH (U.K.) using a Burkard spore trap.

It was reported that ascospores of *Didymella* were highest (42,496) on July 30, 2002 which correlated with high asthma admissions. Further, it was observed that the most important asthma admissions occurred after a thunderstorm with increased fungal spore concentrations. *Alternaria alternata* is known to cause severe asthma and *Didymella* and *Alternaria* have been implicated in thunderstorm asthma in U.K.

FORENSIC AND CRIMINAL APPLICATION

Palynological findings used as corroborative evidence to secure prosecutions in criminal cases is well known, but, for obvious reasons, such forensic evidence on police files is not widely publicized (Faegri and Iversen 1966), Suspected art forgeries, especially by those of the masters, who characteristically worked in the open air when painting, may be subjected to forensic tests to determine whether or not they are genuine. A minute sample of pigment can be removed from the canvas/board, and analyzed for the pollen present. Their presence can confirm, or otherwise, whether or not the airborne pollen spectrum observed in the sample belonged to or originated in the area where the painting was actually undertaken.

The application of pollen analysis in determining the date and location of bodies drowned in peat bogs or buried in ice following an avalanche has proved to be of critical value. The role of pollen analysis in resolving the problems of the approximate date in centuries of bodies deposited in peat bogs and subsequently brought to light, was revealed in a series of discoveries made in Denmark in the 1950s – culminating in the discoveries of Tolland and Grauballe Men. For over 200 years the remains of some 35 bog people have been recorded from Danish bogs alone (Glob 1971).

In 1991 a male body became exposed in the ice on an alpine ridge on the Italian-Austrian border. The body of the 'Ice Man' was subjected to intensive and extensive forensic examination (Spindler 1994).

In both the bog people and the 'Ice Man' respectively, pollen analysis played an important role, where indicator species among the total pollen spectra were able to be used to reconstruct the contemporary landscape, and alongside other dating techniques (e.g., C_{14}), to fix approximate dates of death. Both pollen analysis and radio carbon dating in the Danish examples were found to correlate exactly (0-400 years A.D. or 0-1600 years B.C.) corresponding with the early Iron Age to the Roman Iron Age. In the case of the Tyrolean 'Ice Man', (subsequently named Otzi), some 2,222 pollen grains were recovered from the vicinity of his body. A high count of pollen of Pine (*Pinus*), Alder (*Alnus*) and Meadow grass, (*Poa*); a medium count for Mugwort (*Artemisia*), whilst only individual counts for Birch (*Betula*), Hazel (*Corylus*), Spruce (*Picea*), Elm (*Ulmus*) and Beech (*Fagus*) along with Ivy (*Hedera*), Fern (*Pteridium*), Nettle (*Urtica*) and Plantain (*Plantago*) were recorded and which indicated that their main dispersal period was over. It follows therefore that the ice in which the 'Ice Man' rested was formed between late summer and early autumn-September/October (Spindler 1994). Otzi's' last supper before being killed, following evidence that he was murdered, was described from his stomach contents: various seeds along with conifer pollen.The details of this case have been described in Chapter 20 in this book.

Another example of the application of pollen analysis was only recently revealed, which proved the time of year (season) in which a disputed mass execution of prisoners of war had occurred during the closing years of World War II European Theatre, when Soviet armies were overrunning Germany. The mass execution was shown by the pollen present on the clothing of the exhumed remains to have been carried out by the invading army, not the retreating one, i.e., the Red Army (Radford 1999).

The mutilated torso of a coloured boy aged between 5 and 6 years, was found in the River Thames, London, (U.K.) in September 2001. His death is believed to have been the result of a ritual killing. The identification and origin of the boy, named 'Adam' had proved difficult. The principal clue the police had was from pollen grains found in his stomach, which were of a type not found in the northern hemisphere, suggesting that he arrived in the U.K. from Africa a few days before his death as reported by the correspondent of the Scotsman, on 10-07-02, subsequent police and other enquiries have focussed on a town in northern Nigeria to attempt to identify a missing boy of that age and time.

There are ongoing long-term historic studies on the medication and treatment rendered to pilgrims passing over Soutra in the upland moorland of the Scottish Borders country (U.K.) where a mediaeval hospice was an important outpost of healing and short term respite. Investigations involving the prominent role played by pollen analysis has shed valuable light on mediaeval herbal remedies and nursing practices of the time (Moffat 1989). These studies have linked mediaeval medicine with archaeological excavations of the site.

FORECASTING

The application of aerobiology, especially with regard to allergenic pollen and spores, to forecasting the onset of the pollen seasons, has obvious implications for community and individual health and also for the national and local economy (Gonzalez and Mensaque 1996). The combination of meteorological information (i.e., rainfall and temperature) allows from an estimation of the potential daily pollinosis symptoms during the grass pollen season. Forecasting is at present an inexact science, but is receiving increasing attention in Europe, the U.S.A. and Australia, where years of cumulative pollen/spore and meteorological data allow for computerized models to be constructed (Jones *et al.*, 1994). In areas where less stable climates exist, for example, the U.K., forecasting requires a long period over which data can be analyzed and statistically correlated to produce even-estimated forecasts.

The utilization of parameter data from meteorological, aerobiological, phenological and phytogeographical studies, is another good example of the value of interdisciplinary work related to allergology.

MELISSOPALYNOLOGY

Honey production is an ancient and important aspect of human nutrition and health. Bee keeping is a widespread activity both commercially and privately throughout the world. The blending of different honey types is a commercial activity, bee-keeping and honey production by individuals on a small scale is often described as a 'cottage industry'.

Pollen analysis of honey to determine its type is a labour intensive process. The honey type is determined after counting and evaluating the percentage present of each pollen type. A correction factor is applied to the evaluated figure which enables the honey to be accurately typed as, e.g., 'mixed floral'; 'free blossom'; 'heather'; and oil (seed rape).

Sometimes honey samples can be mislabelled, the error may be due to the assumption that the vegetation in the proximity of the hive(s) is the source of the honey's pollen. After analysis of two such honeys, one labelled 'Orange Blossom' (from Mexico), the other 'Pine' (from Turkey) were found to contain predominantly *Euphorbia* pollen in the former and *Eucalyptus* pollen in the latter!

Pollen present in the honey can possibly be implicated in triggering a type of hay-fever condition, a situation which occurred in a Scottish doctor's surgery around Christmas, 2000 (personal communication).

ARCHELOGICAL APPLICATION

Pollen analysis has a long history as a tool of the archeocologist, particularly with regard to post glacial climatic history in the northern hemisphere by means of pollen analysis of peat deposits, reconstruction of ancient vegetation communities has been possible (Godwin 1975). This area of application and its significance in human evolution and cultural development is now widely recognized (Wood 1968).

Marine acrhaeologists excavating sunken vessels dating back to classical times have recently found pollen to be a useful indicator of the location of the shipyards where the vessels' were constructed. Up to the present this aspect of their sub-marine investigations and discoveries has proved to be a problem. Pollen circulating in the air at the time of the vessels' construction were trapped in the seams and joints of the wooden framework. These pollen long preserved *in situ* over millennia provide a new tool and technique for their extraction and identification. The plant communities represented by the pollen brought to light have enabled archaeologists to match the data obtained with present day representatives of the early plant communities and so pin point the actual locality of the ancient shipyards concerned.

Another example of the role of pollen in submerged fresh water lakes has come from investigations of Bronze Age Cranogs–the dwellings built on stilts sunk into the bottom of lakes, and linked to the land by stilt walkways. (Dixon, unpublished), submerged fractured pottery containing residues of substances involved in food preparation, have revealed solid amorphous organic material containing pollen and spores dating from the time when the broken containers were discarded into the lake. The bracken (*Pteridium aquilinum*) spores and their sporangia with stalks intact, enabled the season of the year concerned, to be determined, as late summer-autumn.

ECOLOGICAL APPLICATION

As pollen analysis has developed over the past half century, so the emphasis in application has changed. In the mid-20th Century, palynology was a minor aspect of study; the emphasis being more concerned with the cytology of pollen development rather than a study in its own right. Plant ecology as developed by Tansley (1953) made only passing reference to pollen analysis, then only emerging as a significant tool, but subsequently underwent important changes with the recognition and identification of communities (plant sociology) and the concomitant concept of **indicator species** of habitat. Pollen analysis of peat and lake deposits became a useful tool in the reconstruction of past vegetation history. This aspect was also of interest and relevant to biogeographers (Tivy 1971; Watts 1971; Cox and Moore 1993). From this point onwards as interglacial and post-glacial vegetation studies became more widely undertaken, the community nature, including indicator species, became more emphasized. The concept of "fossil assemblages" developed and the whole study of vegetation analysis with pollen analysis at the core, took on both a prehistoric and an ecological stance (West 1968; Birks and Birks 1980). The condition and nature of plant fragments alongside pollen *in situ* has also played a part in discoveries relating to fossil remains of man.

AEROBIOLOGY AND METEOROLOGY

Under various global warming scenarios, continuous pollen monitoring could provide early signals that pollen production and dispersal is occurring earlier in the season. Newnham (2001) has pointed out that the North-South lag in onset and peak in grass pollen season in New Zealand indicates that in this country, as elsewhere, routine monitoring of the onset of the pollen season in the lower latitude localities could also help with the accurate forecasting of the pollinosis seasons at higher latitudes. This inference could have important implications for the diagnosis of allergic rhinitis and other respiratory disorders. Thus, it is clear that forecasting of pollen seasons and climate changes have a direct bearing on environmental health. The strong correlation between the onset of pollen season and late Winter, early Spring, temperature has been used as a basis for developing regression models to predict timing and severity of *Betula* (Birch). Pollen seasons from the London area (Emberlin *et al.*, 1993) form the basis of other predictive studies of the onset of pollen season elsewhere.

Many patients had experienced an increase in allergic symptoms during windy days. Hence, Montoyuroy *et al.* (1992) wanted to compare meterological data with clinical data to substantiate the point made above. Wind is a fundamental element of the climate. The south wind, a foehnlike one is special for its thermic and hygrometric characteristic resulting from its particular formation. This south wind is warm, dry and turbulent , often very strong. In order to explain the influence of the south wind on the temporary aggravation of the respiratory allergic symptoms, a release and dispersal of pollen grains was investigated. It was suggested that some of these pollen are not responsible for the real allergies but they can be an irritant.

The influence of rainfall (precipitation) on pollen concentration in the atmosphere has been investigated by many aerobiologists. The general conclusion drawn by these workers indicates a simple and negative relationship i.e., the rainfall will wash out airborne pollen from the atmosphere.

It has been the experience of many allergy sufferers that they feel better during actual rainfall, obviously due to a lower pollen count as the airborne pollen are washed out and settle on the ground. It is said that pollen are to some extent hygroscopic and act as nuclei around which water vapour is accumulated giving rise to raindrops which fall on the ground. However, Norris-Hill and Emberlin (1993) found their observations contrary to the widely accepted view which observations resulted from a four-year study of pollen dispersal in North-Central London (U.K.). On several occasions during their work, they recorded high pollen counts during rainfall. These unusual phenomena may produce unexpectedly high pollen counts and reveal an unusual mechanism of pollen dispersal.

It is possible that besides rainfall, other important meteorological factors such as wind speed, play a significant role. Showery or stormy weather is usually characterized by strong winds independently influencing pollen dispersal. High velocity winds may enable more pollen to become airborne resulting in the increase in pollen counts even during the rainfall. In addition, the size of the pollen grains is also an important factor during washout of pollen from the atmosphere. Large pollen grains are removed from the air at a faster rate than small pollen grains. The small pollen grains of *Urtica* (16 μm in diameter) are recorded at a higher concentration during precipitation than are the larger pollen grains of grasses (25-35 μm in diameter) and *Tilia* pollen grains (ranging from 16 x 33 to 21 x 41 μm). This was the pattern of washout of pollen observed in most rainfall events by Norris-Hill and Emberlin (1993).

No such work has been done in India and elsewhere. It is worth determining the pollen count of the commonest airborne pollen before, during and after rainfall. This can be correlated with wind direction and wind speed in different eco-geographic regions. One more weather dimension can be added to this experimentation, i.e., cloud pattern and density. It would be worth correlating these weather parameters with allergic symptoms in patients.

CHAPTER 20

Forensic Palynology

Forensic palynology is a branch of applied palynology, which deals with use of pollen evidence in detecting crimes and capturing perpetrators. The pollen of living plants or fossil plants store in them a fund of information which is applicable in many fields. In this connection pollen morphology and pollen analysis are the most importent aspects of palynology. The literature on forensic palynology abounds with numerous cases in which a single pollen grain evidence has solved criminal mysteries.

Pollen grains are very small particles ranging from 5-200 μm in size, varying in shape, structure, colour and ornamentation pattern. Pollen grains are like dust particles, which become airborne and are easily carried by wind. They are often sticky and are easily stuck to clothes or any part of the body. The wall of the pollen is highly resistant to physical and chemical forces. Pollen as old as 300 million years have been preserved in rocks.

The most important palynomorphs which are useful in forensic investigation include pollen grains, pteridophytetic spores, fungal spores and other fungal remains. Pollen and spores are present virtually everywhere. They settle on to soils, vegetation, and inanimate objects. These palynomorphs are breathed into the nasal passage and they become trapped in hair, fur and feather. Forensic palynology has been very useful for detection of criminals in various kinds of crimes such as burglary, abduction, rape, assault, forgery, drug trafficking, murder, genocide, homicide,illegal fishing and terrorism.

Palynology finds a number of applications in forensic science because of the various assets that pollen and spores have. The deposition of pollen and spores on to people or objects are three-dimensional. Any part of the body can accumulate pollen and spores. As pollen and spores are very minute particles which cannot be seen with the naked eye, the offender will not be aware of the presence of palynomorphs and will, therefore, be less likely to try to clean contaminated clothing, vehicles, and other objects. Pollen and spores are very difficult to remove from certain materials. There have been several instances where conviction has been possible even when the offender washed his woollen suit in a washing machine. It appears that even washing in biological washing powder does not remove pollen and spores from clothing altogether.

It is advisable that the forensic palynologist has botanical, ecological and preferably, other biological training. Each pollen type has distinguishing pollen morphological features, which help in distinction of different plants. On account of this, pollen and spores are often referred as **"Nature's Finger Prints of Plants"**.

Bryant and Mildenhall (1995) have reviewed very well, various aspects of forensic palynology and described new ways to catch criminals with the help of evidence from pollen and spores. They also listed a number of examples to substantiate the use of forensic palynology in different types of crimes, by giving a number of examples taken from the literature, newspaper reports and television shows.

In their article, they summarized that there are still many misconceptions about the types of information that forensic pollen samples can provide. Often police and other investigators regard palynological results, only as tools that can be used to 'convict' a suspect. The forensic palynology samples are useful tools that can point investigators in the 'right' direction, or narrow the number of suspects, or even eliminate a person as a crime suspect. In spite of this shortcoming, forensic palynology, in a supporting role, can become a powerful tool for the forensic scientist.

In one of the cases, the clothing of the victim and suspect was covered with Southern Birch (*Nothofagus menzsii*) pollen well outside of its normal season. Perhaps the pollen was derived from the direct contact with the ground and with the plants in the area. In such cases, the knowledge of air spora of the area will not help much. In this context, Mildenhall (2006) suggested that palynology should be used as a multidisciplinary approach in solving crimes so that different aspects of forensic geology, botany and zoology are used in a holistic way to provide law enforcement agencies with essential information.

A classical example of application of forensic palynology in solving criminal mystery is from Vienna, Austria. In 1959, a crime was committed by a young man who had killed a person and buried the dead body in a swampy sandy area away from the site of the actual crime. The police were able to arrest a suspect, a young man in this connection, who on interrogation, tried to prove his innocence and stated that at the time of murder he had been climbing a steep sandy mountain. The boots worn by him were handed over to Dr. Wilhelm Klaus, a palynologist working at the time with the geological survey. Though the suspect had cleaned his boots, the palynologist was able to scrap less than a gram of dirt from these boots. Dr. Klaus carried out the pollen analysis of this dirt and was able to recover as many as 1,200 pollen grains. He identified most of these pollen grains, which belonged to *Salix* sp., (Willow), *Picea* (Spruce) and a plant called *Filipendula* (Meadow Sweet) and a 20 million year-old hickory grain. The pollen analysis indicated the source of the dirt not from high dry land but a river basin, because of the ancient *Carya* (Hickory) grain located at a specific swampy outcrop 20 km north of Vienna.

With these findings, the suspect was interrogated again and was informed that he had lied earlier and was actually near a place called Danube. The shocked suspect later confessed his crime and led the police to the scene of the crime pinpointed by pollen evidence to a shallow grave; in damp ground from where the police uncovered the body of the murdered man.

Earlier a well-known Zurich criminologist named Max Frei who specialized in a number of cases that hinged on such minute pollen clues incorporated many cases, some of which are enumerated below:

1. In a criminal case the suspect had indicated that he had not used his pistol since the last March in committing a murder. However, a palynologist was able to identify *Hamamelis* sp. (Witch Hazel) and *Betula* (Birch) pollen on the newly greased gun barrel, which was in the possession of the murderer.
2. A critical palynological observation revealed a grain of *Cedrus atlantica* (Atlantic cedar) pollen embedded in the ink of a signed, dated document proved it a forgery. It was revealed by the palynologist that the forged documents, in this case a 'will', was written in October (as indicated by the cedar pollen in the dried ink of the document) and not in June as claimed by the forger.
3. The body of a murdered woman was transported to a different area away from the crime site in Sweden, where the body was found. Pollen analysis of the sample recovered from the murdered woman's clothes brought to light this finding.

APPLICATION OF FORENSIC PALYNOLOGY IN SOLVING THE DISPUTE REGARDING THE AUTHENTICITY OF THE 'SHROUD OF TURIN'

Forensic palynology has been of great assistance in solving a recent dispute on the authenticity of the 'Shroud of Turin'. One of the most controversial and mysterious disputes in recent historical times has been about the authenticity of the 'Shroud of Turin'. This shroud was supposedly to be the piece of cloth, used to cover Jesus Christ's body after he was crucified and bled to death. The cloth piece bore bloodstains absorbed from the body. It is argued that the original cloth, considered as most sacred was preserved in a silver casket safely deposited in a church in Turin, which was destroyed by a fire. Later, an attempt was made to produce this cloth arguing that this was the original. It was also argued that this cloth remained intact in spite of the fire and was shifted to safety and preserved.

In order to solve the dispute and prove authenticity of the cloth, several agencies performed analysis by biochemical tests including C^{14} radio dating technique. It is said that the cloth was later handed over to a forensic palynologist, who carefully recovered dust from it, which on analysis, yielded a number of pollen grains. On a microscopic examination, it was proved that this was the authentic cloth as the area from where it was recovered was known for the vegetation producing pollen stuck to the cloth.

It was Max Frei, the famous forensic expert who found 49 different taxa of pollen grains trapped in the fibres of the cloth. Comparisons of the shroud's pollen spectrum with pollen from other regions of Israel and the Western Mediterranean revealed similar types. Pollen types recovered from the shroud included desert-type plants (Xerophytes) that still grow in Israel. Other types were represented in nearby areas of Turkey, and a few additional types of plants commonly occurring in the western Mediterranean region. In addition, some of the pollen including *Fagus* (Beech) pollen recovered from the shroud were of types found mostly in central Europe. The conclusion reported by Max Frei was that the majority of the pollen recovered from the shroud represented plants from regions in Israel, the nearby western Mediterranean and Turkey (Wilson 1978). Subsequent to Max Frei's original pollen study, the origin and suspected use of the 'Shroud of Turin' have often been questioned. More recent scientific studies cast serious doubts on the authenticity of the shroud and on the pollen that was reportedly recovered from the shroud.

ROLE OF FORENSIC PALYNOLOGY IN DETECTING ILLEGAL DRUG TRAFFICKING

Wind-pollinated plants of *Cannabis* (Marijuana) produce as many as 70,000 small pollen per anther. These small pollen are known to settle from the air at the rate of 2 cm per second. They are easily dispersed in air during their flowering season. In the summer of 1995, European newspapers reported 'clouds' of *Cannabis* pollen drifting across the Mediterranean from source areas in Morocco, where local farmers reported growing a bumper crop of Marijuana.

It was also reported that inhalation of the pollen grains does not cause hallucination as they do not contain any of the hallucinogenic cannonboids.

Case 1. A person was arrested with a large supply of Marijuana in his possession, but refused to say how or where he had obtained it. Police wanted to know if the drug was from some locally grown source, or if it was imported. If it were imported, might it be part of a larger shipment linked to organized crime? A pollen analysis of the Marijuana sample revealed pollen and spores that were similar to the ones found near where the suspect was arrested. Thus, it was reasoned that the sample was probably from local, homegrown sources (Stanley 1991).

Case 2. Over the course of several weeks, three different people were arrested with fairly large quantities of Marijuana in their possession. Each of the suspects was arrested in a different nearby city and each claimed that he did not know either of the other two suspects. However, an examination of the samples revealed that all three were packaged in a similar manner and might have come from the same source. A pollen analysis of each sample confirmed that all three were from a single large shipment and that the geographical origin was Southeast Asia, thus suggesting a large distribution network run by organized crime.

Case 3. A shipment of 500 g of Cocaine hydrochloride was seized in New York (City). The cocaine was sent for forensic pollen analysis in hopes that the pollen might provide clues about its origin and shipment route. There were three distinctly different types of pollen present in the sample. One group of pollen represented tropical plants currently growing in regions of Bolivia and Columbia, which undoubtedly became mixed with the sample when the coca leaves were being picked and then processed to form coca paste. A second group of pollen consisted of types from Jack Pine (*Pinus* sp.) and Canadian Hemlock (*Tsuga* sp.) trees. As these two trees commonly grow together only in a few regions of North America, it meant that after arriving in the United States, the Cocaine must have been 'cut' and then packaged in one of the following locations: a) northern Michigan or Wisconsin, b) the mountains along the Canadian border of northern New York, or c) in the mountain regions of upper New Hampshire or Maine. The third and final group of pollen in the sample came from weed plants that commonly grow in vacant lots throughout downtown New York (City) and the Manhattan Island. In summary, the pollen data showed that the Cocaine originated and was processed in South America, then smuggled into some northern area of the United States where it was probably 'cut' and packaged. From there it was sent to New York (City) where it was probably 'cut' again, and was being prepared for distribution when it was seized (Stanley 1992).

Some more case examples related to forensic palynological analysis for the detection of crimes are quoted below:

Ex. 1. When the wooded crates reached their final destination, the company found only sacks of dirt. Investigators determined that somewhere en route the machinery had been stolen and the crates had been refilled with sacks of soil. Samples of the dirt were sent to a forensic laboratory and examined for pollen. The types of pollen and spores recovered from the soil showed a geographical association with plants normally found in South Africa. One of the ports where the ship stopped was Capetown, South Africa. Several months later, the missing machinery was discovered in a South African warehouse.

Ex. 2. A shipment of Scotch whisky was shipped overseas. After it arrived, the shipment was unpacked, but only limestone rocks were present. As both the county of shipment origin and the country of destination had ample deposits of limestone, authorities were unsure where the whisky had been illegally removed. Rumours soon spread that the whisky had been removed by local workers after it arrived at its final destination. To confirm or deny these rumours, local authorities requested that samples of the limestone be examined in hopes of discovering the source region of those rocks. The resulting analysis revealed an array of fossil palynomorphs (dinoflagellates and acritarchs) that linked the rocks to the country where the whisky had been produced, not the country where it had been sent. This forensic test proved that the whisky had been removed before the shipment had been sent, not after it had reached its final destination.

Ex. 3. A French antique chest was sold at auction. Later, the chest was tested to see if it was authentic. One of the tests was a forensic examination of the dirt and dust trapped inside each of the drawer's small keyholes. To accomplish this test, the lock was removed and the dirt and wood shavings trapped in the lock's corners were examined. The pollen analysis revealed none of the pollen types that one would expect to find in that region of France, where the antique chest was purportedly made or used. Although the pollen evidence did not prove the antique was a fake, it created doubt and led the owner to pursue an additional test, which eventually revealed that the item was indeed, a fake.

Ex. 4. A classical and complicated case in which forensic palynology has been extremely useful was thoroughly described by Graham (2001).

The case consists of the crash of a private airplane in Ruidoso, New Mexico (U.S.A.) in December 1989. The crash was blamed on a pellet of biological material found in the fuel control unit of the aircraft. Pollen analysis of the material (pellet) revealed the biological contents, pollen, trichomes and insect remains. Subsequent investigation revealed that the pellet was actually a nest of the bee *Ahmeadiella* built after the crash, while the wreckage was in an outdoor storage facility.

The pilot and his wife were on their way to a vacation home near Ruideo from San Diego, California. They were flying a Beechcraft Super Kingair F90. The weather turned very rough with blowing snow. The plane lost the Non-Directional Beacon and was flying off course above the clouds. After it emerged from the clouds, it was heading straight down at full power, to a fiery crash in the mountains. The wreckage of the plane was transported back to Sierra Blanca Regional Airport in the spring of 1990 and remained there uncovered. When the attorneys from Philadelphia visited the site, a mass of biological material about 4 mm in diameter was found in a metal tube referred as B2 elbow which was part of fuel control unit of the plane.

A law suit was filed at San Diego on behalf of the two children of the dead couple. It was argued by the plaintiff's attorney that the engine had ingested biological contaminants from the air which blocked the fuel line causing the plane crash.

A team of scientists was given the following tasks:

1) To identify the various components of the biological material;
2) To determine how it got into the fuel line;
3) To assess whether it was a pre-crash or post-crash accumulation.

Walter Lewis, Professor of Biology, Washington University, St.Louis (U.S.A), who carried out the pollen analysis of the biological mass indicated the presence of pollen of *Grindela* (ASTERACEAE), *Melilotus* (FABACEAE) and *Sphaceralcea* (MALVACEAE). Transmission Electron Microscopy (TEM) of pollen present revealed that some of the pollen still retained the intine and protoplast. Germination experiments on these pollen revealed the emergence of pollen tubes from the apertures of *Grindelia* and *Melilotus.* It was noteworthy that these pollen are produced by insect-pollinated plants, flowering during May and October, and they were found growing in and around the outdoor storage facility of the planes wreckage.

Lewis *et al.* (1990) also carried out an aeropalynological survey in the area and were able to trap only windborne pollen of *Juniperus,* POACEAE, *Morus, Quercus, Pinus,* AMARANTHACEAE/CHENOPODIACEAE and *Ambrosia.* As expected, no pollen of entomophilous plants were present. As mentioned earlier, the plane was flying in winter, during snow when the pollen production was almost nil.

The most important aspect in this civil case was to explain clearly the origin of the biological mass in the B2 elbow of the fuel control unit. The entomologists were also summoned to investigate the biological mass. After a thorough analysis including a SEM examination, revealed the presence of a nest of *Ashmeadiella* bees, which is usually constructed in tubular structures. The nest in this case was constructed by gathering trichomes of *Sphaceralcea* along with a pellet of pollen serving as food for the developing insects. Analytical chemists also gave the opinion that it is impossible that contamination could accumulate at the air pressures operating in the system. The case assumed importance as a large amount of money was involved in addition to the potential awards, insurance ramifications, decisions of the multinational corporations involved regarding the manufacture of small aircrafts and the future of the aviation industry. After a thorough scientific investigation and arguments in the court, the verdict was 10-2 for the defence indicating that the plane crash was not due to contamination of biological mass in fuel control unit, as it was proved to be a post-crash phenomenon.

In an unusual aspect of forensic palynology, Emberlin (2006) has indicated that the pollen found in the nasal cavities of bodies can be used as evidence for the season or even the location of death of the person. It is observed that during the inhalation of air, different types of airborne pollen get deposited in the nasal cavity, especially in the turbinates, and are expelled with the mucous stream created by the ciliary action on the nasal mucosa. During this stream course, the majority of pollen particles will be moved towards the back of the nose where they will enter the throat and be swallowed in a normal living human being. However, in a dead body, the pollen deposited in the nose will remain in the same locality.

In order to understand the whole mechanism, Emberlin carried out some experiments in live subjects. For this purpose, a nasal lavage was collected from each of 11 volunteers in May 2005. They had travelled from within 30 km of the National Pollen and Aerobiology Unit of the University of Worcester (U.K.). The sediments from the lavage were treated and permanent slides were prepared, examined and the pollen identified and counted.

On the same day, airborne pollen were studied using a Burkard pollen trap.She found that, in general, there was a correlation between the results from the Burkard trap and those from the lavages. Hence, this methodology can be adopted to find out the relative time of year and location of the person prior to his death.

Forensic palynology has been extensively used for the last 30 years or so. During this period more than 200 case histories have been established where palynology has been used. It is usually experienced that the pollen numbers decrease very rapidly from source. Thus, the abundance of a specific pollen can give a clue to how close a criminal action was to a parent plant.

Mildenhall (2006) described a classical case in New Zealand, termed as 'Operation Lollipop', which was assisted by forensic palynology. The rape case involved a 14 year-old girl who was grabbed from the street, threatened, and raped near some bushes of *Coprosma* (RUBIACEAE) and taken to another place and raped again. The victim was released later along a remote roadway. *Coprosma* is a very common shrub related to coffee plants in New Zealand.

The suspect confessed having raped the girl, confirmed by DNA, but claimed that he did it at his house. A series of samples from the victim's jeans and jacket were collected and analyzed palynologically. *Coprosma* pollen grains, alongside fungal hyphae and spores were encountered in the above samples. In addition, swabs taken from the victim's vulva and natal cleft also yielded *Coprosma* pollen and other pollen from the scene. Control samples from the alleged scene also contained abundant pollen of *Coprosma.*

The courts accepted this circumstantial evidence as confirming the girl's story and the suspect criminal was sentenced for kidnapping and rape. Mildenhall thus concluded that this case demonstrated the importance of mapping pollen distribution on clothing in order to provide evidence for confirming a scenario of crime.

Forensic palynology appears to be a technique few people know about and even fewer utilize.The full potential of forensic palynology remains untapped and ignored in most countries, except New Zealand and the U.S.A., where forensic palynology is widely accepted and routinely used to gather evidence in civil and criminal cases.

Some of the reasons why forensic palynology is not more widely used might be a result of some or all of the problems listed below:

There are few pollen analysts in the world who have had forensic training or experience and there are even fewer who might be willing to work on forensic pollen samples. Instead, the samples are useful tools that can point investigators to the 'right' direction, or narrow the number of suspects, or perhaps even eliminate a person as a prime suspect. Nevertheless, even in this type of supporting role, forensic palynology can become a powerful tool for the forensic scientist.

Evidence obtained from pollen by pollen-analytical methods has been used extensively for the detection of crimes and criminals in western countries, particularly in Austria, Sweden, Switzerland, New Zealand and the United States of America. Forensic palynological work in India is almost nil. There is an urgent need for the police department to consider establishing forensic palynological laboratories by the State governments, and also include a basic curriculum of forensic palynology in the forensic science departments of the Indian medical colleges.

CHAPTER 21

Applications of Fossil Pollen Studies

A: PALYNOLOGY AS AN AID FOR THE EXPLORATION OF COAL

Coal is one of the most important fossil fuels, which nature has produced and deposited in the sedimentary rocks in different parts of the world. Coal is a rich source of carbon, and is often referred as 'fossilized' sunlight on account of its value as a source of thermal energy. It is a well-known fact that the economic development of any nation depends on fossil fuel deposits and their exploration for various economic and industrial applications.

Coal is primarily a plant product. The source material in the formation of coal comes from organic matter derived from plant debris produced by rich vegetation. Hence fossil plants, both megafossils and microfossils are well preserved in coal seams and coal-bearing rocks (underlying and overlying sedimentary rocks) such as sandstones, shales and limestones.

Pollen grains and spores produced by plants are most ideally suited for their preservation in coal on account of the following assets they possess:

1. Pollen and spores are very minute in size and can be easily carried by transporting agents such as wind and water to the site of sedimentation;
2. The outer wall of pollen and spores is most resistant and durable on account of the presence of sporopollenin, which is a complex proteinaceous polysaccharide;
3. The pollen and spores are characterized by highly variable pollen morphological features, which help in their precise identification;
4. Fossilized pollen and spores can be easily recovered from the sediments including coal by using simple laboratory techniques;
5. On account of variability, durability, restricted distribution in time and space, pollen and spores can be used efficiently in the palynostratigraphic correlation of coal seams.

A variety of microfossils such as pollen grains of pteridosperms, other types of gymnosperms, ptederophytic spores, different types of fungal remains including spores occur in coal. However, the most abundant among them are the pteridospermous pollen and pteridophytic spores. Some of the commonly occurring pollen and spores in coal have been already illustrated and described by Agashe (1995; 1996).

B: APPLICATIONS OF FOSSIL POLLEN STUDIES IN THE EXPLORATION OF OIL

Petroleum is another very precious fossil fuel which nature has taken millions of years to produce. It is often referred to as "liquid gold" or "golden fluid". It is refined from crude oil deposited in sediments deep in the earth's crust. The word 'petroleum' is derived from the Latin words: *petros* meaning rock, and *olium* meaning oil. It may occur in the rocks of all geologic ages ranging from Cambrian to the Pliocene.

The site of production and the site of deposition of oil may or may not be the same. Oil, once produced, is deposited in the pores or interstices of fine and coarse grained sedimentary rocks such as shales, sandstones or limestone, otherwise called 'reservoir' rocks.

Oil is essentially a mixture of hydrocarbons derived from organic matter. It is believed that planktonic animals and plants are the main sources for the generation of hydrocarbons. Terrestrial plant remains, rich in lipid contents, form a potentially good source material for the production of oil. It has been conclusively proved that oil is of biogenic origin and it has been formed by the metamorphosis of organic remains of plants and animals that lived in the remote past. Paleopalynology is a branch of palynology, which deals with the study of fossil microorganisms such as pollen grains, spores, marine and fresh water algae, Foraminifera and other planktonic forms. These have significant applications in the exploration of oil. The importance of palynology in the exploration of fossil fuel was realized as early as 1936, when the Royal Dutch Shell group studied the pollen and spore successions to solve the problem of correlation of Tertiary lignites and lignitic clays in Mexico.

Paleopalynology plays a significant role in the exploration of oil in five different ways:

1. Stratigraphy
2. Palaeogeographic reconstruction
3. Reconstruction of palaeoenvironment
4. Study of oil source rock
5. Maturation of organic matter

Pollen and spores are ideal bioparticles in exploration of oil because of the following assets:

1. On account of their minute size and the chemically resistant nature of walls, pollen and spores are not destroyed or broken during drilling operations, high temperature and pressures. Hence, pollen and spores can be easily recovered from the rocks in a very complete condition; even the oldest Precambrian metamorphosed sediments have yielded identifiable palynofossils.
2. The relative abundance of pollen and spores, especially in sub-surface samples, allows the use of quantitative methods in their study.
3. The wind-dispersed pollen and spores are easily deposited in all sedimentary environments and therefore they are free from the restrictions of the basin environments to which most organisms are subjected. This characteristic feature permits them better possibilities for finer time stratigraphic correlations, inspite of varying basin conditions. The external morphology of pollen and spores is highly variable which helps in their identification and they are restricted to within a short geological time, which makes them ideal index fossils for local and regional correlations.
4. The occurrence of pollen and spores in fresh water and marine environments enables correlation of standard marine sequences of rocks with non-marine sequences. A small quantity of rock sample is enough to yield a large amount of pollen and spores.
5. The colouration of pollen and spores indicates the highest temperature to which the host rock was subjected subsequent to its formation.
6. Invariably, an artificial system of classification is employed for pollen and spores recovered from oil-bearing rocks.

The methodologies involved in the extraction of pollen and spores from rock samples for microscopic examination have already been described in Chapter 9. Further, the microscopic observations lead to two types of analysis:

Qualitative Analysis

This involves the correct identification of pollen and spores. Histograms or curves for zonation and correlation are drawn for different samples.

Quantitative Analysis

This involves the estimation of quantities in terms of numbers of different types of pollen and spores identified during the qualitative analysis. The quantitative analysis is useful in interpreting the relative abundance of different groups of pollen and spores and concurrently their sources in vegetation of the remote past.

The fossil pollen and spores recovered from rock samples are immensely useful in the reconstruction of the palaeoenvironment favourable for the formation and accumulation of oil. They also help in the biostratigraphic correlation of 'oil-bearing' strata. This is achieved by plotting 'isobotanical lines', that is, the lines, which join localities of nearly equal quantities of similar fossil pollen and spores, and by comparing histograms compiled after carrying out qualitative and quantitative analyses.

Hoffmeister (1954) in his classical research work on the determination of ancient shorelines by means of pollospores, used the technique of plotting isobotanical lines. After studying the pollospore contents of rock samples from several oil wells of Seminole in the State of Oklahoma, (U.S.A.) Hoffmeister (1960) clearly demonstrated that the isobotanical lines ran very close and parallel to the oil fields. It has also been scientifically proven that the ideal locations for the formation and deposition of oil are situated in the continental shelf area or near the shorelines. The sediments deposited under shelf conditions generally yield abundant pollen and spores of the terrestrial plants in association with the marine microplanktons, such as hystericosphaerids and dinoflagellates.

It is a well-known fact, which has been scientifically proven, that pollen rain progressively lessens in a seaward direction. Pollen assemblages are limited in 'near shore', marine and lacustrine sediments. Hence, it is possible to determine the distance and direction of the ancient shorelines by the types and quantities of the microfossils as pollen and spores will decrease in density in a seaward direction with a corresponding increase in the marine forms **(Fig. 21.1)**.

It is therefore very evident that the location of the 'paleoshore lines' bears special significance in oil exploration and that palaeopalynology has been proved to be a valuable tool to geologists and oil explorers in this task.

Microfossil Contents of Oil-bearing Beds

A variety of microfossils belonging to different groups of plants, animals and intermediate types occurs abundantly in the oil-bearing sediments of different geological ages. The microfossils of plant origin include pollen grains, spores, fungal filaments, bacteria, dermal appendages, cuticles, vascular elements and remains of various algal groups such as chlorophyceae, charophyceae, desmids and diatoms. Among the animal microfossils occurring in the oil-bearing sediments, the most prominent are foraminifera, microforaminifera, ostracods, discoasters, chitinozoa, radiolaria, scolicodonts and otoliths, whereas the intermediate types, that is without definite ancestory, include members of the coccolithorids, dinoflagellates, silicoflagellates, hystricosphaerids and acritarchs. These

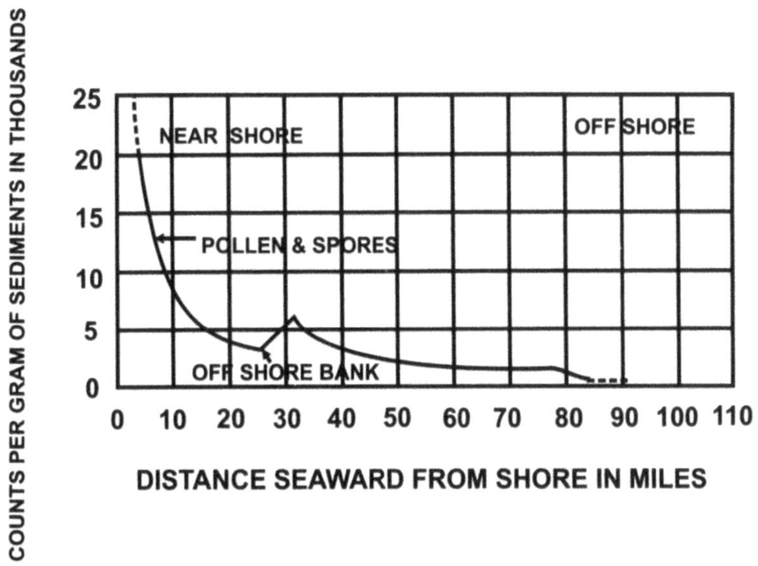

Fig. 21.1 Graph showing decrease of pollen and spore density in a seaward direction (From Hoffmeister, 1954).

forms are also considered to be intermediate between plants and animals by virtue of their mixed characters. A brief description of some of these microfossils has been given in the Chapter 8 on Morphology of Microfossils.

Thus, it can be concluded that palaeopalynology, an applied branch of palynology, has tremendous potential in the exploration of oil.

C: POLLEN ANALYSIS IN THE RECONSTRUCTION OF PAST VEGETATION

There are several reasons why microfossils such as pollen and spores have proved to be valuable as indicators of environmental conditions in the past. They are preserved much more easily than other parts of the plants on account of their structural chemistry. The pollen and spore walls are constructed of a complex chemical substance called sporopollenin. This material is actually a polymer of carotenoids and carotenoid esters. Pollen and spores are generally preserved in considerable abundance in peat deposits and lake sediments. Due to their small size, they can be carried easily for some distance from their source. The pollen and spores are highly variable in their structure and sculpturing.

On account of the above-mentioned assets, pollen grains and spores have proved extremely useful to ecologists, palaeoecologists, medical scientists, medical microbiologists, criminologists and agriculturalists.

Various applications of pollen analysis are listed below:

1. To trace the history of plant groups and species
2. To trace the history of plant communities and their habitats
3. Dating deposits
4. To study climatic history
5. To follow the course of man's influence upon his environment
6. To study the pollen content of the atmosphere and its effect upon human and plant health
7. To study the pollen contents of honey
8. To employ knowledge of pollen and spores in the detection of crime

Some of the above applications have been discussed in various chapters. In the present chapter only the role of pollen analysis in the reconstruction of past vegetation and palaeoclimate will be discussed.

Almost a century ago, a Swedish botanist von Post (1916) published the first pollen diagram showing the occurrence and type of pollen extracted from a peat core.

The virtually indestructible nature of the complex exine of pollen grains, combined with the anaerobic conditions of cumulative layers of dead plant material over long periods of time, enables palynologists to interpret past vegetation.

Peat is primarily produced in those areas of the world, where due to high precipitation, impervious underlying rock formation and on the whole, cold temperatures, dead plant remains can successively be buried and remain more or less intact and recognizable. Most of such areas occur in high latitudes bordering glaciers or in areas of high elevation below the permanent snow and ice zones. Residual pollen trapped and buried occurs at each layer of plant remains and provides a record of the vegetation, which produced it at the time of burial.

The method of illustrating a sequence of pollen extracted from a core has essentially remained unchanged since von Post's diagram was published in 1916. The two familiar axes of a graphical presentation where the x axis represents in most cases time, and the y axis usually records details which were changed for the presentation of pollen occurrence in a peat core: The x axis comprises a series of vertically aligned graphs of each taxon and the y axis concerns depth of core marked into matters or a sub division depending on the depth to which the core descended, as well as a vertically-aligned key to the different soil types within and underlying the peat .

Pollen is quantified at each level examined on a percentage basis of all pollen present at that particular depth. At the end of the core examination a graphic picture is produced which shows the changes in taxa, number and type, as the core is read from top (the surface) to the bottom (often the bed rock).

The changes in the pollen percentages affect the changing environmental conditions prevailing as one sampled level (depth) passes to the next below. The distance between one sampled level and another is determined by the researcher, based in part, on the overall length of the core, and the duration or consistency of a particular deposit – usually in centimeters.

Interpretation of the vegetation extant at any particular level derived from the pollen type and the quantity present, needs to be carefully considered before conclusions can be drawn. This is all the more important as the researcher is investigating vegetation, some thousands of years in the past – perhaps as distant as the glacial melt and retreat some twelve thousand years B.P.

Plants produce varying amounts of pollen. This variation may be affected by weather conditions operating at any given site, but generally speaking, allowing for such local or regional fluctuation, some plants given optimum conditions, produce prolific quantities of pollen. This is especially true of trees whose pollen is dispersed by wind and whose pollination is anemophilous. An Oak tree (*Quercus* sp.) does not produce prolific amounts of pollen characteristic of Birch (*Betula*) or Pine (*Pinus*). Therefore, any attempt to reconstruct vegetation present at any one level (= time in the past) of a core must relate to modern pollen fecundity. This, of course, assumes that the fecundity of flowering plants with regard to pollen has remained virtually the same time for any given species. Working on this not an unreasonable assumption, it is possible to reconstruct vegetation for any given period of time represented in the core.

The reconstruction of vegetation in the past has undergone an interesting development in recent years with the advent of palaeoecology. The living environment does not consist only of plants, but of an association of interacting organisms ranging from bacteria to large animals all of which profoundly affect and influence change in vegetation.

The earlier designation of plant ecology (Tansley 1946) and animal ecology (Elton 1927) has given way to community studies not only today, but extended to the past time. This is essentially an ecological approach, which collectively draws on both flora and fauna extant at any particular point in time.

Palynology in this area of investigation has become palaeopalynology and is an important contributor to the reconstruction of biological communities. Palaeoecology is based, like contemporary ecology, on the concept of communities or assemblages (= fossil assemblages) of different groups of organisms, which naturally interact to produce ecosystems.

Pollen assemblages are indicative of the vegetation present in any one place at any one time. Assemblages of molluscs, Coleoptera and other arthropods, Diptera larvae (the hard non-decomposable elements), can be identified by an appropriate specialist. As with pollen all these assemblages, individually or collectively can act as environmental indicators. Palynology has therefore moved into an ecological phase. Peat is not the only substrate to reveal information about past vegetation communities. Cores extracted from lake bottom deposits, likewise show a succession of change depending on the age of the lake.

There are some time markers observable in pollen cores some 4,000 years ago in northern Europe, for example a noticeable drop in the percentages of pollen of Elm (*Ulmus* spp.). This coincides with the period when Neolithic farmers arrived and selectively felled elms as fodder for their pigs. This elm decline has become a date marker in peat deposits. A more recent (c 2,000 years B.P.) occurred in lime (*Tilia* spp.), because of which and its interpretation is a subject of debate.

The pollen assemblages determined by researchers have bestowed a new and promising area for palynological investigations into the vegetation of the interglacial and postglacial periods.

Applications of Pollen Analysis

The use of pollen data and climatic model simulations provide opportunities to gain better understanding of climate change impact on vegetation (Cheddadi *et al.*, 2001). Reconstruction of various climatic variables for the 6000 B.P. period using pollen data from the African pollen database has been done.

For pollen analytical investigation of Holocene deposits in the Bengal basin, Barui (2003) collected subsurface peat samples from different localities scattered from north to south Kolkata, India, from the vertical sections exposed during the construction of the metro railway project from 1978-1985. In addition, he also collected samples from Hoogli and Howrah belonging to the lower Bengal basin. These samples dated by C^{14} dating technique indicated their age ranging from 7030 to 730 B. P. The dominant fossil spore types extracted from these samples were *Heritiera* along with *Sonneratia, Bruguiera, Suaeda, Phoenix* and *Nypa,* most of which originated from mangrove communities. The fossil pollen assemblage indicated the existence of swampy halophytic vegetation in and around Kolkata about 7030 B.P., which, to some extent resembles the present-day vegetation of the Sundarbans. In addition these samples also yielded large number of fern spores belonging to *Acrostichum, Pteris, Dryopteris* and *Polypodium.* This finding was very significant because these ferns are not largely represented in the present-day vegetation of the Sundarbans.

Pollen Analysis and Reconstruction of Past Vegetation

One of the prime objectives of Quaternary palynological investigation is to reconstruct past vegetation and decipher the paleoclimate. A thorough knowledge of the present vegetation and modern pollen rain spectrum, vegetation patterns and environmental conditions are essential, for this, they are closely linked and interdependent. A change in vegetation in a region is a reliable index of environmental change. With this in mind, Sharma (2001) carried out excellent work in the eastern Himalaya such as, Darjeeling and Sikkim. The Himalayan region is very unique on account of lofty mountain ranges supporting subtropical, temperate and alpine vegetation.

Pollen analysis of recent moss-cushions from these regions revealed the predominance of *Alnus* followed by *Quercus, Betula, Carpinus, Juglans, Rhododendron, Ulmus, Celtis* and *Pinus*. The pollen spectra constructed by analysis of samples from deforested areas of these regions indicate the dominance of non-arboreal taxa such as grasses, CARYOPHYLLACEAE, ASTERACEAE, CHENOPODIACEAE/AMARANTHACEAE.

Pollen Analytical Work at the French Institute, Pondicherry, India

Pollen analytical work concerning both the modern and fossil pollen is being carried out in the well-established Palynology and Ecology Laboratory of the French Institute in Pondicherry, India. The vegetation history of the recent past, that is, the historic period, has been investigated, using palynological records from two sides in the Western Ghats in Kerala. One of the core samples was collected in the Pulicat Lake along the east coast of South India. The sample is dated approximately from 1800 to 1400 B.P. The pollen analysis of the sample indicated the dominance of mangrove pollen and the appearance of *Casuarina* pollen indicating large scale planting of *Casuarina equisetifolia*.

Substantial work was also carried out on the Nilgiri Hills, particularly with reference to pollen analysis and their significance in different climatic regimes.

References

Abrol, D.P. 1993. Insect pollination and crop production in Jammu and Kashmir. *Curr. Sci.* 65: 265-269.

Acosta-Castellanos, S. and Palacio-Chavez, R. 2001. *Aerobiologia*. 8: 277-284.

Acosta-Castellanos, S. and Palacio-Chavez, R. 2001. Plants of apicultural interest in the plume hidalgo zone. Ceaxaca, Mexico; In: D.K. Goodman and R.T. Clarke (eds), Proceedings of the IX International Palynological Congress, Houston, Texas, U.S.A., 1996; American Association of Stratigraphic Palynologists Foundation, pp. 459-469.

Agarwal, M.K., Jones, R. and Yunginger J.W. 1982a. Immunochemical and physiochemical *Alternaria* extracts: a model for standardization of mold allergen extracts. *J. Allergy Clin. Immunol.* 70: 432-436.

Agarwal, M.K., George, P., Menon, M.P.S. and Shivpuri, D.N. 1973. Further studies on the modified rapid method for the preparation of antigenic extracts. *Asp. Allergy Appli. Immunol*. 6: 88-92.

Agarwal, M.K., Shivpuri, D.N., Mukerji, K.G. and Shali, P.L. 1968. Species specific antigens in Fungi. *Aspects Allergy Appl. Immunol.* 2: 69.

Agarwal, M.K., Singh, K., Chanda, S., Nair, P.K.K. and Baldeva, V.S. 1973. Aerobiology and allergens–Methods of research in allergy and related disciplines. *Bulletin 1, Indian College of Allergy and Applied Immunology*. New Delhi.

Agashe, S.N. (ed) 1994. Recent Trends in Aerobiology, Allergy and Immunology. (A collection of plenary lectures and contributory articles) Oxford & IBH Pub. Co. New Delhi. p. 315.

Agashe, S.N. 1979. Circadian rhythms in air-spora with particular reference to some fungal spores and pollen. *The Botaninque* 10: 1-4.

Agashe, S.N. 1989. Airborne entomophilous pollen: Potential Source of Allergy. In: S.T. Tilak (ed) *Environmental Ecology and Aerobiology*, Today and Tomorrow's Printers and Publishers, New Delhi. pp. 153-157.

Agashe, S.N. 1994. Importance of updating pollen calendars in the assessment of allergy. In: S.N. Agashe (ed), *Recent Trends in Aerobiology, Allergy and Immunology*. Oxford & IBH Pub. Co. New Delhi. pp. 43-53.

Agashe, S.N. 1995. Paleobotany: Plants of the past, their evolution. paleoenvironment and application in exploration of fossil fuels. Science Publishers, Inc. USA. p. 359.

Agashe, S.N. 1996. Aerobiology from Laboratory to patients. Presidential address, 8th National Conference of the Indian Aerobiological Society, Bharati Vidyapeeth's Yeshvantrao Mohite College, Pune.

Agashe, S.N. 1997. Melittopalynological work for apiculture development in Karnataka, with particular reference to the contributions by the Palynology Laboratary, Department of Botany, Bangalore University. *Indian Bee Journal* 59: 36-38.

Agashe, S.N. 2000. Role of Aerobiological monitoring in practice of allergy. In: Rajendra Prasad (ed) *Advances in Allergy and Asthma.* Shyam Publications, Lucknow. pp. 99-100.

Agashe, S.N. 2003. Public awareness of allergens. Abstract 1, *36th Annual Convention of College of Allergy, Asthma and Applied Immunology, Chennai, India.* mentioned in *Indian J. Allergy Asthma Immunol.* 17: 33.

Agashe, S.N. 2005. Pollen Calendar: A bridge between aerobiologists and allergists. 13th National Conference on Aerobiology, Silver Jubilee celebrations of Indian Aerobiological Society, Institute of Science, Nagpur, India.

Agashe, S.N. 2006. Palynology and its applications. Oxford & IBH Publishing Co. Ltd. New Delhi. pp. 257.

Agashe, S.N. 2007. Impact of aeroallergens as biopollutants on human health. *Proc. Worksop and Exhibition on Air and Water Pollution, organized by the Environmental Association of Bangalore.* pp. 55-59.

Agashe, S.N. and Abraham, J.N. 1988. Pollen Calendar of Bangalore City: Part I. *Indian J. Aerobiol.* 1: 35-38.

Agashe, S.N. and Abraham, J.N. 1990. Pollen Calendar of Bangalore City: Part II. *J. Palynol.* 91: 279-304.

Agashe, S.N. and Anand, P. 1982. Immediate type of hypersensitivity to common pollen and molds in Bangalore city. *Asp Allergy Appl. Immunol.* 15: 49.

Agashe, S.N. and Chatterjee, M. 1987. Aircraft sampling of the upper airspora. *Advances in Aerobiology, Experimentia supplementum* (Birkhavser verlag, Basel). 51: 411-414.

Agashe, S.N. and Elfadil, A.G. 1989. Atmospheric biopollutant monitoring in relation to meteorological parameters. *Grana* 28: 97-104.

Agashe, S.N. and Philip, E. 1990. Correlation of incidence of bronchial asthma and high pollen counts in Bangalore–A retrospective study. *Ann Convention of ICAI.* Abstracts.

Agashe, S.N. and Rangaswamy, B.E. 1997. Melittopalynological studies of honey samples from Bangardka, Dakshina Kannada District, Karnataka. *Indian Bee Journal* 59: 8–11.

Agashe, S.N. and Rangaswamy, B.E. 2002. Bee floral calendar of Dakshina Kannada and Udupi Districts of Karnataka, India. *Geophytology* 30(1 and 2): 47-59.

Agashe, S.N. and Rangaswamy, B.E. 2002. Chemical characterization of *Apis cerana* and *Apis dorsata* honey from Dakshina Kannada District, Karnataka, India. *Indian Bee Journal* 63 (3 and 4): 81-88.

Agashe, S.N. and Scinthia, M.J.D. 1995. Pollen spectrum of pollen loads collected from Thally block, Dharmapuri District, Tamil Nadu. *Journal of Palynology* 31: 207-212.

Agashe, S.N. and Soucenadin, S. 1992. Pollen productivity in some allerginically significant plants in Bangalore. *Ind. J. Aerobiol.* (Special Volume): 63-67.

Agashe, S.N. and Sudha, P. 1997. Studies on seasonal periodicity of airborne fungal spores in Bangalore city. In: S.N. Agashe (ed) *Aerobiology*. Proceedings of the 5th International Conference on Aerobiology, Bangalore 1994. Oxford & IBH Pub. Co. Pvt. Ltd. New Delhi. pp. 31-37.

Agashe, S.N. and Vidya, M.P. 2000. Biopollution in the Bangalore urban environment and its implications on public health. *Proc. Nat. Symp. on Environment. Bangalore University*. pp. 40-42.

Agashe, S.N. and Vinay, P. 1975. Aeropalynological studies of Bangalore City. Part I. Pollen morphology of *Parthenium hysterophorus*. Linn. *Curr. Sci.* 216-217.

Agashe, S.N. and Vinay, P. 1980. Aerobiological studies of Bangalore City, Part II: A preliminary report. *Advances in Pollen Spore Research* 5: 185-193.

Agashe, S.N., Bapat, B.N., Bapat, H.B. and Philip, E. 1994. Aerobiology of *Casuarina* pollen and its significance as a potential aeroallergen. *Aerobiologia* 10: 123-128.

Agashe, S.N., Nagalakshamma, K.V., Chatterjee, M. and Anand, P. 1983. Aeromycoflora of Bangalore (A preliminary study). *Asp. Allergy Appl. Immnunol.* 15: 49-51.

Agashe, S.N., Philip, E. and Anuradha, H.G. 1997. Aerobiological analysis of intramural environment and atopic allergy in Bangalore (India) population. In: S.N. Agashe (ed) *Aerobiology*. Proceedings of 5th International Conference on Aerobiology, Bangalore 1994. Oxford & IBH Pub. Co. Pvt. Ltd., New Delhi. pp. 169-175.

Agashe, S.N., Rangaswamy, B.E. and Mamlakatoi, K. 2002. Monitoring environmental biopollution and its relevance in public health. *Proceedings of National Symposium on Environment-11*. S. Sadasivan *et al.* (eds). pp. 252-255.

Agashe, S.N., Rangaswamy, B.E. and Mamlakatoi, K. 2003. Monitoring environment of airborne pollen and its significance in public health. *Prospects and Problems of Environment*. M.N. Madhyastha (ed). Daya Pub. House, New Delhi. pp. 85-90.

Agashe, S.N., Sudha, P. and Kanaka Rao, K. 1995. Studies in Circadian Periodicity of Fungal Aeroallergens at Vimanpura, Bangalore. *Botany towards 200A. D.* pp. 38-43.

Agnihotri, M.S. and Singh, A.B. 1971. Observations on pollinosis in Lucknow with special reference to offending factors. *Aspects of Allergy Appl. Immnol.* 5: 135-141.

Agnihotri, M.S. and Singh, B.P. 1975. Pollen production and allergenic significance of some grasses around Lucknow. *J. Palynol.* 11 (1 and 2): 151-154.

Agnihotri, M.S., Mishra, S.R. and Misra, U.K. 1980. Study of fungi in bedrooms of asthmatics. (A preliminary report). *Asp. Allergy Appl. Immnunol.* 13: 11-13.

Ahmed, A., Khan, K.A., Ahmed, V.U. and Qazi, S. 1986. The antibacterial activity of juliflorine isolated from *Prosopis juliflora. Planta Medica* 4: 285.

Akihama, T., Omura, M. and Kozaki, I. 1978. Further information of freeze-drying for deciduous fruit tree pollen. In: T. Akihama, and K. Nakakama (eds). Long-term presentation of favourable germplasm in arboreal crops. *Fruit Tree Res Station.* Japan. pp. 1-7.

Al-Doory, Y., Domson, J.F. and Best, J. 1982. Further studies on the airborne fungi and pollens of the Washington, D.C., metropolitan area. *Ann. Allergy* 49: 265-269.

Allit, U. 2006. Slightly amusing story. 8th International Congress on Aerobiology, Neuchatel, Switzerland. Abstracts, p. 81.

Allitt, U. and Stern, M.A. 1997. Asthma and Didymella ascospores following a thunderstorm in Leicestor (England) In: S.N. Agashe (ed) *Aerobiology* Proceedings of 5th International Conference on Aerobiology Bangalore 1994. Oxford & IBH Pub. Co. Pvt. Ltd. New Delhi. pp. 321-326.

Amici, C.B. 1924. Observations microscopiques sur deverses espices de plantes. *Ann. Sci. Nat. Bot.* 2: 471-482, 211-248.

Amoako, J. and Pickard, R.S. 2001. The identification of Nectar sources for the African honeybee, Apis mellifere adonsonii, by the examination of Ghanian honey. In: D.K. Goodman and R.T. Clarke (eds), *Proceedings of the IX International Palynological Congress, Houston, Texas, U.S.A.* 1996; American Association of Stratigraphic Palynologists Foundation, pp. 471-482.

Anand, P., Agashe, S.N. and Manjunath, K. 1981. Aerobiological studies at Bangalore city and scope in immunotherapy. *Proc. Nat. Conf. Env. Bio.* 117.

Andersen, A.A. 1958. New sampler for the collection, sizing and enumeration of viable airborne particles. *J. Bacteriol.* 76: 471-484.

Andrews, H.N. 1961. Studies in Paleobotany. John Wiley and Sons. New York, p. 487.

Anon, J.B. 1994. Emergencies in the allergy office. In: Richard L. Mabry (ed) *Skin Endpoint Titration.* Thieme Medical Publishers, Inc. New York. pp. 46-54.

Anupama, K. and Achyuthan, H. 2003. Late Quaternary environment change. Emerging Issues. French Institute, Pondicherry, India. P. 288.

Anupama, K. 2003. Modern and Paleoenvironmental Studies: An overview of the different projects in the framework of the research programme at the French Institute, Pondicherry. Late Quaternary environment change. Emerging Issues. French Institute, Pondicherry. India. pp. 179-186.

Ariano, R., Panzani, R.C. and Amedeo, J. 1991. Pollen allergy to *Mimosa (Acacia floribunda)* in a Mediterranean area: an occupational disease. *Ann. Allergy* 66 (3): 253.

Ariatti, A. and Comtois, P. 1993. Louis Pasteur: the first experimental aerobiologist. *Aerobiologia* 9: 5-14.

Arobba, D., Troise, C., Voltolini, S. and Negrini, A.C. 1997. Trend of airborne concentrations of *Ambrosia* pollens in Genoa (Italy) during the period 1981-1993. In: J. Charpin and R. Suriyach (eds) *Atlas Europeen des pollens allergisants. (Atlas of European allergic pollen).* Paris. pp 193-200.

Asthma Research Society, Bangalore. 1979. The report of Asthma Research Project for the year 1976-78. p. 96.

Atluri, J.B. and Subba Reddi, C. 2000b. Self incompatibility, nectar resorption and cross pollination in the Indian laurel-*Terminalia tomentosa* (Combretaceae). *Int. J. Environ. Sci.* 26: 27-36.

Atluri, J.B., Veeranna, M. and Subba Reddi, C. 1993. Pollination cycles in Cyperaceae. *Ind. J. Aerobiol.* 6: 19-25.

Audo, Y., Méon, H., Duisit, D., Pontier, J.L. and Déchamp, C. 2006. Comparison between an important terrestrial ground truth identifying short ragweed in different crops with remote sensing data. *8th International Congress on Aerobiologia, Neuchâtel, Switzerland., Abstract volume*, p. 256.

Avetissian, N.M. 1950. Méthod simplifiée d'acetolyse dans la préparations des pollens. *J. Bot. URSS.*, T. 35: 385-386 (Transl. No. 854 by S. Ketchian, C.E.P.D. Paris).

Bapat, B.N. and Bapat, H.B. 1994. The patient of Atopy: A more complete approach. In: S.N. Agashe (ed) *Recent trends in Aerobiology, Allergy and Immunology*. Oxford and IBH Pub. Co., New Delhi. pp. 182-195.

Bapat, B.N., Godbole, S.H., Vartak, V.D. and Wagle, P.M. 1978. Allergenicity of *Parthenium hysterophorous. Indian J. Med. Res.* 68: 1007-1011.

Barss, M.S. and Williams, G.L. 1973. Palynology and nannofossil processing techniques. *Geological Survey of Canada*, Paper, 33: -26: 1-25.

Barui, N.C. 2003. Pollen analytical investigation of Holocene in Bengal basin, India. *Late Quaternary environment change Emerging Issues*. French Institute, Pondicherry, India. pp. 173-176.

Bastos, E.M. 1995. Espectro polinicio de mel produzido esse algumas areas autropicas de Minas Gerasis, Revista Brasileira de Biologia, 55: 789-799.

Batten, D.J. 1981. Palynofacies, organic maturation and source potential for petroleum. In: J Brooks (ed), *Organic maturation studies and fossil fuel exploration*. Academic Press, London. pp. 201-223.

Bawa, K.S. 1980. Evolution of dioecy in flowering plants. *Annu. Rev. Ecol. Syst.* 11: 15-39.

Beaumont, F., Kauffman, H.F., Sluite, H.T. and de Vries, K. 1985. A volumetric aerobiological study of seasonal fungus prevalence inside and outside dwellings of asthmatic patients living in Northeast Netherlands. *Ann. Allergy*. 53: 486-492.

Belli, E. and Tosco, U. 1962. Folia, Allergol. *Immunol. Clin.* 7: 104.

Bharadwaj, D.C. 1957. The palynological investigations of the Saar Coals. Part I. Morphology of Sporae Dispersae. *Palaeontographica*, Bd. 101, Abt. B, if. 5-6, pp. 73-105.

Bhat, K.G. 1998. Effect of honey on bacteria isolated from burn wound infection: an in vitro study. *Biomedicine* 18 (2) 68.

Birks, H.J. and Gibson, W. 1988. Etudie du contenu polliniques des maieres fecalesd' herbivores, selectionnes *Inst. Fr. Pondicherry, Trv. Sec. Sci. Tech.* LXXXV, 301-308.

Birks, H.J.B. and Birks, H.H. 1980. Quaternary Palaeoecology. Arnold. London.

Blackley, C.H. 1873. Experimental Researches on the Causes and Nature of Catarrhus aestivus (Hay Fever or Hay Asthma). Baillere, Tindal and Cox, London.

Blackley, C.H. 1959. Experimental Researches on the cause and Nature of Catarrhus Aestivus. Dawson, (First printed by Baillllierc, Tindall and Cox in 1873).

Blackmore, S. and Ferguson, I.K. 1986 (eds). *Pollen and spores: Form and Function*. Academic Press, London.

Boehm, G. 1987. Aerobiology its past and its future. In: G. Boehm and R.M. Leuschner (eds) *Advances in Aerobiology. (Experientia Supplementum 51)* Birk hauser Verlag, Basel 3-8.

Bolick, M. 1981. Mechanics as an aid to interpreting pollen structure and function. *Review of Palaeobotany and Palynology* 35: 61-79.

Bostock, J. 1819. Case of periodical affection of the eyes and chest. *Medico-chirurgical Trans.* 10 (i): 161-165.

Bostock, J. 1820. On catarrhus Aestivus or summer catarrh. *Medico-Chirurgial Trans.* 12: 437-446.

Brewbaker, J.L. 1957. Pollen cytology and self-incompatibility systems in plants. *J. Hered.* 48: 271-277.

Bricchi, E. Frenguelli, G., Mincigrucci, G., Fornaciari, M. and Romano, B. 1995. Time linkages between pollination and onset of different taxa over an 11-year period in Perugia, Central Italy. *Aerobiologia* 11: 57-61.

Brooks, J. and Burns, M.W. 1968a. The post tetrad ontogeny of the pollen wall and the chemical structure of sporopollenin with older kerogen and new evidence for the possible biological source of chemicals in sedimentary rocks. *Nature*, 220: 678-679.

Brooks, J. and Shaw, G. 1971. *Pollen development and physiology*. In: J. Heslop Harrison (ed), Butterworths, London. p. 99.

Brooks, J. and Shaw, G. 1971. Recent developments in the chemistry, biochemistry, geochemistry and post-tetrad ontogeny of sporopollenins derived from pollen and spores exine. In: J. Heslop-Harrison (ed), *Pollen: Development and Physiology*, Butterworths, London, pp. 99-114.

Brooks, J. and Shaw, G. 1973. The role of sporollenin in palynology. In: *Problems in palynology*, 3rd International Palynological Conference, Novesibirsk, Nauka, Moscow, USSR. p. 80.

Brooks, J. and Shaw, G. 1977. Recent advances in the chemistry and geochemistry of pollen and spore walls. In: Sir J.C. Bose Memorial Lecture, Bose Research Institute, Calcutta, India. *Proc. Bose Res. Int.* 40.

Brown, C.A. 1960. Palynological Techniques, Pub. C.A. Brown, Baton Rouge. p. 188.

Brown, G.K. and Perking, R.M. 1969. Experiments with aircraft methods for pollinations dates. *Date Growers Inst.* Rep. 46: 35-40.

Bryant, V.M. Jr., Mildenhall, D. and Jones, J. 1990. Forensic palynology in the United States of America. *Palynology* 14: 193-208.

Bryant, V.M., Jr. 1974. The role of Coprolite analysis in *Archaeology. Bulletin of Texas Archaeological Society* 45: 1-28.

Buch, H. 1986. Pollen forecast. In: B. Nuchel Petersen *et al.* (eds), *Pollen and Mould spore counts in Denmark. Description of methods*. Pharmacia, Hillerod, Denmark, pp. 330-338.

Buller, H.R. 1909. The production, liberation and dispersal of the spores of Hymeomycetes. Vol. 1. Collins, London.

Caeiro, E., Branadao, R. and Lopes, M.L. 2006. Aeropalynology and pollen sensitization in Evora (Alentejo region, Portugal). 8th International Congress on Aerobiology, Neuchâtel, Switzerland, 2006. Abstracts, p. 286.

Caeiro, E., Branadao, R., Lopes, M.L., Morais de Almeida, M., Gaspar, A., Leitao, M.T., Nunes, C. and Carmo, S. 2006. The Portuguese Aerobiology Network. In: *Airborne Pollen Results (2002-2005)*. 8th International Congress on Aerobiology, Neuchâtel, Switzerland, 2006. Abstracts, p. 205.

Canright, J.E. 1953. The comparative morphology and relationships of the Magnoliaceae-II, significance of the pollen. *Phytomorphology* 3: 355-365.

Caramiello, R.V., Polini, V., Siniscalco, C. and Mercalli, L. 1989. A pollen calendar from Turin 1981-1988 with reference to geography and climate. *Grana* 29: 239-249.

Caulton, E. 2001. Up on the roof with a passion for pollen. *Biologist* 48 (2): 83-86.

Caulton, E. 1988. Rabbit faeces as a source of pollen for dietary and habitat studies. *Journal of Biological Education* 22 (1): 37-40.

Caulton, E. 2003. Pollen and Spore Count-2003. Annual Report of Scottish Centre for Pollen Studies, Scotland, U.K. pp. 6-11.

Caulton, E. 2006. Aerobiological aspects of pollen contamination of bats *(Chiroptera)*. 8th International Congress on Aerobiology, Neuchâtel, Switzerland, 2006. Abstracts, p. 43.

Caulton, E. and Gibson, W. 1988. Etude du contenu pollinique des matieres d' herbivores selectionnes. Inst. Fr. Pondicery, Tran, sci, tech. LXXXV, 301-308.

Caulton, E., Carmichael, R., Minebois., E. and Vaschetti, C. 1997. Calendar of allergenic pollens in Scotland, 1988-1997. Published as a leaflet, Scottish Centre for Pollen Studies, Napier University, Edinburgh, U.K.

Caulton, E., Keddie, S. and Dyer, A.F. 1995. The incidence of airborne spores of bracken *(Pteridium aquilinium* (L.) Kuhn in the rooftop airstream over Edinburgh, Scotland, U.K. In: R.T. Smith and J.A. Taylor (eds) *Bracken: An environmental issue. J.A. International Bracken group* special publication 2 (17): 82-89.

Caulton, E., Leigh-Brown, S. and Ash Kenani, T. 2005. Pollen in the faeces of bats (Chiroptera) as an indicator of foraging habitats. *The Naturalist* 130: 27-31.

Cecchi, L., Malaspina, T. Albertini, R., Zanaca, M., Ridolo, E., Usberti, I., Morabito, M., Dall'Agilo, S. and Orlandini, S. 2007. The contribution of long-distace transport to the presence of *Ambrosia* pollen in central northern Italy. *Aerobiologia* 23: 145-151.

Cecchi, L., Morabito, M., Domeneghetti, Crisci, A., Onorari, M. and Orlandini, S. 2006. Long distance transport of ragweed pollen as potential cause of allergy in Central Italy. *Annals of Allergy, Asthma and Immunology* 96: 86-91.

Celenk, S., Kemal Altunoglu, M., Canitez, Y., Malyer, H., Sapan, N. and Ones, U. 2006. Aeropalynologic investigation of the atmosphere of Istanbul, Turkey-2005. 8th International Congress on Aerobiology, Neuchâtel, Switzerland, 2006. Abstracts, p. 192.

Chaloner, W.G. 1970. The evolution of miospore polarity. *Geoscience and Man* 1: 47-56.

Chanda, S. and Caulton, E. 1999. David Douglas Cunningham (1843-1914) a biographical profile. *Aerobiologia* 15: 255-258.

Chanda, S. and Mandal, S. 1980. Aerobiology in India with reference of Upper respiratory tract allergy and organic environmental pollution. 1st International Aerobiological conf Munich. pp. 288-306.

Chaturvedi, M. 1973. An analysis of honeybee pollen loads from Banthra, Lucknow, India. *Grana* 13: 139-144.

Comtois, P. 1995. The experimented research of Charles H. Blackley, *Aerobiologia* 11: 63-68.

Comtois, P. 1998. Ragweed (*Ambrosia sp.*): Phoenix of allergophytes. Satellite Symposium Proceedings: Ragweed in Europe. 6th International Congress on Aereobiology, Perugia, Italy, ALK- Abello, Denmark. pp. 3-5.

Corden, J.M. and Stepalska, D. 2006. *Alternaria* and *Didymella* spore concentrations in Derby, UK and Cracow, Poland–Can these Fungal spores concentrations have a role in thunderstorm asthma. 8th International Congress on Aerobiology, Neuchâtel, Switzerland, 2006. Abstracts, p. 102.

Corden, J., Millington, W., Bailey, J., Brooks, M., Caulton, E., Emberlin, J., Mullins, J., Simpson, C. and Wood, A. 2000. Regional variations in *Betula* pollen 1993-1997. *Aerobiologia* 16: 227-232.

Cour, P. 1974. Nouvelles techniques de détection des flux et des retombées polliniques: etude de la sédimentation des pollen déposés à la surface du sol. *Pollen et Spores* 16: 103-141.

Cox, C.B. and Moore, P.D. 1993. Biogeography–an ecological and evolutionary approach. (5th edit.) Blackwell, London.

Cresti, M. 2001. In memorium Gamal El-Ghazaly. International Association of Aerobiology News letter. Issue No. 54-55.

Cunnigham, D.D. 1873. Microscopic Examinations of Air. Government Printer, Calcutta.

D'Amato, G., Spieksma, F. Th. M. and Bonini, S. 1991. Allegenic pollen and pollinosis in Europe. Blackwell, Scientific Publishers, Oxford, London, Vienna.

D'Echamp, C. 1995. L'Ambroisie, Un nouveaux fleau Edit Verso, AHUN.

D'Echamp, C. and Meon, M. 2002. Ambroisies pollutants Biologiques. Arppam-Edit. Lyon.

Dahl, Å. Orville. 1952. The comparative morphology of the Icacinaceae VI. The pollen. *Journ. Arn. Arbor*. 33: 252-286.

Dahl, A., Strandhede, S.O. 1996. Predicting the intensity of the birch pollen season. *Aerobiologia* 12: 97-106.

Davet, A., Clot, B., Guibet, S., Van Eck, F., Gaudart, M. and Gumowski, P.I. 2006. Increasing incidence of ragweed (*Ambrosia artemisiifolia*) pollen allergy in the Geneva area: the prespective from a clinical allergy department. 8th International Congress on Aerobiology, Neuchâtel, Switzerland, 2006. Abstracts, p. 309.

Debasis Jana and Bera, S. 2004. *Sonneratia apetala* buch-ham as major nectar source for honeybees during mid-mummer days in Sunderbans, West Bengal, India. *Phytomorphology* 54 (1 and 2): 51-57.

Deshpande, S.U. and Chitaley, D.D. 1976. Pollen Calendar of Nagpur, India. Rev. Paleobot. Palynol. 22: 256-262.

Devadoss, H.K.P., Gideon, V.A. and Pankajavalli, T. 1997. Pollen survey from suburbs of Madras. In: S.N. Agashe (ed) *Aerobiology* Proceedings of 5th International Conference on Aerobiology, Bangalore 1994. Oxford & IBH Pub. Co. Pvt. Ltd. New Delhi. pp. 39- 48.

Dixit, A.B., Lewis, W.H. and Wedner, H.J. 1997. Indoor fungal aerospora of buildings damaged by the U.S. Mississippi river flood of 1993. In: S.N. Agashe (ed) *Aerobiology* Proceedings of 5th International Conference on Aerobiology, Bangalore 1994. Oxford & IBH Pub. Co. Pvt. Ltd. New Delhi. pp. 129-135.

Dixon, P.M., Mcgorum, B. and Caulton, E. 1992. A study of pasture associated respiratory disease in horses. Index of Equine Research in the British Isles and Iceland. British Equine Veterinary Association Trust. 8, 22.

Domingnez-Vilches, E. 1994. Aerobiology as a useful tool in Forensic Sciences. The control of Adulterations in Illegal Drugs. In: S.N. Agashe (ed) *Recent Trends in Aerobiology, Allergy and Immunology.* Oxford & IBH Pub. Co. New Delhi. pp. 55-73.

Doyle, J. and 'O'Leary, M. 1935a. Pollination in *Pinus. Sci. Proc. R. Dublin Soc. NS,* 21: 181-190.

Doyle, J. 1945. Development lines in pollination mechanisms in the Coniferales. *Sci. Proc. R. Dublin Soc. NS.* 24: 43-62.

Driessen, M.N.B., Van Herpen, RMA., Morelands, RPM and Spieksma, F.Th.M. 1989. Prediction of the start of the grass pollen season for the western part of The Netherlands. *Grana* 28: 37-44.

Ducker, S.C., Pettitt, J. and Knox, R.B. 1978. Biology of Australian seagrasses: pollen development and submarine pollination in *Amphiobolis antaractica* and *Thalassodendron ciliate* (Cymodoceaceae). *Aust. J. Bot.* 26: 265-285.

Dumon, H., Charpin-Kadouch, C., Maurel, G., Filipo, R., Queralt, J., Garans, G. and Charpin, D. 2006. Mold species identified in flooded dwellings. 8th International Congress on Aerobiology, Neuchâtel, Switzerland, 2006. Abstracts, p. 110.

Dunbar, W.P. 1903. On the cause and specific cure of Hay fever. R. Oldenbourg, Munich and Berlin.

Durham, O.C. 1946. The volumetric incidence of atmosphereic allergens. IV. A proposed standard method of gravity sampling, counting and volumetric interpolation of results. *J. Allergy.* 17: 79-86.

Dutkiewicz, J. 1997. Airborne Bacteria and Endotoxin as occupational hazards. 1997. In: J. Charpin and R. Suriyach (eds) *Atlas Europeen des pollens allergisants. (Atlas of European allergic pollen).* Sandoz. Parix. pp. 193-200 and pp. 513-525.

Edmonds, R.L. 1979. Aerobiology, Dowden, Hutchinson and Ross, Inc. Strousburg, Pennsylvania.

El-Ghazaly, G., El-Ghazaly, P.K., Larsson, K.A. and Nilsson, S.N. 1993. Comparison of airbone pollen grains in Huddinge and Stockholm. *Aerobiologia* 9: 53-68.

El-Ghazaly, G., Satti, E.B. and Tsuji, S.I. 1997. Intra-specific pollen polymorphism in *Mimosa pigra* L. (Mimosaceae). *Grana* 36: 279.

El-Ghazaly, G. and Fawzy, M. 1988. Pollen calendar of Alexandria (Egypt), 1981-1982. *Grana* 27: 85-87.

Elton, C.S. 1927. Animal ecology. Sidgwick and Jackson. London.

Emberger, L. 1968. Les Plantes Fossiles dans leur rapports avec Vegetaux Vivants, 2nd Edit., Paris. p. 758.

Emberlin, J. 1992. Pollen U.K. The British Pollen Monitoring Network. *Aerobiologia* 8: 47-49.

Emberlin, J. 1994. The effects of patterns in climate and pollen abundancy on allergy, *Allergy* 49: 15-20.

Emberlin, J. 2006. Pollen deposition in nasal cavities; comparisions from the analysis of 22 samples of nasal lavage taken at Worcester, UK during May 2005, and simultaneous pollen counts from a Burkard spore trap. 8th International Congress on Aerobiology, Neuchatel, Switzerland, 2006. Abstracts, p. 65.

Emberlin, J., Jones, S., Bailey, J., Caulton, E., Corden, J., Dubbels, S., Evans, J., Mcdonagh, N., Millington, W., Mullins, J., Russel, R. and Spencer, A. 1994. Variation in the start of grass pollen season at selected sites in the United Kingdom 1987-1992. *Grana* 33: 94-99.

Emberlin, J., Mullins, J., Corden, J., Millington, W., Brooke, M., Savage, M., and Jones, S. 1997. The trend to earlier Birch pollen seasons in the U.K.: A biotic response to changes in weather conditions? *Grana* 36: 29-33.

Emberlin, J., Norris-Hill, J. and Bryant, R.H. 1990. A tree pollen calendar for London. *Grana* 29: 301-309.

Emberlin, J., Savage, M. and Jones, S. 1994. Annual variation in grass pollen seasons in London 1961-1990. Trends and forecast models. *Clin. Exp. Allergy* 23: 911-918.

Emberlin, J., Savage, M. and Woodman, R. 1993. Annual variations in the concentration of Betula pollen in the London area, 1961-1990. *Grana* 32: 359-363.

Engstorm, I. and Nilsson, S. 1979. Pollen calendar for Stockholm somradet. *Lakartidn* 42: 3624-3625.

Erdtman, G. 1934. Uber die verwendung von Essigsäurenhybrid bei Pollenuntersuchungen, sv. *Bot. Tidsk.*, Bd. 28. H, 2: 354-358.

Erdtman, G. 1940. An introduction to pollen analysis. Natham Marschroninca Botanica, Co., U.S.A.

Erdtman, G. 1952. Pollen morphology and plant taxonomy of angiosperms. Almquist and Wiksell, Stockholm. p. 553.

Erdtman, G. 1956. "LO-analysis" and "Welcker's rule". *Svensk bot Tidskr.* 50: 1-7.

Erdtman, G. 1957. Pollen and Spore Morphology/Plant toxonomy. Vol. I-IV.

Erdtman, G. 1960. The acetolysis method. A revised description. *Svensk Botanisk Tidskrift* 54: 561.

Erdtman, G. 1969. Handbook of Palynology, Morphology, Toxonomy, Ecology (An Introduction to the study of pollen grains and spores) Munksgaard, Copenhagen, Denmark. p. 486.

Erdtman, G. and Straka, H. 1961. Cormophyte spore classification. *Geol. Foren. Stockh. Forhandl*, 83, HF. 1, pp. 65-78.

Faegri, K. and Iversen, J. 1964. Textbook of pollen analysis, 2nd Edit. Hafner Publishing Co., New York. p. 265.

Faegri, K. and Iversen, J. 1975. *Textbook of Pollen analysis*. Hafner Pub. Co. New York, p. 295.

Faergri, K. and Iversen, J. 1988. Textbook of pollen analysis IV Edit. John Wiley and Sons. New York.

Falk, R.H. 1980. *Scanning electron microscopy*. SEM Inc., AMF O'Hare, Chicago, IL. 60666, USA. p. 78.

Feinberg, S.M. 1935. Mould allergy. Its importance in asthma and hayfever. *Wisconsin. M.J.* 34: 254.

Felix, C.J. 1961. An Introduction to Palynology. In: Andrews book: "Studies in Paleobotany". An Introduction to Palynology, pp. 436-462.

Frankland, A.W. and Aalberse, R.C. 1987. Silver Birch (*Betula*) pollen allergy and fresh fruit allergy. *Clin. Ecology* 5: 2.

Fraser, M.A., Caulton, E. and McNeil, P.D. 2001. Examination of faecal samples as a method of identifying pollen exposure in the dog. The Veterinary Record. 149: 424-426.

Frenguelli, G., Bricchi, E., Romano, B. and Mincigrucci, G. 1989b. A predictive study on the beginning of the pollen season and daily variations. *Proceedings of the Joint Canadian and Pan-American Symposium on Aerobiology and Health.* Otawa. pp. 11.

Frenguelli, G., Bricchi, E., Romano, B., Mincigrucci, G., Fornaciari, M. and Goretti, G. 1989c. Pollen forecast of some taxa in central Italy. *Proceedings of the International Symposium on Pollinosis in Mediterranean Area*, Naples, pp. 148. 1991.

Frenguelli, G., Bricchi, E., Romano, B., Mincigrucci, G. and Spieksma, F. Th. M., 1989a. A predictive study on the beginning of the pollen season for Gramineae and *Olea europea* L. *Aerobiologia* 5: 64-70.

Frenguelli, G., Spieksma, F. Th. M., Bricchi, E., Romano, B., Mincigrucci, G., Nikkels, A.H., Dankaart, W. and Ferrant, F. 1991. The influence of air temperature of the starting dates of the pollen season of *Alnus* and *Populus*. *Grana* 30: 196-200.

Frinking, W.D. 1987. Philip Herries Gregory FRS. DSc. (1907-1986) *Advances in Aerobiology*. Birkhauser Verlag, Basel 9-10.
Fulton, J.D. 1966. Microorganisms of the upper atmosphere. III. Relationship between altitude and micropopulation. *Appl. Microbiology* 14: 237-240.
Funkhouser, J.W. and Evitt, W.R. 1959. Preparation techniques for acid-insoluble microfossils. *Micropaleontology* 5: 369-375.
Ganeshan, S., Rajasekharan, P.E. and Shashikumar, S. 2004. Pollen cryobank at IIHR, Bangalore. *Technical Report* from the Pollen Storage Laboratory, Division of Plant Genetic Resources, Indian Institute of Horticultural Research, Hessaraghatta, Bangalore, India.
Garg, A. and Nair, P.K.K. 1993. Honeybee pollen load as a bioindicator of bee pasturage in Bhimatal area of Western Himalayas. *Indian. J. Palynol.* 29: 89-107.
Gjebrea, E. and Hyso, E. 2006. Pollen calendar of Vlora (Albania) during 2002-2005. 8th International Congress on Aerobiology, Neuchâtel, Switzerland, 2006. Abstracts, p. 200.
Glob, P.V. 1971. The Bog People. Faber and Faber, London.
Godwin, H. 1968. The origin of the exine. *New Phytol.* 67: 667-676.
Godwin, H. 1975. History of the British Flora (2nd Edit.) Cambridge University Press, Cambridge, U.K.
Goldberg, C., Buch, H., Moseholm, L. and Weeke, E.R. 1988. Airborne pollen records in Denmark. 1977-1986. *Grana* 27: 209-217.
Gonzâlez Minero, F.J. and Fernândez-Mensaque, P.C. 1996. Prediction of the beginning of the olive full pollen season in the south-west Spain. *Aerobiologia* 12: 91-96.
Govind, K. 2004. *Botanical, Clinical and Immunological Studies of Intramural Fungal Spore Aeroallergens of Kerala*. Ph.D. Thesis submitted to MGM University, Kottayam, Kerala, India.
Graham, A. 2001. Forsenic palynology: The Rusdiso plane crash. In: D.K. Goodman and R. Clarke (eds). Proceedings of the IX International Palynological Congress, Houston, Texas, U.S.A., 1996; American Association of Stratigraphic Palynologists Foundation, pp. 609-612.
Gravesen, S. 1979. Fungi as a cause of allergic disease. *Allerg*. 34: 135-154.
Gray, J. 1965a. Palynological techniques. In: B. Kummel and D. Raup (eds), *Handbook of paleontological techniques*; W.H. Freeman & Co., San Francisco. U.S.A. pp. 471-481.
Gray, J. 1965b. Techniques in Palynology: extraction techniques. In: B. Kummel and D. Raup (eds), *Handbook of paleontological techniques*; W.H. Freeman & Co., San Francisco. USA. pp. 530-587.
Green, B.T., Dattman, M., Panula, E.Y., Rutherfold, S. and Simpson, R. 2004. Atmospheric Poaceae pollen frequencies and associations with meteorological parameters in Brisbane, Australia: a 5-year period, 1994-1999. *Int. J. Biometeorology* 40: 172-175.

Gregory, P.H. 1961. Microbiology of the Atmosphere. First Edit. Leonard Hill Books, Aylesbury, p. 373.

Gregory, P.H. 1973. Microbiology of the Atmosphere. Second Edit. Leonord Hill Books, Aylesbury. p. 373.

Gregory, P.H. 1983. Aerobiology: past, present and future. *International Aerobiology News letter* 17: 9-16.

Gregory, P.H. 19951. Deposition of airborne *Lycopodium* spores on cylinders. *Ann. Appl. Biol.* 38: 357-376.

Gregory, P.H. and Hirst, J.M. 1952. Spore content of the atmosphere near the ground. *Nature* 170: 475.

Gupta-Battacharya, S. and Chakraborty, P. 2006. *Phoenix sylvestris* pollen allergy: a two year aerobiological study and immunotherapy follow-up in seasonal allergic patients from an agricultural area of West Bengal, India. 8th International Congress on Aerobiology, Neuchâtel, Switzerland, 2006. Abstracts, p. 296.

Gupta, S. and Chanda, S. 1991. Aerobiology and some chemical parameters of *Pathenium hysterophorus*. *Grana* 30: 497-503.

Hallsdótter, N. 1999. Birch pollen abundance in Reykjavik, Iceland. *Grana* 38: 368-373.

Harrington, J.B., Gill, G.C. and Earr, B.R. 1959. High efficiency pollen samplers for use in clinical allergy. *J. Allergy* 30: 357-375.

Hasselman, H. 1919. Iakttagelser over skogsträdpollens spridnings förmåga-*Meddel. Stat. Skogs förs* anst. 16: 27-53.

Hawksworth, D.L., Kirk, P.M., Sutton, B.C. and Pegler, D.N. 1961. Ainsworth and Bisby's Dictionary of the Fungi. Commonwealth Mycological Institute, Kew, U.K.

Heslop-Harrison, J. 1968. Pollen wall development. *Science* 161: 230-237.

Heslop-Harrison, J. 1971. Pollen: Development and Physiology. Appleton Century-Crofts, Butterworth, London.

Heslop-Harrison, J. 1975. The physiology of the pollen grain surface. *Proceedings of the Royal Society, B,* 190: 275-299.

Heslop-Harrison, J. 1979a. An Interpretation of the hydrodynamics of pollen. *American Journal of Botany* 66: 737-743.

Heslop-Harrison, J. 1979b. Pollen walls as adaptive systems. *Annals of Missouri Botanical Garden,* 813-829.

Heslop-Harrison, J., Heslop-Harrison, Y., Knox, R.B. and Howlett, B. 1973. Pollen wall proteins: gametophytic and sporophytic fractions in the pollen walls of the Malvaceae. *Ann. Bot.* 37: 403-412.

Heslop-Harrison, Y., Reger, B.J. and Heslop-Harrison, J. 1984. The pollen-stigma interaction in the grasses. 8. The stigma ('silk') of *Zea mays* L, as host to the pollens of *Sorghum bicolor* (L) Moench and *Pennisetum americanum* (L.) Leike. *Acta Bot. Neeri* 33: 205-227.

Hills, L.V. and Sweet, A.R. 1972. The use of "Quaternary O" in megaspore palynological preparations. *Review of Palaeobotany and Palynology* 13: 229-231.

Hirst, J.M. 1952. An automatic volumetric spore trap. *Ann. Appl. Biol.* 39: 257-265.

Hirst, J.M. 1952. Spore liberation and dispersal. In: Holten *et al.* (eds.) *Plant Pathology Problems and Progress (1908-1958)* Madison, USA. pp. 529-535.

Hirst, J.M. 1992. Biography of an aerobiologist P.H. Gregory 24. 7. 1907 – 9.2. 1986. *Aerobiologia* 8: 209-218.

Hjelmroos-Koski, M.K. and Solomon, G. 2006. Fungi succession and airborne concentrations of indoor and outdoor fungi in New Orleans, LA after hurricane Katrina. 8th International Congress on Aerobiology, Neuchâtel, Switzerland, 2006. Abstracts, p. 109.

Hodges, D. 1975. The pollen loads of the Honey Bee (2nd Edit) IBRA, London.

Hodgkinson, R.L. 1991. Microfossil processing–a damage report. *Micropaleontology* 37: 320-326.

Holman, R.M. and Brubaker, F. 1926. On the longevity of pollen. *Univ. Calif. Publ. Bot.* 13: 179-204.

Huang, Tseng-Chieng and Lin, Chia-Chin. 2001. Aeropollen of the Pingtung area, South Taiwan. In: D.K. Goodman and R.T. Clarke (eds), Proceedings of the IX International Palynological Congress, Houston, Texas, U.S.A, 1996; American Association of Stratigraphic Palynologists Foundation, pp. 571-577.

Hugg, T., Valtonen, A. and Rantio-Lehtimäki. 2007. Pollen concentrations inside private cars during the Poaceae and *Artemisia spp.* pollen season-a case study. *Grana* 46: 110-117.

Hyde, H.A. 1952. Grass pollen in Great Britain. *Acta Allerg.* 5: 98-112.

Hyde, H.A. 1969. Aeropalynology in Britain-an outline. *New Phytol.* 68: 579-590.

Ikuse, M. 1956. *Pollen grains of Japan*. Kirokawa Publishing Co. Tokyo. p. 103.

Ingold, C.T. 1971. Fungal spores – their liberation and dispersal. Clarendon Press, Oxford, p. 302.

Iszhizaka, K., Iszhizaka, T. and Hornbook, M.M. 1966. Physiochemical properties and antigenic structure of reagenic antibodies. *J. Immunol.* 97: 750-785.

Iwanami, Y., Sasakuma, T. and Yamada, Y. 1988. Pollen: Illustrations and Scanning Electromicrographs. Kodansha, Tokyo and Springer-Verlag, Tokyo. p. 198.

Jäger, S. 2000. Ragweed *(Ambrosia)* sensitization rates correlate with the amount of inhaled airborne pollen. A 14-year study in Vienna, Austria. *Aerobiologia* 16: 149-153.

Jäger, S. 1989. Trends in the frequency of different pollen types in Vienna from 1976-1989. *Aerobiologia* 5: 9-16.

Jäger, S. 1991. Allergenic significance of *Ambrosia* (Ragweed). In: G. D'Amato, F. Th. M. Spieksma and S. Bonini (eds), *Allergenic pollen and pollinosis in Europe*. Blackwell scientific publ. Oxford. London, Vienna. pp. 125-127.

Jäger, S. 1998. Global Aspects of Ragweed in Europe. 6th International Congress on Aerobiology, Perugia, Italy, ALK-Abello, Denmark. pp. 6-10.

Jäger, S. 2006. The trouble of threshold values for allergy forecasts. 8th International Congress on Aerobiology, Neuchâtel, Switzerland, 2006. Abstracts, p. 22.

Jain, A. and Shivanna, K.R. 1990. Membrane state and pollen viability during storage in organic solvents. In: S.K. Sinha, P.V. Sane, S.C. Bhargava and P.K. Agarwal (eds). *Proc. Intl. Cong. Plant Physiology*. Soc. Plant Physiol. Biochem, New Delhi, pp. 1342-1349.

Janaki Bai, A. and Subba Reddi, C. 1982. Airborne pollen grains of Visakhapatnam. A combined field and air sampling study. *Proc. Indian Acad. Science* (Plant Science) 91: 329-350.

Járai-Komlódi, M. 1988. Ragweed in Hungary. *Satellite Symposium Proceedings: Ragweed in Europe. 6th International Congress on Aerobiology, Perugia, Italy*. ALK ABELLO, Denmark. pp. 33-38.

Jarzen, D.M. and Nichols, D.J. 1996. Pollen. In: J. Jansonius and D.C. McGregor (eds) *Palynology: principles and applications;* American Association of Stratigraphic Palynologists Foundation, 9 (1): 261-291.

Jayanth, K.P. 1997. Biological control of *Parthenium* weed in India – Progress and prospects. In: S.N. Agashe (ed) *Aerobiology*, Proceedings of 5th International Conference, Bangalore 1994. Oxford & IBH Pub. Co. Pvt. Ltd. New Delhi. pp. 109-115.

Jelks, M. 1987. Allergy plants: that cause sneezing and wheezing worldwide publications, Singapore.

Johri, B.M. and Vasil, I.K. 1961. Physiology of pollen. *Bot. Rev.* 27: 325-381.

Jones, G.D. and Bryant, V.M. Jr. 1996. Melissopalynology (Chapter 23 D) In: Jansonius, J. and McGregor, D.C. (eds) *Palynology: Principles and Applications*. American Association of Stratigraphic Palynologist Foundation. 3: 933-938.

Jones, G.D. and Bryant, Vaughan M., Jr. 2001. Alcohol dilution of honey. In: D.K. Goodman and R.T. Clarke (eds.). Proceedings of the IX International Palynological Congress, Houston, Texas, U.S.A., 1996; American Association of Stratigraphic Palynologists Foundation, pp. 453-458.

Jonker, F.P. 1974. Reflections on pollen evolution. In: P.K.K. Nair (ed) *Advances in Pollen spore Research*. Today and Tomorrow's Printers and Publishers, New Delhi. pp. 52-63.

Juhász, M. 1998. History of ragweed in Europe. Satellite Symposium Proceedings: Ragweed in Europe. 6th International Congress on Aerobiology, Prugia, Italy, ALK-Abello, Denmark, pp. 11-14.

Käpylä, M. 1984. Diurnal variations of tree pollen in the air of Finland. *Grana* 23: 167-176.

Kasliwal, R.M., Sethi, J.P. and Soagi, J.C. 1959. Studies in the atmospheric pollen a daily census of pollen at Jaipur: 1957-58. *Ind. J. Med.* Res. 47: 515-521.

Kaushal, P.S and Kaushal, C.K. 1984. Spore morphology in some hepatics of western Himalayas. *J. Palynol.* 20 (2): 139-148.

Kessler, R. and Harley, M. 1994. The Hidden sexuality of flowers. Papadakis. pp. 1-264.

Keynan, N., Waisel, Y., Shomer-Ilan, A., Goren, A. and Brener, S. 1991. Annual variations of airborne pollen in the coastal plain of Israel. *Grana* 30: 477-480.

Khandelwal, A. and Mittre, V. 1973. Palynology of *Holoptelea integrifolia* Panch. *Palaeobotanist* 22: 127.

King, J.R. 1965. The storage of pollen particularly by the freeze-drying method. *Bull. Torrey Bot. Club.* 92: 270-287.

Knox, E. 1939. Spores of Bryophyta compared with those of Carboniferous age. *Trans. Bot. Soc. Edinburgh.* 32: 477-487.

Knox, F.M. 1950. The spores of *Lycopodium, Phylloglossum, Selagenella* and *Isoetes* and their value in the study of microfossils of Palaeozoic age. *Transactions and Proceedings of the Botanical Society of Edinburgh.* 35: 209-357.

Knox, R.B. 1979. *Pollen and Allergy.* Institute of Biology Studies in Biology No. 107. Edward Arnold, London.

Knox, R.B. 1980. Pollen and allergy. Arnold Heinemann (India) Pvt. Limited.

Knox, R.B., Heslop-Harrison, Y. 1975. Pollen wall proteins: Localization of gametophytic and sporophytic fractions. In: J.G. Duckett and P.A. Racey (eds). *The Biology of the male gamete. Biol. J. Linn. Soc.* 7 (suppl. 1): 177-187.

Knox, R.B., Willing, R.R. and Ashford, A.E. 1972. Pollen wall proteins: Role as recognition substances in interspecific incompatibility in Poplars. *Nature* 237: 381-383.

Komlodi, M. and Medzihradszky, Z. 1997. Five years airborne pollen monitoring in Hungary. (1989-1993). In: S.N. Agashe (ed) *Aerobiology,* Proceedings of 5th International Conference, Bangalore 1994. Oxford & IBH Pub. Co. Pvt. Ltd. New Delhi. pp. 389-400.

Komlodi, M. and Miklos Juhasz. 1993. *Ambrosia elatior* (L.) in Hungary (1989-1990). *Aerobiologia* 9: 75-78.

Kosanke, R.M. 1950. Pennsylvanian spores of Illinois and their use in correlation. III. *Geol. Surv. Bull.* 74: 1-128.

Kramer, C.L. 1979. Sources and characteristics of airborne materials;Factors in the production, release and variability of biological particles. Fungus, moss and fern spores. In: R.L. Edmonds (ed) *Aerobiology, The ecological systems approach*. US/IBP synthesis series 10, Dowden pp. 24-41. Hutchinson & Ross, Inc., Strondsburg, Pennsylvania, U.S.A.

Kremp, G.O.W. 1968. Morphologic Encyclopedia of Palynology. Arizona University Press, Tucson. U.S.A. p. 411.

Kundu, S., Chanda, S. and Kundu, I. 1985. Clinical investigation of some allergenic pollen grains from Calcutta. – *Trans. Bose Res. Inst. Calcutta* 48 (1): 21-28.

Lacey, J. 1994. Whither Aerobiology. In: S.N. Agashe (ed) *Recent Trends in Aerobiology, Allergy and Immunology*. Oxford & IBH Pub. Co., New Delhi. pp. 31-41.

Lacey, J., Williamson, P.A.M., Janaki Bai, A. and Mandal, S. 1997. Detection of airborne allergens. In: S.N. Agashe (ed) *Aerobiology* Proceedings of 5th International Conference, Bangalore 1994. Oxford & IBH Pub. Co. Pvt. Ltd. New Delhi. pp. 607-615.

Lacey, M. and McCartney, A. 1994. Measurement of airborne concentrations of spores of bracken (*Pteridium aquilinum*). *Grana* 33: 91-93.

Lacey, M.E. 1997. Painting as an aid in identifying fungal spores. In: S.N. Agashe (ed) *Aerobiology*. Proceedings of 5th International Conference, Bangalore 1994. Oxford & IBH Pub. Co. Pvt. Ltd. New Delhi. pp. 7-12.

Lacey, M.E. and West, J.S. 2006. The air-spora. Springer, Dordrecht.

Lakhanpal, R.N. and Nair, P.K.K. 1958. Survey of atmospheric pollen in Lucknow. *J. Sci. Industr. Res.* 17C: 80-87.

Lakhanpal, R.N. and Nair, P.K.K. 1960. Atmospheric pollen survey at Almora. *J. Sci. Industr. Res.* 19C: 51-53.

Latif, A.A., Al-Kazimi, K.A., Mohamed, D.E. and Al-Omany, M. 1989. Evaluation of sensitivity and specificity of skin prick test in correlation with specific IgE for nine inhalant allergens in Kuwait. Abstract. *8th Medical Conf. Kuwait Medical Assoc.* 124. Min. Pub. Health, Kuwait.

Lee, S. 1978. A factor analysis study of the functional significance of angiosperm pollen. *Systematic Botany* 3: 1-19.

Legge, A.P. 1976. Hive insects and pollen dispensers for tree fruits. *Bee World*, 57: 159-167.

Leuschner, R.M. 1974. Pollen Calendars for Switzerland: In: J. Charpin and R. Suriyach (eds) *Atlas Europeen des pollens allergisants. (Atlas of European allergic pollen)*. Sanduz. Paris. pp. 193-200.

Leuschner, R.M. 1978. Registrierte Luftpollen von Ambrosia L. als Hinweis auf ein Vorkommen dieser Adventivpflanze. *Bauhinia*, 6: 2, 265-271.

Leuschner, R.M. and Jenkins, J. 1996. Ruth M. Leuschner: a Biographical sketch. *Aerobiologia* 12: 141-147.

Levetin, E., Davis, E., Lehrer, S.B. and Horner, E. 2006. Post Katrina changes in the Airborne Fungal Spore Load in New Orleans. 8th International

Congress on Aerobiology, Neuchâtel, Switzerland, 2006, Abstracts, p. 111.

Levetin, E. and Buck, P. 1980. Hay fever plants in Oklohoma. *Ann allergy* 45: 26-32.

Lewis, W.H. 1984. Pollen allergy. In: P.E. Korneblat and H.J. Wedner (eds) *Allergy: Theory and practice* Grune and Startton, Orlando, FL. pp. 353-370.

Lewis, W.H. 1991. Distribution and Incidence of North American Pollen Aeroallergens. *Am. J. Otolaryngol* 12: 205-226.

Lewis, W.H., Dixit, A.B. and Wedner, H.J. 1987-1988. Reproductive biology of *Parthenium hysterophorus* (Asteraceae). *J. Palynol.* 23-24: 73-82.

Lewis, W.H., Dixit, A.B. and Wedner, H.J. 1990. Aeropollen of herbaceous plants at Corpus Christi, Texas. *Aerobiologia* 6: 141-146.

Lewis, W.H., Vinay, P. and Zenger, V.B. 1983. Airborne allergenic pollen of North America. The John Hopkins University Press. Baltimore, U.S.A. P. 254.

Licorish, K., Novey, H., Kozak, P. and Wilson, A. 1984. Studies of *Alternaria* and *Penicillium* extracts and spores in the pathogenesis of asthma. *J. Allergy Clin. Immunol.* 73 (suppl.): 117.

Lieux, M.H. 1981. An analysis of Mississippi (U.S.A.) honey: Pollen, color and moisture. *Apidologie.* 12: 137-158.

Lieux, M.H. 1978. Minor honeybee plants of Louisiana indicated by pollen analysis. *Economic Botany* 32: 418-432.

Litwin, R.J. and Traverse, A. 1989. Basic guidelines for palynoform extraction and preparation from sedimentary rocks. In: R.M. Feldman, R.E. Chapman and J.T. Hannibal (eds) Paleo-techniques; *Paleontological Society, Special Publication,* 4: 87-98.

Lloyd de Brul, D. 1974. Biological form and function. *The New Encyclopedia Britanica.* 7: 542-547.

Louveaux, J. and Vorwohl, G. 1970. Methods of Melissopalynology. *Bee World* 51: 125-131.

Louveaux, J. and Vorwohl, G. 1978. Methods of Melissopalynology by International Commission of Bee Botany of IUBS. *Bee World* 59: 139-157.

Luz, C. Fernandes P., da and Barth, Orturd Monika. 2001. Melissopalynological observations in a mangrove area next to Rio de Janeiro, Brazil. In: D.K. Goodman and R.D. Clarke (eds) Proceedings of the IX International Palynological Congress, Houston, Texas, U.S.A., 1996; American Association of Stratigraphic Palynologists Foundation. pp. 489-492.

Mackay, T.W., Wathen, C.G., Sudlow, M.F., Elton, R.A. and Caulton, E. 1992. Factors affecting asthma mortality in Scotland. *Scottish Medical Journal* 37: 005-007.

Madan, P., Lamba, L.C. and Asseja, K.R. 1982. The most suitable medium for trapping fungi from air. *Science and Culture* 48: 77-78.

Madhusoodanan, P.V. and Jyothi, P.V. 1993. Palynology in relation to Taxonomy of South Indian Cheilanthoid Ferns. *J. Econ. Tax. Bot.* 17: 105-107.

Mahadevan, V. and Chandy, K.C. 1959. Preliminary studies on increase in cotton yield due to honey bee pollination. *Madras Agri. J.* 46: 23-36.

Majumdar, M.R. and Barui, N.C. 2006. Indoor Mould Exposure, a potential source of Occupational Health Hazard among the workers of Potato cold stores in West Bengal, India. 8th International Congress on Aerobiology, Neuchâtel, Switzerland, 2006. Abstracts, p. 308.

Makovcová, S., Zlinská, J., Mikolás, V., Salát, D., Krio, M. and Kúpele Strbské Pleso, A.S. 1998. Ragweed in Slovak Republic. *Satellite Symposium Proceedings: Ragweed in Europe, 6th International Congress on Aerobiology, Perugia, Italy*. ALK ABELLO. pp. 27-28.

Mandrioli, P. 1997. Aerobiology Today. In: S.N. Agashe (ed) *Aerobiology*. Proceedings of 5th International Conference, Bangalore 1994. Oxford & IBH Pub. Co. Pvt. Ltd. New Delhi. pp. 3-6.

Markgraf, V. and Antoni, H.L.D. 1978. Pollen flora of Argentina, The University of Arizona Press, Tucson, Arizona. p. 208.

Mascarnehas, J.P. 1975. The biochemistry of Angisperms pollen. *Bot. Rev.* 41: 259-314.

May, K.R. 1945. The cascade impactor: an instrument for sampling coarse aerosols. *J. Sci. Instrum.* 22: 187-195.

Mehra, P.N. 1976. Evolution of Spore through the Ages. Today and Tomorrow's Printers and Publishers, New Delhi.

Menon, V.K. 1967. A preliminary note on the pollen analysis of the post glacial deposits from Pykara, Ootacamund. *Palynol. Bull.* II, III: 100.

Meyer, G., Homer, P. and Warren, R. 1983. Airborne fungi – a resurvey. *Ann. Allergy* 51: 26-29.

Mezei, G., Komlodi, M.J., Dzibradxzky, M. and Cserhati, E. 1997. Pollen allergy in Hungary based on a five years comparative study. In: J. Charpin and R. Suriyach (eds) *Atlas Europeen des pollens allergisants. (Atlas of European allergic pollen)*. Sandoz. Paris. pp. 193-200 and pp. 369-374.

Middleton's Allergy Principles and Practice. 2003 Vol. 1, 6th edit, Mosby.

Mildenhall, D.C. 2006. Forensic Palynology: Operation Lollipop. 8th International Congress on Aerobiology, Neuchâtel, Switzerland, 2006. Abstracts, p. 66.

Mildenhall, D.C. 2006. Pollen and crime. 8th International Congress on Aerobiology, Neuchâtel, Switzerland, 2006. Abstracts, p. 17.

Moar, N.T. 1985. Pollen analysis of New Zealand honey. *New Zealand J. Agri. Res.* 28: 39-70.

Moar, N.T. 1993. Pollen grains of New Zealand Dicotyledonous plants Manaaki Whenna Press Lincoln, Canterbury, New Zealand.

Mondal, A.K. and Mandal, S. 1998. Pollen production in some terrestrial angiosperms. *Current Science* 74: 906.

Monserrat Roses-Codinachs, Suarez-Cervera and Jesus Marquez. 1992. An aerobiological study of pollen grains and fungal spores of Barcelona (Spain). *Aerobiologia* 8: 255-265.

Montouroy, C., Martin, M.H., Boillean, Y. and Colbert, R. 1992. Foehnlike wind, pollination and allergic symptoms in French Basque country. *Aerobiologia* 8: 277-284.

Morrow Brown, H. 1978. Aerobiological studies based in Derby III, A comparison of simultaneous pollen and spore counts from the east coast, Midlands, and West coast of England and Wales. *Clin. Allergy* 8: 611-619.

Morrow Brown, H. 1992. The relationship of aerobiological data to seasonal allergic symptoms. *Immunology and Allergy practice* 14: 318-329.

Morrow Brown, H. 1994. The relationship between Aerobiology and Allergology (A Tribute to Charles Blackley and a Review of Personal Experience). In: S.N. Agashe (ed) *Recent trends in Aerobiology, Allergy and Immunology*. Oxford & IBH Pub. Co. New Delhi. pp. 1-30.

Mullenberg, M.L., Rogers, C., Maestretti, R., Santiago, T. and Chew, G.L. 2006. Airborne fungal spore concentration in New Orleans homes following Hurricane Katrina. 8th International Congress on Aerobiology, Neuchâtel, Switzerland, 2006. Abstracts, p. 107.

Muller, J. 1970. Palynological evidence of early differentiation of angiosperms. *Biol. Rev.* 45: 417-450.

Nagarajan, C.R., Rao, V.N.M. and Sheshu, K.A. 1972. Standardisation of haemocytometer method for pollen output estimation *J. Palynol.* 8: 22-26.

Nair, P.K.K. (ed) 1964. Advances in Palynology. National Botanical Garden, Lucknow, India.

Nair, P.K.K. and Rastogi, K. 1963. Pollen production in some allergenic plants. *Current Science* 32: 566.

Nair, P.K.K., Joshi, A.P. and Gangal, S.V. 1986. Airborne pollen, spores and other plant materials of India–A survey. C.S.I.R. Centre for Biochemicals, Delhi and National Botanical Research Institute, Lucknow, India.

Nakagawa, T., Yasuda, Y. and Tabata, H. 1996. Pollen morphology of Himalayan *Pinus* and *Quercus* and its importance in palynological studies, Himalayan area. *Rev. Paleobotany and Palynology* 91: 317-329.

Nautiyal, D.D. and Midha, M. (Now Sahney, M.). 1984. Aerobiology of Allahabad 1973-1979. *Phyta Monograph* 2: 1-100.

Nayar, B.K. and Devi, S. 1963. Spore morphology of Loxogramme. *Grana Palynol.* 4: 388-392.

Nayar, B.K. and Devi, S. 1964. Spore morphology of Indian ferns II and III, Aspleniaceae, Blechnaceae and Polypodiaceae. *Grana Palynologica* 5 (2, 3): 222-246 and 242-395 (#3).

Nayar, J. and Ramanujam, C.G.K. 1989. Pollen calendar of a semi-urban area of Secundrabad. *Ind. J. Aerobiol.* 2: 7-12.

Nayar, T.S. 1990. Pollen Flora of Maharashtra State, India. Today & Tomorrow's Printers & Publishers, New Delhi. p. 155.

Nevalainen, A., Rantiala, S., Hyvarinen, A. Reponen, T., Husman, T. and Kalliokoski, P. 1994. Exposure to fungal spores in mouldy houses - effect of remedial work. In: Agashe, S.N. (ed) *Recent Trends in Aerobiology, Allergy and Immunology*. (A collection of plenary lectures and contributory articles) Oxford & IBH Pub. Co. New Delhi. pp. 99-107.

Newnham, R.M. 2001. Aeropalynology and global warming. In: D.K. Goodman and R.T. Clarke (eds), Proceedings of the IX International Palynological Congress, Houston, Texas, U.S.A., 1996; American Association of Stratigraphic Palynologists Foundation, pp. 567-570.

Nikkels, A.H., Terstegge, P. and Spieksma, F.Th.M. 1996. Ten types of microscopically identifiable airborne fungal spores at Leiden, The Netherlands. *Aerobiologia* 12: 107-112.

Nilsson, S. (ed) 1983. Atlas of Airborne Fungal spores in Europe, Springer-Verlag, Berlin. p. 139.

Nilsson, S. and Palberg-Gotthard, J. 1982. Pollen calendar for Huddunge (Sweden), 1977-1981. *Grana* 21: 183-185.

Nilsson, S. and Praglowski, J. 1974. Pollen and spore incidence and phenology in the Stockholm area during 1972. *Grana* 14: 78-84.

Nilsson, S., Praglowski, J. and Nilsson, N. 1977. Atlas of Airborne Pollen Grains and spores in Northern Europe. Almquist and Wiksell International, Stockholm, Sweden.

Noon, L. 1911. Prophylactic inoculation for hay fever. *Lancet*. 1: 1572.

Norris-Hill Jane and Emberlin, J. 1993. The incidence of increased pollen concentrations during rainfall in the air of London. *Aerobiologia* 9: 27-32.

Oberle, G.D. and Goertzen, K.L. 1952. A method for evaluating pollen production of fruit varieties. *Proc. Amer. Soc. Hort. Sci*. 59: 263-265.

Oeggel, K.D. 2001. Pollen analysis of the Iceman's colon content: In: D.K. Goodman and R.T. Clarke (eds), Proceedings of the IX International Palynological Congress, Houston, Texas, U.S.A., 1996; American Association of Stratigraphic Palynologists Foundation, pp. 511-516.

Ogden, E.C., Raynor, G.R., Hayes, J.V., Lewis, D.M. and Haines, J.N. 1974. Manual for sampling airborne pollen, Lewis and Hames. Hafner Press, New York, 182p.

Olmo, H.P. 1942. Storage of grape pollen. *Proc. Amer. Soc. Hort. Soc*. 41: 219-224.

Ong, E.C., Singh, M.B. and Knox, R.B. 1995. Seasonal distribution of pollen in the atmosphere of Melbourne: An airborne pollen calendar. *Aerobiologia* 11: 51-55.

Org, E.K., Abramson, M., Farish, J., Singh, M.B. and Knox, R.B. 1997. Grass pollen, rainfall and Asthma. Relationship between Environmental allergen load and pollen counts. In: S.N. Agashe (ed) *Aerobiology* Proceedings of 5th International Conference, Bangalore 1994. Oxford & IBH Pub. Co. Pvt. Ltd. New Delhi. pp. 347-356.

Panchaksharappa, M.G. and Tirlapur, U.K. 1984. Storage of pollen grains in organic solvents for preserving pollen viability. *J. Palynol.* 20 (2): 132-134.

Panicker, Cherriyan, K.T. 2002. *A study of Biochemical, Morphological and Immunological features of pollen in Mimosaceae*. Ph.D. Thesis submitted to the University of Kerala, India.

Penel, V. and De Clercq, G. 2006. Personilized pollen counts: using the pollinic distribution in patient hair. 8th International Congress on Aerobiology, Neuchâtel, Switzerland, 2006. Abstracts, p. 144.

Penel, V. and De Clereq, G. 2006. Study on bees from the Drome valley in France in mid-September: Bee pollen may contain a high quantity level of *Ambroisa* pollen. 8th International Congress on Aerobiology, Neuchâtel, Switzerland, 2006. Abstracts, p. 302.

Perkins, W.A. 1957. The rotorod sampler. 2nd Seminnual Rept. aerosol. Lab., Dept. Chemistry & Chem. Enging., Stanford Univ. CMh., 186: 66.

Peters, A.G. 1998. Ragweed in Switzerland. 6th International Congress on Aerobiology, Perugia, Italy, Satellite Symposium Proceedings: Ragweed in Europe, pp. 16-19.

Peterson, S., Maples, C.G. and Lane, N.G. 1983. Use of surfactants in disaggregation of argillaceous rock. *Micropaleontology* 29: 467-469.

Phadke, R.P. 1962. Physico-chemical composition of major unifloral honeys from Mahabaleshwar (Western Ghats). *Indian Bee J.* XXIV: 59-65.

Praglowaski, T. and Blackmore, S. 2003. In Memorium: Siwert Nilsson. 1933-2002. *Grana* 42 (2): 65-71.

Pugalmaran, M. and Vittal, B.P.R. 1997. Aeromycological survey of indoor environment in leather godowns. In: S.N. Agashe (ed) *Aerobiology* Proceedings of 5th International Conference, Bangalore 1994. Oxford & IBH Pub. Co. Pvt. Ltd. New Delhi. pp. 143-147.

Punt, W. 1986. Functional factors influencing pollen form. *Pollen storage Pollen and spores: form and Function* The Linnean Society of London. pp. 97-101.

Radford, T. 1999. Pollen points to grave KGB skullduggery. The Guardian, 01 October (Home News). London. p. 9.

Ramalingam, A. 1968. The construction and use of simple air sampler for routine aerobiological surveys. *Envir. Health* 10: 61-67.

Ramanujam, C.G.K. and Kalpana, T.P. 1994. Summer pollen source for honeybees in the coastal belt of East Coast Godavari District, A.P., India. *Geophytology* 23: 249-252.

Rangaswamy, B.E. 2001. *Melittopalynological Investigation of Karnataka with particular reference to Dakshina Kannada and Udupi Districts*. Ph.D. Thesis submitted to the Bangalore University, Bangalore, Karnataka, India.

Rao, A.R. and Vimal, K.P. 1952. Preliminary observations on the plant microfossil contents of some lignites from Warkalli in Travancore. *Curr. Sci.* 21: 302-305.

Rao, M., Prakash, O. and Rao, P.V. 1985. Reaginic allergy to Parthenium pollen: evaluation of skin test and radioallergosorbent test. *Clin. Allergy.* 15: 449-454.

Rao, S.P., Atluri, J.B. and Subba Reddi, C. 2001. Intermittent mass blooming, midnight anthesis and rockbee pollination in *Pterocarpus santalinus* (Fabaceae). *Nord. J. Bot.* 21: 271-276.

Rao, M. and Lakshmi, K. 1996. African milk bush. *Bee World* 77: 12-14.

Ravindran, P., Gopi, T.V., Prakashkumar, R. and Nair, P.K.K. 1991. Airborne Aspergilli concentration in Kerala. *New Botanist* 18: 151-156.

Renault, B. 1896. Études des Gites minéraux de la France. Bassin Houller et Permien d'Autun et d'Épinac. Fasc. IV *Flore Fossile,* 2 partie. Paris. p. 578. Atlas.

Ribeiro, M., Cunha, M. and Abreu, L. 2007. Improving early-season estimates of olive production using airborne pollen multi-sampling sites. *Aerobiologia* 23: 71-78.

Richardson, M.J. 1966. The occurrence of airborne *Didymella* spores in Edinburgh. *Mycological Research* 100: 213-216.

Rocha-Estrada, A., Alvarado-Vazquez, M.A., Torres-Cepeda, T.E. and Foroughbakhch-Pournavab, R. 2006. Airborne pollen flora in the metropolitan area of Monterrey, Nuevo Leon, Mexico. 8th International Congress on Aerobiology, Neuchâtel, Switzerland, 2006. Abstracts, p. 195.

Rosoanaivo, P. and Ratsimamanga-Urveg, S. 1993. *Biological evaluation of plants with reference to the Malagasy flora.* Monograph for the IF-NAPRECA workshop on Bioassays, Antavanarivo, Madagascar. p. 72.

Rothwell, G.W. 1972. Evidence of pollen tubes in Paleozoic pteridosperms. *Science* 175: 772-774.

Rowley, J.R. and Skvarala, J.J. 1976. Surface coating of germinal apertures of pollen and evolution of apertures. 34th *Annual Proceedings of the Electron Microscopy Society of America*: 42-43.

Rybnícek, O. 1988. Ragweed in the Czech Republic. *Satellite Symposium Proceedings: Ragweed in Europe. 6th International Congress on Aerobiology, Perugia, Italy.* ALK ABELLO, Denmark. pp. 29-32.

Saar, M., Gudzinskas, Z., Ploompuu, T., Linno, E., Minkiene, Z. and Motiekaityté, V. 2000. Ragweed plants and airborne pollen in the Baltic States. *Aerobiologia* 16: 101-106.

Salvaggio, J. and Aukrust, L. 1981. Mold induced asthma. *J. Allergy Clin. Immunol.* 68: 327-346.

Savitsky, V.D. and Martynyuk, O.A. 2001. Polyplicate pollen – its prevalence and peculiarity of evolution in the subclasss Ranunculidae and

Caryophyllidae: A review. In: D.K. Goodman and R.T. Clarke (eds), Proceedings of the IX International Palynological Congress, Houston, Texas, U.S.A., 1996; American Association of Stratigraphic Palynologists Foundation, pp. 593-596.

Saxena, M.R. 1993. Palynology. Oxford and IBH Pub. Co. Pvt. Ltd, New Delhi, p. 127.

Schulze, Franz. 1855. Über das vorkommen wohler-haltener cellulose in Brraunkohle und steinkohle. *Ber. K. Acad. Wiss. Berlin*. pp. 676-678.

Schwanitz, G. 1967. Untersuchungen zur postmeiotischen MiKrosporogenese II. Vergleichende Analyse der Pollentwicklung sub. Und emers bluehender Arten. *Pollen et Spores* 9: 183-209.

Seetharamaiah, A.M., Viswanath, B. and Subba Rao, P.V. 1981. Atmospheric survey of pollen of *Parthenium hysterophorus*. *Ann. Allergy* 47: 192-196.

Seligar, A.K. 1998. Ragweed in Slovenia. 6th International Congress on Aerobiology, Perugia, Italy, Satellite Symposium Proceedings: Ragweed in Europe, pp. 39-41.

Seligmann, J., Donosky, L., Shapiro, D. and Allan, R.A. 1982. Allergy: New Insights, Newsweek, 38-43.

Severova, E.E. and Polevova, S.V. 2001. Three years of pollen monitoring in Moscow: Preliminary pollen calendar for 1993-1995. In: D.K. Goodman and R.T. Clarke (eds), Proceedings of the IX International Palynological Congress, Houstan, Texas, U.S.A., 1996; American Association of Stratigraphic Palynologists Foundation, pp. 579-583.

Sharma, C. 2001. Modern pollen rain compared to vegetation in the Himalayas. In: D.K. Goodman and R.T. Clarke (eds), Proceedings of the IX International Palynological Congress, Houston, Texas, U.S.A., 1996; American Association of Stratigraphic Palynologists Foundation, pp. 557-565.

Sharma, D. 1981. Pollen morphology of some cultivars of *Carthamus tinctorius* L. (Compositae). *Indian J. Bot.* 4 (1): 70-79.

Sharma, R., Deval, R., Priyadarshi, V., Gaur, S.N. and Singh, A.B. 2006. Sensitization pattern to indoor fungal allergens in asthmatic/allergic children in Delhi-Clinical and immunological aspects. 8th International Congress on Aerobiology, Neuchâtel, Switzerland, 2006. Abstracts, p. 300.

Sheldon, J.M., Lovell, R.G. and Mathews, K.P. [eds]. 1967. *A manual of Clinical allergy*.

Shesahgiri, T.N. 1985. Melittopalynological studies of Bangalore and its environs. M. Phil. Dissertation, Bangalore University, Bangalore.

Shivanna, K.R. 2003. Pollen Biology and Biotechnology. Oxford & IBH, New Delhi. p. 301.

Shivanna, K.R. and Rangaswamy, N.S. 1992. Pollen Biology: A Laboratory Manual. Springer-Verlag, Berlin. Heidelberg, New York.

Shivanna, K.R. and Johri, B.M. 1985. The Angiosperm Pollen: Structure and Function. Wiley Eastern, New Delhi.

Shivpuri, D.N. 1962. Comparative evaluation of sensitivity of diagnostic antigen tests in patients of respiratory allergy. *Ind. J. Chest diseases* 4: 102-108.

Shivpuri, D.N. 1964. Aeropalynology and its significance in allergy. In: P.K.K. Nair (ed) *Recent Advances in Palynology,* National Botanical Garden, Lucknow. pp. 420-430.

Shivpuri, D.N. 1980. Clinically important pollen, fungal and insect allergens for nasobronchial allergy patients in India. *Asp. Allergy App. Immunol.* 13: 19-23.

Shivpuri, D.N. and Dua, K.L. 1963. Studies in pollen allergy in Delhi area. Part IV: Clinical investigations. *Ind. Jour. Med. Res.* 51: 68-74.

Shivpuri, D.N. and Singh, K. 1971. Studies in yet unknown allergenic pollens of Delhi State Metropolitan (Clinical aspects). *Ind. J. Med. Res.* 59: 1411-1419.

Shivpuri, D.N., Vishwanathan, R. and Dua, K.L. 1960. Studies in pollen allergy in Delhi area. Pt. I. *I. Jour. Med. Res.* 48: 15-20.

Sihag, R.C. 1994. Influence of environmental factors on the pollination activity of bees. *Environ. Ecol.* 2: 149-152.

Sikoparija, B., Radisic, P., Pejak, T., Simic, S. and Boza, P. 2006. *Ambrosia spp.* Pollen in Southern Pannonian valley. 8th International Congress on Aerobiology, Neuchâtel, Switzerland, 2006. Abstracts, p. 48.

Singh, A.B. 1984. Pollen production in common allergenic plants of Delhi and their atmospheric prevalence. *Science and Cult* 50: 65-67.

Singh, A.B. 1987. Airborne pollen types of allergenic significance in India. *Advances in Aerobiology* 61-64.

Singh, A.B. and Chatterjee, M. 1990. Airborne viable fungi in Library. Before and after agitation of books. *J. Aerobiol.* 3: 32-38.

Singh, A.B. and Singh, A. 1994. Pollen allergy: A Global scenario In: S.N. Agashe (ed) *Recent trends in Aerobiology, Allergy and Immunology.* Oxford & IBH Pub. Co. New Delhi. pp. 143-170.

Singh, S.P. 1965. Role of tapetum. *Palynological Bulletin* 1: 11.

Sitholey, R.V. 1965. Professor Birbal Sahni's contribution to Palynology. *Palynol. Bulletin* 1: 5-6.

Smart, I.J. and Know, R.B. 1979. Aerobiology and grass pollen in the city atmosphere of Melborne, quantitative analysis of seasonal and diurnal changes. *Aust. J. Bot.* 27: 317-331.

Smith, E.G. 1984. Sampling and Identifying Allergenic Pollens and Molds. Blewstone Press, San Antonio.

Sobolik, K. 1988. The importance of pollen concentration values from coprolites: an analysis of Southwest Texas samples. *Palynology* 12: 201-214.

Spieksma, F.Th.M. 1984. Comparative European counts-uniformation of the data. In: K. Wilen-Jensen and S. Gravesen (eds) *Atlas of Mould in Europe causing Respiratory Allergy*. ASK Publishing, Copenhagen. pp. 62-65.

Spieksma, F.Th.M. 1988. Retrospective and predictive systems of information on airborne pollen concentrations in relation to hay fever. *Aerobiologia* 4: 8-11.

Spieksma, F.Th.M. 1990. Allergenic plants in different countries. In: D.V.M. Paola Falagiani (ed) *Pollinosis*, CRC Press, Inc., Boca Raton, Florida. pp. 19-38.

Spieksma, F.Th.M., D'Amato, G., Mullins, J., Nolard, N., Wachter, R. and Weeke, E.R. 1989. City spore concentrations in European Economic Community (EEC). VI. Poaceae (Grasses), 1982-1986. *Aerobiologia* 5: 38-43.

Spieksma, F.Th.M., Nikkels, B.H. and Dijkman, J.H. 1995. Seasonal appearance of grass pollen allergen in natural, paucimicronic aerosol of various size fractions: Relationship with airborne grass pollen concentration. *Clin. Exp. Allergy* 25: 234-239.

Spindler, K. 1994. The man in the ice. Weidenfield and Nicholson, London.

Stanley, R.G. and Linskens, H.F. 1974. Pollen: Biology, Biochemistry and Management. Springer, Heidelberg, New York.

Stix, E. and Ferretti, M.L. 1974. Grundwerte des Pollen and Sporengehalt der Luft 1907-1970. DfG. Komm, Erforsch. Luftvernur Luftvernur. Mitt Xi, Bonn-Bad Godesberg, pp. 1-53.

Stix, E. 1971. Vorkmen von Pollen and spores in dex Luft. In: E. Landes and H. Zielger (eds), *Allergien durch Pollen and Spores*, DFG Komm. Erforch. Luftvernur Luftverunr, Mitt. VI, Bonn-Bad Godesberg, pp. 46-73.

Stix, E. 1974. Pollen calendars of three locations in Western Germany. In: J. Charpin and R. Suriyach (eds) *Atlas Europeen des pollens allergisants. (Atlas of European allergic pollen)*. Sandoz, Paris. pp. 85-94.

Subba Reddi, C. and Reddi, N.S. 1985. Relation of pollen release to pollen concentrations in air. *Grana* 24: 109-113.

Subba Rao, P.V. 1985. Sensitivity of Parthenium rhinitis patients to ragweed pollen allergens. *Ann. Allergy* 55: 247.

Surange, K.R., Singh, P. and Srivastava, P.N. 1953. Megaspores from the West Bokaro coalfield (Lower Gondwanas) of Bihar. *Palaeobotanist* 2: 9-17.

Tansley, A.G. 1946. Introduction to Plant Ecology. George Allen & Unwin Ltd. London.

Tansley, A.G. 1953. The British Isles and their vegetation. Cambridge University Press, Cambridge, U.K.

Teranishi, H., Katoh, J., Inadera, H., Yoshida, Y. and Hayashi, S. 2006. Pollen count of Japanese Cedar (*Crytomeria japonica*) by an automatic pollen monitor and by the volumetric method, in Toyama, Japan. 8th International Congress on Aerobiology, Neuchâtel, Switzerland, 2006. Abstracts, p. 128.

Thanikaimoni, G. 1966. Pollen morophology of the genus *Utricularia*. *Pollen et Spores* 8: 265-284.

Thanikaimoni, G. 1986. Pollen apertures: form and function. In: *Pollen and spores: form and function*. pp. 119-136. The Linnean Society of London, London.

The airborne fungal spores in New Orleans. 8th International Congress on Aerobiology, Neuchâtel, Switzerland, 2006. Abstracts, p. 111.

Thibauden, M. 1998. Ragweed in France in Lyon, France 1987-1997. In: 6th International Congress on Aereobiology, Perugia, Italy, ALK- Abello, Denmark. p. 15.

Thomson, J.D. 1986. Pollen transport and deposition by bumble bees *Erythronium*: Influences of floral nectar and bee grooming. *Journal of Ecology* 74: 329-341.

Tilak, S.T. 1989. Airborne pollen and fungal spores. Vijayanthi Prakashan, Aurangabad, India. p. 316.

Tilak, S.T. and Kulkarni, R.L. 1970. A new air sampler. *Experientia* 26: 443-444.

Tissot, C., Chikhi, H. and Nayar, T.S. 1994. Pollen of wet evergreen forests of the Western Ghats, India. French Institute, Pondicherry, India.

Tivy, J. 1971. Biogeography–a study of plants in the ecosphere. Oliver and Boyd, Edinburgh.

Tomlinson Harvard Forest, P.B. 1966. Functional morphology of saccate pollen in conifers with special reference to Podocarpaceae. In: Pollen and Spores Museum National D'Histoire Naturelle Vol. VIII - No. 2. AOUT Harvard University, Petersham, Massachusetts. p. 1366.

Traverse, A. 1955. Occurrence of the oil-forming alga *Botryococcus* in lignites and other Tertiary sediments. *Micropaleontology* 1: 343-349.

Traverse, A. 1988. Paleopalynology. Allen and Unwin Inc. U.S.A.

Triago, M.M., Melgar, M., Sinclair La Rosa, N. and Meinardus-Hager, G. 2006. Two years of aerobiological study of the atmosphere of Münster (northwestern Germany). 8th International Congress on Aerobiology, Neuchâtel, Switzerland, 2006. Abstracts, p. 208.

Tripathi, D.M., Gupta, S.M. and Vas, C.J. 1982. Aerobiological Survey of Bombay Part II (Pollen and Spore Calendar). *Asp. Allergy. Appl. Immunol.* 15: 43-48.

Tschudy, H. 1958. A modification of the schulze digestion method of possible value studying coals. *Grana Palynologica*, n. s. 1: 34-38.

Tschudy, R.N. and Scott, R.A. 1969. Aspects of Palynology. John Wiley and Sons. New York. p. 509.

Tyldesley, J.B. 1973. Long range transmission of the tree pollen to Shetland. (i) Sampling and trajectories. (ii) Calculation of pollen deposition. (iii) Frequencies over the past hundred years. *New Phytol.* 72: 175-181, 183-190, 691-697.

Ueno, J. 1911. Studies on pollen grains of Gymnospermae. Concluding remarks to the relationships between Coniferae. *J. Inst. Polytech/Osaka City Univ.* Ser. D. 11: 109-136.

Uthayashankar, R., Prakash, O. and Subba Rao, P.V. 1985. Sensitivity of *Parthenium* rhinitis patients to ragweed pollen allergens. *Ann. Allergy* 55; 247. (Abst.)

Van Campo, M. 1966. Variations polliniques intra florales. *Adansonia* 6: 55-64.

Van Leewen. 1924. Bronchial asthma in relation to climate. *Proc. Roy. Soc. Med.* 17: 19.

Vasil, I.K. 1961 Studies on pollen storage of some crop plants. *J. Indian. Bot. Soc.* 41: 178-196.

Vijay, H.M. 1994. Advanced studies on *Alternaria alternata* mould allergens with reference to standardization. In: S.N. Agashe (ed) *Recent trends in Aerobiology, Allergy and Immunology*. Oxford & IBH Pub. Co. New Delhi. pp. 247-277.

Visser, T. 1955. Germination and storage of pollen. *Meded Landb-Hoogesch.* 55: 1-68.

Vittal, B.P.R. 1974. Obituary: Professor T. Sreeramulu, 1925-1974. *Kavaka.* 2: 93.

Vittal, B.P.R., Udaya Prakash, N.K. and Bhuvaneshwari, S. 2001. Pollen Calendar of Chennai.

Vlietinck, A.J., Van Hoof, L., Totte, J., Lasure, A., Vanden Berghe, D., Rwangabo, P.C. and Mvukiyumwami, J. 1995. Screening of a hundred Rwandese medicinal plants for anti-microbial and antiviral properties. *J. Ethanopharmacol.* 46: 31.

Von Pirquet, C. 1907. Die Allergieprobe zur diagnose der Tuberkulose in Kindersalter. *Wien. Med. Wochenschr.* 57: 1369.

Von Post, L. 1916. Om skogsträdpollen I sydsvenska torfmosslagerföljder. Geol. for Stockholm förhdl. 38: 384-394.

Vyas, M.K., Agarwal. M.K., Shali, P.L. and Mehra, A. 1969. A modified rapid method for the preparation of antigenic extracts (*A. tamerii*). *Aspects of Allergy and Applied Immunology* 3: 71.

Walker, J.W. 1971. Unique type of angiosperm pollen from the family Annonaceae. *Science* 172: 565-567.

Walker, J.W. and Doyle, J.A. 1975. The basis of angiosperm phylogeny. *Ann. Missouri Bot. Garden.* 62: 664-723.

Watts, D. 1971. The principles of biogeography – an introduction to the functional mechanisms of ecosystems. McGraw Hill. London.

Wedner, H.J., Wilson, P. and Lewis, W.H. 1989. Allergic reactivity to *Parthenium hysterophorus* pollen: An ELISA study of 582 sera from United States Gulf Coast. *J. Allergy Clin. Immunol.* 84: 263-271.

West, R.G. 1968. Pleistocene Geology and Biology. Longmans, London.

White, J.M. 1988. Methodology of the exotic spike: Differential settling of palynomorphs during sample preparation. *Pollen et Spores* 30: 131-148.

Whitehead, D.R. 1969. Wind pollination in the angiosperms: evolutionary and environmental considerations. *Evolution* 23: 28-36.

Wilken-Jensen, K. and Gravesen, S. (eds) 1984. Atlas of Moulds in Europe causing Respiratory Allergy. Ask, Copenhagen. p. 110.

Williams, R.R. and Legge, A.P. 1979. Pollen application by the chemical dusting in English apple orchards. *J. Hort. Sci.* 54: 67-74.

Wilson, A.T., Vickers, M. and Mann, L.R.B. 1979. Metabolism in dry pollen –a novel technique for studying anhydrobiosis. *Naturwissenschaften* 66: 53-54.

Wodehouse, R.P. 1935. Pollen Grains, their structure identification and significance in science and medicine. McGraw-Hill, New York. p. 574.

Wodehouse, R.P. 1948. Patterns of allergic sensitization. *Annals of Allergy* 6: 358.

Wodehouse, R.P. 1965. Pollen Grains, their structure identification and significance in science and medicine (Second Edition). McGraw-Hill, New York. p. 574.

Wolters-Arts, M., Lush, W.M. and Marium, C. 1998. Lipids are required for directional pollen tube growth. *Nature* 392: 818-821.

Wood, G.D., Gabril, A.M. and Lawson, J.C. 1996. Chapter 3: Palynological techniques-processing and microscopy. In: J. Jansonius and D.C. McGregor (eds) *Palynology: Principles and Applications.* American Association of Stratigraphic Palynologist Foundation. Vol. 1, pp. 655-659.

Wyman, M. 1872. Autumnal catarrah. Hurd and Honghtan. Cambridge

Yankova, R. 1998. Ragweed in Bulgaria. 6th International Congress on Aerobiology Perugia, Italy 1998. Satellite Symposium Proceedings: Ragweed in Europe. 6th International Congress on Aerobiology, Perugia, Italy. ALK- Abello, Denmark. pp. 42-44.

Yermekova, R.K. 1991. Allergenic pollen and pollinosis in USSR. In: G. D'Amato, F.Th.M. Spieksma and S. Bonini (eds) *Allergenic pollen and pollinosis in Europe,* Blackwell, Scientific Publishers, Oxford, London, Vienna. pp. 217-219.

Zanon, P., Berra, D., Alesima, R., Cirla, A., Corsico, R. and Guidoboni, A. 1998. Spread of ragweed allergy in Lombardy (North Italy). Satellite Symposium Proceedings: Ragweed in Europe, 6th International Congress on Aerobiology, Perugia, Italy, pp. 20-21.

Index

B

D

Gravity Separation Technique 113

N

O

P

T

About the Authors

Shripad N. Agashe M.Sc. (Poona), Ph.D. (Washington), has rich experience of 36 years in Post-Graduate teaching, and, research experience of more than 40 years. He is a author of two books and Chief Editor of three volumes of research publications concerning Palynology, Aerobiology, Paleobotany and Paleopalynology. His recent book entitled *Palynology and its applications* was published in 2006. After retirement as a Professor of Botany at Bangalore University he was selected as Professor Emeritus by the University Grants Commission, New Delhi. He has more than 65 research papers to his credit and has presented more than 125 research papers in conferences and seminars.

He was the Founder President of Indian Academy of Allergy and has held several positions in scientific organizations such as Indian Aerobiological Society, International Association of Aerobiology. He was the Chairman of Organizing Committee and successfully organised 5th International Conference on Aerobiology in August 1994 in Bangalore, India. On account of his significant scientific contributions spanning more than four decades he was honoured as Fellow of Indian College of Allergy, Asthma and Applied Immunology (FCAI), Indian Aerobiological Society (FIAS), Palaeobotanical Society (FPbS), Palynological Society (FPnS), Indian Academy of Allergy (FIAA).

Professor Agashe was the recipient of prestigious Dr. D.N. Shivpuri Oration award in 2003 from the Indian College of Allergy, Asthma and Applied Immunology and Lifetime Achievement Award of International Asthma Services, Colorado, USA in 2007.

Eric Caulton B.Sc. (Hons.), Dip. Ed., M.Sc., Ph.D., C.Biol., F.I.Biol., F.L.S., established the Scottish Centre for Pollen Studies in 1987 and has been its Director since then. He was a Senior Lecturer in the School of Life Sciences, Napier University, Edinburgh, Scotland, U.K. prior to retirement, since which time he has been an Honorary Research Fellow.

For almost twenty years, he has been involved in the monitoring of allergenic pollen and spores in South-East Scotland, providing a daily pollen count for sufferers of seasonal Rhinitis (hay fever) and Asthma.

Eric Caulton is a founder member of the British Aerobiology Federation and contributes data to the European Aeroallergy Network. He is the author of numerous publications relating to Aerobiology, Ecology, Mycology, Environmental Pollution and Wildlife conservation, including papers on pollen in animal faeces to determine diet and foraging habitats.